A MATLAB®
COMPANION
TO COMPLEX
VARIABLES

TEXTBOOKS in MATHEMATICS

Series Editors: Al Boggess and Ken Rosen

PUBLISHED TITLES

ABSTRACT ALGEBRA: AN INTERACTIVE APPROACH, SECOND EDITION
William Paulsen

ABSTRACT ALGEBRA: AN INQUIRY-BASED APPROACH
Jonathan K. Hodge, Steven Schlicker, and Ted Sundstrom

ADVANCED LINEAR ALGEBRA
Hugo Woerdeman

APPLIED ABSTRACT ALGEBRA WITH MAPLE™ AND MATLAB®, THIRD EDITION
Richard Klima, Neil Sigmon, and Ernest Stitzinger

APPLIED DIFFERENTIAL EQUATIONS: THE PRIMARY COURSE
Vladimir Dobrushkin

COMPUTATIONAL MATHEMATICS: MODELS, METHODS, AND ANALYSIS WITH MATLAB® AND MPI,
SECOND EDITION
Robert E. White

DIFFERENTIAL EQUATIONS: THEORY, TECHNIQUE, AND PRACTICE, SECOND EDITION
Steven G. Krantz

DIFFERENTIAL EQUATIONS: THEORY, TECHNIQUE, AND PRACTICE WITH BOUNDARY VALUE PROBLEMS
Steven G. Krantz

DIFFERENTIAL EQUATIONS WITH MATLAB®: EXPLORATION, APPLICATIONS, AND THEORY
Mark A. McKibben and Micah D. Webster

ELEMENTARY NUMBER THEORY
James S. Kraft and Lawrence C. Washington

EXPLORING LINEAR ALGEBRA: LABS AND PROJECTS WITH MATHEMATICA®
Crista Arangala

GRAPHS & DIGRAPHS, SIXTH EDITION
Gary Chartrand, Linda Lesniak, and Ping Zhang

INTRODUCTION TO ABSTRACT ALGEBRA, SECOND EDITION
Jonathan D. H. Smith

PUBLISHED TITLES CONTINUED

INTRODUCTION TO MATHEMATICAL PROOFS: A TRANSITION TO ADVANCED MATHEMATICS, SECOND EDITION
Charles E. Roberts, Jr.

INTRODUCTION TO NUMBER THEORY, SECOND EDITION
Marty Erickson, Anthony Vazzana, and David Garth

LINEAR ALGEBRA, GEOMETRY AND TRANSFORMATION
Bruce Solomon

MATHEMATICAL MODELLING WITH CASE STUDIES: USING MAPLE™ AND MATLAB®, THIRD EDITION
B. Barnes and G. R. Fulford

MATHEMATICS IN GAMES, SPORTS, AND GAMBLING—THE GAMES PEOPLE PLAY, SECOND EDITION
Ronald J. Gould

THE MATHEMATICS OF GAMES: AN INTRODUCTION TO PROBABILITY
David G. Taylor

MEASURE THEORY AND FINE PROPERTIES OF FUNCTIONS, REVISED EDITION
Lawrence C. Evans and Ronald F. Gariepy

NUMERICAL ANALYSIS FOR ENGINEERS: METHODS AND APPLICATIONS, SECOND EDITION
Bilal Ayyub and Richard H. McCuen

ORDINARY DIFFERENTIAL EQUATIONS: AN INTRODUCTION TO THE FUNDAMENTALS
Kenneth B. Howell

RISK ANALYSIS IN ENGINEERING AND ECONOMICS, SECOND EDITION
Bilal M. Ayyub

TRANSFORMATIONAL PLANE GEOMETRY
Ronald N. Umble and Zhigang Han

TEXTBOOKS in MATHEMATICS

A MATLAB®
COMPANION
TO COMPLEX
VARIABLES

A. David Wunsch

University of Massachusetts Lowell

CRC Press
Taylor & Francis Group
Boca Raton London New York

CRC Press is an imprint of the
Taylor & Francis Group an **informa** business

A CHAPMAN & HALL BOOK

CRC Press
Taylor & Francis Group
6000 Broken Sound Parkway NW, Suite 300
Boca Raton, FL 33487-2742

© 2016 by Taylor & Francis Group, LLC
CRC Press is an imprint of Taylor & Francis Group, an Informa business

No claim to original U.S. Government works

Printed on acid-free paper
Version Date: 20160315

International Standard Book Number-13: 978-1-4987-5567-2 (Paperback)

Library of Congress Cataloging-in-Publication Data

Names: Wunsch, A. David.
Title: A Matlab companion to complex variables / A. David Wunsch.
Description: Boca Raton : Taylor & Francis, 2016. | Series: Textbooks in mathematics ; 41 | "A CRC title." | Includes bibliographical references and index.
Identifiers: LCCN 2015044810 | ISBN 9781498755672 (alk. paper)
Subjects: LCSH: Functions of complex variables--Data processing. | MATLAB.
Classification: LCC QA331.7 .W865 2016 | DDC 515/.9028553--dc23
LC record available at http://lccn.loc.gov/2015044810

Visit the Taylor & Francis Web site at
http://www.taylorandfrancis.com

and the CRC Press Web site at
http://www.crcpress.com

Contents

Acknowledgments

In writing this book, I consulted with a number of very generous people. One is my brother, Carl Wunsch, formerly of MIT and now at Harvard. He is a physical oceanographer and knows his fluid dynamics and helped me with mine. His son, Jared, who is in the mathematics faculty at Northwestern University, answered some questions about fractals without indicating that I should have been able to figure these out for myself. He is a good mathematician and a tactful teacher.

Kenneth Falconer, who is in the mathematics department at the University of St. Andrews, has written a very graceful introduction to the subject of fractals published by Oxford University Press. He has kindly answered some of my e-mail inquiries on this subject. And equally generous has been Professor Bodil Branner of Denmark, the first woman to head the Danish Mathematical Society, who replied to my e-mails. I can recommend her essay, The Mandelbrot Set, which appeared in the book *Chaos and Fractals* published by the American Mathematical Society.

Michael F. Brown ably assisted me on the last two editions of my book on complex variables and kindly found time to help me with this text. He is a good mathematician and a careful reader.

Finally, MathWorks, publisher of MATLAB®, provides a team of consultants to help authors who are writing MATLAB-based books. These people have responded quickly to my e-mails and have almost always managed to correct my work and supply helpful advice.

MATLAB® is a registered trademark of The MathWorks, Inc. For product information, please contact:

The MathWorks, Inc.
3 Apple Hill Drive
Natick, MA 01760-2098, USA
Tel: 508-647-7000
Fax: 508-647-7001
E-mail: info@mathworks.com
Web: www.mathworks.com

Introduction

I hope the reader will enjoy using this book. It is for someone with at least a beginner's knowledge of MATLAB® who is learning the branch of advanced calculus called "functions of a complex variable." Complex variable theory is the calculus of functions dependent on variables that can assume complex numerical values. I want you to discover that MATLAB is your friend when you are learning this type of mathematics.

One of the daunting aspects of complex numbers (and variables) is that simple arithmetic operations that are easily done in one's head, if they involve real quantities, become tedious when the numbers are complex. You can ask a class of third graders for the product of the numbers 3, 4, and 5, and someone with the answer will raise his or her hand in less than 5 seconds. Ask a college student for the product of the numbers $(1 + 3i)$, $(1 + 4i)$, $(1 + 5i)$ and he or she will, after a little squirming, probably resort to pencil and paper, although the problem is solvable in one's head with some serious concentration. If you were to ask me this question as I'm typing this sentence, I would open the MATLAB window on my computer and type in the product. Here we see a glimpse of the utility of MATLAB in complex arithmetic.

If you glance at the table of contents, you will see how MATLAB can be your companion in such staples of complex variable theory as conformal mapping, infinite series, contour integration, and Laplace and Fourier transforms. Fractals, the most recent interesting topic involving complex variables, cries out to be treated with a language such as MATLAB, and you might want to begin this book at its end, The Coda, which is devoted entirely to this visually intriguing subject. However, I must add that as you progress through the book, the MATLAB skill required increases gradually, and leaping to the end is not for everyone.

Sometimes while working with MATLAB and complex algebra, you may be puzzled at what you find. For example, if you ask a class of high school students to compute $\left(5^3\right)^{1/3}$, someone will quickly call out "five." He or she will know to multiply the exponents together. Now asking the same class for $\left((-1+i)^2\right)^{1/2}$, a student will follow the same logic and produce a correct answer: $-1 + i$. But another student, using MATLAB, will say $1 - i$. Both answers are correct, but why did MATLAB choose this one? This book answers such questions and many similar ones.

This textbook does not purport to present MATLAB as a substitute for a knowledge of the functions of a complex variable any more than MATLAB can be used as a replacement for an actual understanding of elementary calculus or linear algebra. This is also not a text from which one learns the elements of MATLAB, although if you already know a little of the language, it will expand your knowledge. Some books that will get the reader started in the elements of MATLAB programming are listed as references [1,2]. MATLAB is not without constraints, assumptions, limitations, irritations, and quirks, and there are subtleties involved in performing the calculus of complex variable theory with this language that will be made evident here. Without knowledge of these subtleties, the engineer or scientist who is attempting to use MATLAB for solutions of practical problems in complex variable theory suffers the real risk of making major mistakes. This book should serve as an early warning system about these pitfalls.

This book should be read as a companion to standard texts on functions of a complex variable. Throughout what follows, we refer to two of the author's favorites. Not surprisingly, one volume is his own *Complex Variables with Applications* (3rd edition) published by Addison-Wesley in 2005. We refer to this book as "W" in the text. In most cases, the section numbers referred to apply to the second edition as well. There is a Spanish translation available for those who prefer that language: *Variable Compleja con Aplicaciones*. The reference book in the Schaum's outline series, *Complex Variables*, 2nd edition by M. Spiegel et al. is an old favorite of mine, and although it is more handbook than textbook, it is remarkably well done, and I will refer to it with the letter "S." Notice that the section numbers that I refer to apply to the second edition only. *Complex Variables and Applications* by R. V. Churchill and colleagues is very well written, but the book has had so many editions that it is difficult to refer the student to any particular section.

References

1. Hunt, B., Lipsman, R., and Rosenberg, J. *A Guide to MATLAB for Beginners and Experienced Users*, 3rd edition. New York: Cambridge University Press, 2014.
2. Hahn, B. and Valentine, D. *Essential MATLAB for Engineers and Scientists*, 5th edition. New York: Academic Press, 2013.

A. David Wunsch
University of Massachusetts Lowell
Lowell, Massachusetts

A Note to the Reader

To keep the cost of this book reasonable, all figures are rendered in black and white and shades of gray. Some of the codes provided here will produce only these colors, while other programs will produce color plots on your screen even though the plots are in grayscale in your book. In the solutions manual I have not avoided color.

MATLAB® is upgraded at least once a year with new releases. Its capabilities change and for the most part are improved. This book is based on R2015a, the release available through most of the year 2015.

Corrections to this book as well as to the solutions manual will be posted at the author's web page hosted by the University of Massachusetts Lowell. Here is the URL: http://faculty.uml.edu/awunsch/wunsch_complex_variables/faculty.htm.

The author invites corrections and comments for his work. An email address can be found at the above website where you will also find solutions to the odd numbered problems. To assist readers with data entry in MATLAB, code text can be found for download at https://www.crcpress.com/A-MatLab-Companion-to-Complex-Variables/Wunsch/9781498755672.

Instructors using my book in the classroom can also receive a copy of a complete solutions manual, for all problems, if they write to me on their college stationary at the ECE Department, University of Massachusetts Lowell, Lowell, MA 01854, USA.

1

Complex Arithmetic

1.1 The Rectangular Form

A complex number z can be stated in the form

$$z = x + iy \tag{1.1}$$

where x and y are real numbers, and there is multiplication between the i and the y. An equally valid representation is

$$z = x + yi \tag{1.2}$$

since complex numbers obey the commutative law of multiplication so that the order of multiplying y with i is immaterial. Right away we must deal with an idiosyncrasy of MATLAB®. The statement $z = 3 + 4i$ can be entered in the MATLAB command window with the result given in the following example. Note that >> preceding a line of code indicates that the expression was *entered from the keyboard* into the command window of MATLAB.

Example 1.1

```
>> z=3 + 4i
z = 3.0000 + 4.0000i
```

which is exactly what we hoped for: MATLAB has returned 3.0000 + 4.0000i. However, entering $z = 3 + i4$ in the command window will result in an error message. The value i must appear as the second factor in the multiplication if i is to be interpreted by MATLAB as a multiplicative factor in $z = x + yi$. The practice is only valid if i is *preceded* by an explicitly stated *real* (not complex) number. A symbol cannot be used for that number. This entire convention can be overlooked provided we employ the MATLAB $*$ for multiplication. We will adopt that practice throughout this book even though the authors of MATLAB claim that eliminating $*$ where allowed will speed up calculations; the advantage is often slight.

Note that in general, all the arithmetical operations that one does with real numbers in MATLAB can be carried out for complex numbers, using the same operators (i.e., the + and − signs for addition and subtraction, the / for division, the ^ for raising a number to a power, and as noted, the * for multiplication). The precise meaning of what the ^ will yield when followed by a fraction will be treated in section 1.3.

To get the magnitude (or absolute value or modulus) of a complex number, we apply the operation **abs** to that number. Thus,

```
>> abs(3 + 4*i)
ans = 5
```

Example 1.2

Multiply $3 + i4$ by $(1 + 2i)$ and add $i2$ to that result. Then find the absolute value of that quantity.

Solution:

```
>> (3 + i*4)*(1 + 2*i) + 2*i
ans = −5.0000 + 12.0000i
>> abs(ans)
ans = 13
```

Electrical engineers often prefer to use j instead of i, and this notation can be used automatically in MATLAB. We just use j instead of i and follow the same conventions as above.

You should not use both i and j in the same code, or you will confuse yourself and whoever reads what you wrote.

Example 1.3

Use j instead of i, and raise $(1 + j3)$ to the −2 power, divide that result by $2 − j3$, and subtract $3 + 4j$ from that result.

Solution:

```
>> (1 + 3*j)^(−2)/(2−j*3)−(3 + 4*j)
ans = −2.9985 − 4.0277i
```

Note that although we used j instead of i, MATLAB returned an answer employing i.

For some purposes, it is more convenient to put $j = −i$, and you should make this statement at the start of your work if that is your preference. (Note that $j^2 = −1$ as before.) Recall that physicists prefer $e^{-i\omega t}$ in lieu of the electrical engineer's $e^{j\omega t}$, where ω is a radian frequency and t is the time.

The real and imaginary parts of a complex number, for example z, are found from the functions **real**(z), **imag**(z), and the conjugate of z is given by

conj(z) as shown by the following sequence of calculations. Note that in the text material of this book, we will designate the conjugate of a quantity by an overbar, as in $\bar{z}$, which in MATLAB would of course be **conj**(z).

Example 1.4

```
>> z=3 + 4*i
z = 3.0000 + 4.0000i
>> x=real(z)
x = 3
>> y=imag(z)
y = 4
>> w=conj(z)
w = 3.0000 – 4.0000i
```

1.1.1 A Caveat on Complex Numbers as Matrix Elements

MATLAB, as its name might suggest, is a computer language for technical computation that is based on matrices (i.e., rectangular arrays of numbers). Sometimes the matrix consists of just a horizontal row of numbers, in a specific order, or a column of numbers in a certain order. Such matrices are known as vectors and more specifically as row or column vectors. The numbers that make up matrices, the elements, can be complex or real numbers, or symbols. When entering complex numbers as the elements, it is useful to state them with parentheses surrounding each complex number. In other words, instead of entering 3 + 4*i, you should enter (3 + 4*i). Otherwise, an inadvertent use of the space bar can result in an error, as shown by the following where we want to enter a row vector a having two elements, one being $3 + i$ and the other $2 - i$. In the first instance, we will do this correctly without parentheses, then incorrectly without parentheses, because we have placed a space where it does not belong, and finally correctly with parentheses even though there is an unnecessary space:

```
>> a=[3 + i   2–i]
a = 3.0000 + 1.0000i   2.0000 – 1.0000i
>> a=[3   +i   2–i]
a = 3.0000   0 + 1.0000i   2.0000 – 1.0000i
>> a=[(3   + i)   (2–i)]
a = 3.0000 + 1.0000i   2.0000 – 1.0000i
```

Comment: Note that in the second instance, the value obtained for a is

3.0000   0 + 1.0000i   2.0000 – 1.0000i

which is a row vector with *three* elements, not the desired two elements. The first element is 3, and the second is $0 + 1.0000i = i$. The error was caused by our having a space after the 3 and before the adjacent + sign.

In the third output for *a*, we got the correct value even though we typed a space between the 3 and the plus sign following it.

Note that you can always tell the number of elements in a row or column vector, let us call it *a*, with the command **length**(*a*). If we apply this operation to the three values of *a* described above, we get 2, 3, and 2, respectively.

Exercises

1. Using MATLAB, determine

   a. $\dfrac{3-4i}{1+i2}+i2$

   b. $\dfrac{2+i}{3+5i}-1-i/2$

   c. $\left(1+i2/3\right)^3-\dfrac{i}{3-4i}$

   d. $\left(1+\dfrac{2}{i3}\right)^{11}$

   e. $\left(\dfrac{2i}{(3+4i)}+\overline{\left(\dfrac{1}{5-4i}\right)}\right)^5$

2. The numerical value of the expression $(1 + i)^{37}$ can be determined without recourse to MATLAB by the following technique: $(1+i)^{37}=(1+i)^{36}(1+i)=\left((1+i)^2\right)^{18}(1+i)$. Now complete the calculation by evaluating $(1 + i)^2$ and noting that *i* raised to any even power is easily computed. Check your answer by using MATLAB to find $(1 + i)^{37}$ directly.

3. Consider the finite series ($N \geq 0$ is an integer) and its sum.

   $$\sum_{n=0}^{N} z^n = \frac{1-z^{N+1}}{1-z} \quad z \neq 1,$$ whose derivation is identical to that given

   in real calculus for the sum of a finite geometric series.

   a. Using MATLAB, sum the series $1+\dfrac{i}{2}+\left(\dfrac{i}{2}\right)^2+\left(\dfrac{i}{2}\right)^3+...+\left(\dfrac{i}{2}\right)^{10}$ directly by using a loop created from the **while** or **for** commands. Use the "long format" of MATLAB as you will need it in part (b).

   b. Verify the answer obtained in (a) by using the closed-form expression for the sum given above, again using the long format so that you can notice any small discrepancy.

1.2 The Polar Form of Complex Numbers

Referring to Figure 1.1, we are reminded that a complex number z can be expressed using the polar coordinates r and θ. Thus,

$$z = x + iy = r\cos\theta + ir\sin\theta = r(\cos\theta + i\sin\theta) \tag{1.3}$$

Sometimes $cis\theta$ is used to mean $\cos\theta + i\sin\theta$, but other times, the notation $\angle\theta$ is used for the same quantity. Thus,

$$z = rcis\theta = r\angle\theta \tag{1.4}$$

Generally, θ is expressed in radians, but one might also use degrees.

Now $|z| = \sqrt{x^2 + y^2} = r$, i.e., the polar coordinate r is simply the magnitude of the complex number z, and we have in MATLAB that $r = \text{abs}(z)$. The quantity θ is called the *angle* or the *argument* of the complex number z.

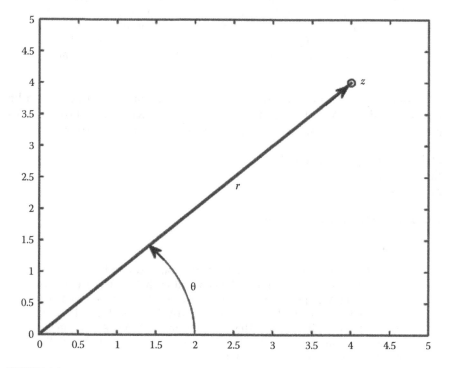

FIGURE 1.1
The polar coordinates r and θ for the point z.

There are some subtleties with θ since it is multivalued. If θ_0 is a possible value of θ, then so is $\theta = \theta_0 + 2k\pi$, where $k = 0, \pm1, \pm2, ...$ If we are working in degrees, the expression becomes $\theta = \theta_0 + k360°$. Given a complex number in rectangular form, MATLAB can be used to yield a value of θ in radians from the function **angle**(z). But which value will it yield? It yields the *principal value of the argument*, the one satisfying $-\pi < \theta \leq \pi$. Note that a few textbooks use a different definition for the principal value and that the value returned by MATLAB would not be in conformity with theirs.*

Example 1.5

Find the polar form of the complex number $-\sqrt{3} - i$.

Solution:

```
>> z= -sqrt(3)-i;
r=abs(z)
r = 2.0000
>> theta=angle(z)
theta = -2.6180
```

Thus, we have shown $z = 2\angle -2.6180...$ in polar form.

A short calculation with MATLAB shows that -2.6180 agrees with $-5\pi/6$ (or -5pi$/6$ in MATLAB) to as many decimal places as MATLAB will display in the standard format. Construction of the vector for $-\sqrt{3} - i$ on a piece of paper and recognizing the resulting triangle with angles 30, 60, and 90 degrees should show where the angle $-5\pi/6$ arises from.

Incidentally, the above rectangular to polar calculation can be done a little more easily in MATLAB as $[th,r] = $ **car2pol**(x,y), which converts complex numbers from Cartesian to polar form. Thus, using numbers from above,

```
>> [th,r]=cart2pol(-sqrt(3),-1)
th = -2.6180
r = 2.0000
```

which agree with what we found.

Example 1.6

Using MATLAB, compute the angles of the complex numbers $-1+$eps$*$i and $-1-$eps$*$i and compare them by computing their difference. Recall that in MATLAB, eps is the smallest possible difference between two floating point numbers. In the version of MATLAB used in this book, eps $\approx 2.22 \times 10^{-16}$. If you type the word eps in the MATLAB command window, you can find out what eps is for your version of MATLAB. We chose here to work in the "long" format.

---

* These books take $0 \leq \theta < 2\pi$.

Solution:

```
>> format long
>> angle1=angle(-1 + i*eps)
angle1 = 3.141592653589793
>> angle2=angle(-1-i*eps)
angle2 = -3.14159265358979
>> angle1-angle2
ans = 6.28318530717959
>> 2*pi
ans = 6.28318530717959
```

Note that although graphical representation of $-1 + i*$eps and $-1 - i*$eps as points in the complex plane will involve neighboring points that are *exceedingly* close to one another, these points lie just above and below the negative real axes. Their principal angles are nearly π and $-\pi$, respectively. The differences in their angles, as computed by MATLAB, are indistinguishable from 2π.

Referring to Figure 1.1, we see that $\theta = \arctan(y/x)$, and we might be tempted to use the function **atan** in MATLAB, which yields the arctangent, in order to obtain θ, the angle of the complex number $z = x + iy$. This is to be avoided; because of the multiplicity of values of the arctangent, the computer might yield the wrong value. If, for example, $x = 1$ and $y = 1$, then **atan**(1/1) is evaluated by MATLAB to be 0.78539816339745 or $\pi/4$, that is a correct value as a sketch of the vector for this complex number shows. However, if $x = -1$, $y = -1$, we have from MATLAB that atan$(-1/-1)$ is again 0.78539816339745, which is incorrect for the angle of this complex number. A correct value, as a sketch of the corresponding vector shows, is $-3\pi/4$ $= -2.35619449019234$. In general, it is better to use the **angle** function of MATLAB to avoid these situations. However, one might use the function **atan2**(x,y) in MATLAB where the real and imaginary parts of the complex number $z = x + iy$ are entered in a particular order into the function. Additionally, the MATLAB function $[th,r]$ = car2pol(x,y) will always yield the correct angle th (or θ).

Given a complex number in the polar form (a magnitude r and an angle θ), we can compute the Cartesian form by asking MATLAB for $r\cos\theta = x$ and $r\sin\theta = y$. Alternatively, we can use a function M file built into MATLAB that will automatically do the two calculations. The function is $[x,y]$ = **pol2cart**(th,r), where the angle θ in radians, here called th, is entered first followed by r. Suppose we wish to convert the complex number $2\angle\pi/6$ to rectangular form. We proceed as follows:

```
>> [x,y] = pol2cart(pi/6,2)
x = 1.73205080756888
y = 1.00000000000000
```

Note that the program does not yield the complex number $z = x + iy$. It simply gives the real and imaginary parts of z. To get z, one must take the two results from above and make that additional computation.

Exercises

1. We know that the polar angle of the product of two complex numbers can be expressed as the sum of the angles of each.

   a. Multiply $(1 + i)$ by $\left(\sqrt{3} + i\right)$ using MATLAB. Find the angle of the product using MATLAB and show that it is the sum of the angles of each factor.

   b. Repeat the above exercise, but use as factors $(1 + i)$ and $-\sqrt{3} + i$. Explain why the angle of each factor, obtained from MATLAB, when added, does not agree with the angle that MATLAB yields for the angle of the product. Compute these angles using MATLAB.

2. When MATLAB is asked for the argument (angle) of a complex number, it always returns the principal value. Use MATLAB to find the principal argument of the following numbers where all angles are in radians except in part (d).

   a. $5cis(197)$

   b. $1/(4\angle79)$

   c. $(1 + 2i)^{17}$

   d. $(5cis(53.3°))^8$ (give answer in degrees)

3. Using the **pol2cart** function in MATLAB described above, convert these numbers to rectangular components where the angles below are in radians:

   a. $3cis(-207)$

   b. $4\angle50$

4. Write a new function, similar to the **pol2cart** function in such a way that if it is given θ in *degrees* and r, it will convert the polar form of the number to rectangular form. Call the new function **pold2cart**. Use it to convert these numbers to rectangular form. Give x and y.

   a. $5\angle60°$

   b. $2\angle(-320°)$

1.3 Fractional Powers of Complex Numbers

First, we make a note on notation. In this book, as in many texts on complex variable theory, the symbol $\sqrt{\beta}$ is applied only when β is a non-negative real number and is intended to mean the non-negative square root. Thus, $\sqrt{16}$ is always 4 and never −4. Similarly, the $\sqrt[m]{\beta}$ is applied when β has this same restriction and m is an integer and refers to the mth root of β but only to the non-negative real value. Thus, $\sqrt[3]{64}$ is 4 and not any of the other two possible values, which are complex.

Suppose you wish to solve the equation $z^4 - 1 = 0$. We have $z^4 = 1$. Notice that our equation has four solutions: ± 1 and $\pm i$. However, if you attempt to solve the given equation using MATLAB by asking it for $z = 1^{1/4}$, which is written in MATLAB as 1^(1/4), you will receive just one of the four answers, in this case, 1. The question arises as to how MATLAB decides which answer to give you when a number is raised to a fractional power and how we might get the rest.

If z is to be raised to a fractional power n/m where m and n are integers, with $m \neq 0$, and where the fraction n/m cannot be reduced, we convert z to the polar form $z = r\angle\theta$, and as shown in the standard texts, we have m values given by

$$z^{n/m} = \left(\sqrt[m]{r}\right)^n cis\left[\frac{n}{m}\theta + 2k\pi\frac{n}{m}\right] \quad k = 0,1,2,\ldots |m| - 1 \tag{1.5}$$

where as we recall, $cis\phi = \cos\phi + i\sin\phi$. This result is consistent with the definition $z^{n/m} = (z^{1/m})^n$. Notice the order of the operations.

In what follows, we will take m as positive and skip the absolute magnitude signs around the m. We can do this because it is easily shown that the set of values of $z^{-n/m}$ is identical to the set of values of $z^{n/(-m)}$. Thus, given the problem of computing $(1 + i)^{2/(-3)}$, we can compute the set of values of $(1 + i)^{-2/3}$ and use a positive m.

Equation 1.5 is equivalent to

$$z^{n/m} = \left(\sqrt[m]{r}\right)^n cis\left[\frac{n}{m}\theta\right]cis\left[2k\pi\frac{n}{m}\right] \quad k = 0,1,2,\ldots (m-1) \tag{1.6}$$

which arises from the identity $cis(\alpha + \beta) = cis\alpha\, cis\beta$; it applies to any numbers α and β. Note that in Equation 1.6, we can generate the same set of values for $z^{n/m}$ if we let k go through any set of m consecutive integers, for example, $1,2,\ldots m$.

The results from Equations 1.5 and 1.6 are m numerically distinct values. When plotted as points in the complex plane, they all lie on a circle having radius $\left(\sqrt[n]{r}\right)^m$ (i.e., they have identical magnitudes). The angular spacing of these values is $2\pi/m$. The separation of these points on the circle is uniform.

If you ask MATLAB for the computation of a fractional power, for example, $(81i)^{3/4}$, you will receive exactly one value, in this case, $10.3325 + 24.9447i$, which in polar form is $27 \angle 1.1781....$ The angle here is $3\pi/8$ radians. According to Equations 1.5 and 1.6, there are three other values for $(81i)^{3/4}$. After some experimentation with MATLAB, you will see how it decides which value of $z^{n/m}$ to give you.

MATLAB Rule for Fractional Powers *The angle* θ *in Equations 1.5 and 1.6 is taken as the principal value, and the value of* k *is set equal to zero. MATLAB returns the result in Cartesian form.*

You should confirm that this convention yields the MATLAB result just given for $(81i)^{3/4}$.

Example 1.7

Use Equation 1.5, taking the MATLAB convention for fractional powers, to determine one value of $(-1 - i)^{2/3}$, and verify that MATLAB does indeed yield this value.

Solution:

We have $-1 - i = \sqrt{2}\angle(-3\pi/4)$, where the principal value of the argument of $(-1 - i)$ is used.
Using Equation 1.5, with $k = 0$ we have

$$(-1-i)^{2/3} = \left[\sqrt{2}\angle(-3\pi/4)\right]^{2/3} = \left(\sqrt[3]{\sqrt{2}}\right)^2 \angle\left(-3\frac{\pi}{4}\frac{2}{3}\right) = \sqrt[3]{2}\angle\left(-\frac{\pi}{2}\right) = -i\sqrt[3]{2}$$

Proceeding with MATLAB, we have

```
>> (-1-i)^(2/3)
ans = 0.0000 - 1.2599i
```

These two answers are numerically identical because the cube root of 2 is $1.2599....$

If we compute a value of $z^{n/m}$ and raise it to the m/n power, do we necessarily get back z? That is, does $(z^{n/m})^{m/n}$ equal z when the inner and outer operations are done in MATLAB? Because MATLAB will choose just one value of the n possible values when raising a number to the m/n power, you cannot say for certain that in MATLAB $(z^{n/m})^{m/n} = z$.

The reader should confirm that in MATLAB we have $\left((-1)^{4/3}\right)^{3/4} = -i$, so the original z (namely, -1) is *not* returned. The result comes about because the principal angle of -1 is π. When MATLAB evaluates $(-1)^{4/3}$ using this angle, and the rule stated above, the result is $\frac{-1}{2} - i\frac{\sqrt{3}}{2}$ whose polar representation, using the principal angle, is $1\angle\left(-2\frac{\pi}{3}\right)$. Raising the preceding to the ¾ power using the MATLAB rule for fractional powers, we get $1\angle\left(-\frac{\pi}{2}\right) = -i$.

For a different problem, MATLAB yields $\left((-i)^{4/3}\right)^{3/4} = -i$. The value of z is returned. One should be able to predict in advance whether the original z is obtained if we compute $(z^{n/m})^{m/n}$. This matter is explored in problem 2 in the Exercises.

1.3.1 Using sqrt

If you want to obtain the square root of a number, z, using MATLAB, you could ask for the computation $z^{\wedge}(1/2)$. The resulting root is the one supplied by the MATLAB rule given above for fractional powers. Alternatively, we might ask for **sqrt**(z), which requires about the same amount of typing. The same root is obtained as if you were using $z^{\wedge}(1/2)$—that is, you get $\sqrt{r}cis(\theta/2)$, where θ is the principal angle of z. So, for example, we have

```
>> format short
>> sqrt(-9)
ans = 0 + 3.0000i
```

We did not get $-3i$, the other square root.

1.3.2 Reminder: A Warning about Square Roots and Fractional Powers in MATLAB

If z is a matrix and you ask MATLAB for **sqrt**(z), then MATLAB will return to you a matrix whose elements are the square root of each element of z. However, if you ask MATLAB for $z^{\wedge}(1/2)$, it will return the square root of the matrix z (provided it is square matrix, like 2×2)—that is, a matrix whose square is z. To get the same matrix as produced by **sqrt**(z), you will need z.$^{\wedge}(1/2)$, where a dot follows the z. Similar precautions must be taken where other exponents are concerned.

1.3.3 Use of roots

Suppose we want to obtain all m values of $z^{n/m}$ through the use of MATLAB. A simple method involves the function **roots**. Let us begin by using **roots** to find $z^{1/m}$, where m is a positive integer. Observe that all m solutions for the unknown w in the equation $w^m = z$ are the m values of $z^{1/m}$. MATLAB has a function called **roots** that will solve the polynomial equation

$$c_1 w^m + c_2 w^{m-1} + c_3 w^{m-2} + \ldots + c_m w + c_{m+1} = 0$$

and yield all m roots. The procedure is to ask MATLAB for **roots(p)**, where **p** is a row vector whose entries, from left to right, are the numerical coefficients $c_1, c_2, \ldots c_{m+1}$. Note that if we take $c_1 = 1, c_2 = 0, c_3 = 0, \ldots c_{m+1} = -z$ in the polynomial equation, we obtain $w^m - z = 0$ whose solutions all satisfy $w = z^{1/m}$.

Suppose for example we wanted all six values of $(1 + i)^{1/6}$. This is the same problem as finding all six solutions of the polynomial equation $w^6 - (1 + i) = 0$. The expression on the left is a polynomial of degree 6. We should see that this polynomial has coefficients of zero for the terms w^5, w^4, ..., w, while the coefficient of w^0 is $-(1 + i)$. Thus, to solve our polynomial sixth-degree equation in w, we proceed as follows:

```
>> coeffs = [1 0 0 0 0 0 -(1 + i)];
>> format short
>> roots(coeffs)
ans =
   -1.0504 - 0.1383i
   -0.6450 + 0.8405i
   -0.4054 - 0.9788i
    0.4054 + 0.9788i
    0.6450 - 0.8405i
    1.0504 + 0.1383i
```

The above are the six values of $(1 + i)^{1/6}$. The following provides a check.

```
>> ans.^6
ans =
    1.0000 + 1.0000i
    1.0000 + 1.0000i
    1.0000 + 1.0000i
    1.0000 + 1.0000i
    1.0000 + 1.0000i
    1.0000 + 1.0000i
```

There is some round-off error, which is hidden because we are using short format, but each of these expressions is, to our degree of approximation, $(1 + i)$. This is correct, as we expect all six values of $(1 + i)^{1/6}$ when raised to the sixth power to yield $(1 + i)$.

Suppose we want all seven values of $(3 + 4i)^{5/7}$. We would proceed as before, calculating $(3 + 4i)^{1/7}$. We would then take all seven of these values and raise them to the fifth power. Here is the procedure:

```
>> coeffs = zeros(1,8);
%the above takes all the coeffs as zero at first
coeffs(1) = 1;
coeffs(8) = -(3 + 4*i);
poly_roots = roots(coeffs);
final_answer = poly_roots.^5
```

The output is

final_answer =

  −3.0852 − 0.6691i

   0.0342 + 3.1567i

   3.0700 − 0.7358i

  −1.4005 − 2.8293i

   1.3388 − 2.8590i

  −2.4467 + 1.9950i

   2.4894 + 1.9414i

Note that there are seven answers, as expected. The first line of our code coeffs=zeros(1,8); establishes a row vector having eight elements. Initially all eight are zero. Then we set the first element equal to one and the last equal to −(3 + 4i).

The above technique can be applied to finding all $|m|$ values of $z^{n/m}$ when m is a *negative* integer. We compute $z^{-n/|m|}$ and proceed as above, first finding the $|m|$ values of $z^{1/|m|}$ and then raising them to the −nth power.

Suppose we were to take each of the values of $(3 + 4i)^{5/7}$ obtained above and, using MATLAB, raise them all to the 7/5 power. Can we expect in each case to recover $(3 + 4i)$? Trying this, we have

```
>> recover = final_answer.^(7/5)
recover =
```

  −2.8772 + 4.0892i

  −2.8772 + 4.0892i

   4.7313 − 1.6171i

  −4.7782 − 1.4727i

  −0.0759 − 4.9994i

  −4.7782 − 1.4727i

   3.0000 + 4.0000i

Notice that only one of these results is the original 3 + 4i. This is because there are five possible values of a number raised to the 7/5th power. Just one of the above yields ultimately to our having $\left((3+4i)^{5/7}\right)^{7/5} = 3+4i$.

Keep the following in mind: Suppose using any method at your disposal you compute *all* $|m|$ values of $z^{n/m}$, where n and m are integers, and $m \neq 0$. Now suppose you raise each value to the m/n power using the MATLAB expression V^(m/n), where V is any one of those m values of $z^{n/m}$. There is no guarantee that you will get back z. A simple example will suffice. We know that $(-1)^{3/2} = (-1)^{1/2} (-1) = \pm i$. From MATLAB, we raise both i and $-i$ to the 2/3rds power:

```
>> i^(2/3)
ans = 0.5000 + 0.8660i
```

>> (–i)^(2/3)
ans = 0.5000 – 0.8660i

In neither case did we recover –1. However, if you computed *all* values of $i^{2/3}$ and $(-i)^{2/3}$, you would find in each case that one of the values obtained is –1.

1.3.4 A Program to Give You Fractional Roots and to Do a Check

Here is a simple MATLAB program that asks you to choose a value of z and a positive integer m. The program then gives you all values of $z^{1/m}$. It does this by solving the polynomial equation in $w : w^m - z = 0$ using **roots**.

We have added some additional checks to our program. Notice that if you were to factor $w^m - z$ into its m factors $(w - w_1) (w - w_2) \ldots (w - w_m)$, then the product of the roots $w_1 w_2 \ldots w_m$ is seen to equal $(-1)^{m-1}z$; the product of the m values of $z^{1/m}$ is thus $(-1)^{m-1}z$. Similarly, you should convince yourself that the coefficient of w^{m-1} obtained when you multiply the m factors is $-(w_1 + w_2 + w_3 + \ldots w_m)$. But since w^{m-1} does not appear in $w^m - z$, we conclude that the sum of the m values of $z^{1/m}$ is zero.

Try the following problem with some numbers:

```
z = input('the complex value of z to be considered')
m = input('input the positive integer m')
v = zeros(1,m+1);%creates a row vector length m+1
%elements are zero.
v(1,1) =1;% the first coeff is 1, the coeff of w^m
v(1,m+1) =-z;% the constant term (the last coeff) is -z
%the following solves the equation w^m-z=0 for w=z^(1/m)
the_roots =roots(v)
prod_roots =prod(the_roots)
%the preceding should equal (-1)^(m-1) times z.
check_prod=(-1)^(m-1)*z
check_roots =sum(the_roots)
%the preceding should be zero to a good approx.
```

As an exercise, the reader should modify the above code so that the user is prompted to enter an integer n and a positive integer m where the fraction n/m is irreducible. The program then computes all m values of $z^{n/m}$ where again z must be supplied.

1.3.5 A Further Caveat with Fractional Powers: The Plot Function

Suppose x is a row vector whose elements are real numbers. Let $w(x)$ be a function defined for these values of x. A commonly used line in MATLAB code is **plot**(*x*,*w*), which will produce a plot in the Cartesian plane of *w* versus *x*. For example, if **x = linspace**(0,10,1000) and *w* = sin(*x*), the command **plot** (*x*,*w*) gives you a nice plot of the sine function based on 1000 data points spaced between 0 and 10.

One must be careful, however. If *w* is a complex function of the real variable *x*, then the command plot (*x*,*w*) yields a plot of *only the real part* of *w* versus *x*. This is demonstrated in the following example.

Example 1.8

Using the MATLAB **plot** function, graph $w = x^{1/3}$ for the interval $-2 \le x \le 2$ and explain your result. Use at least 100 data points.

Solution:

```
x = linspace (-2,2,101);
w = x.^(1/3);
plot(x,w);grid
```

which results in Figure 1.2.

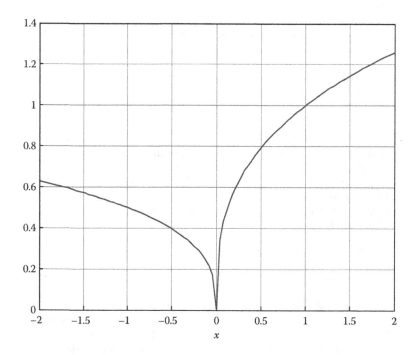

FIGURE 1.2
The plot for Example 1.8.

Notice the lack of symmetry about $x = 0$. In executing the program, MATLAB issues a warning about the imaginary value of the function being ignored in the plotting process.

To explain this result, notice that

$$x^{1/3} = \sqrt[3]{|x|} \angle \left(\frac{1}{3}\theta + \frac{2}{3}2k\pi \right) \; k = 0,1,2$$

MATLAB employs the principal value, taking $k = 0$, and $\theta = 0$ when $x > 0$. Thus, for $x \geq 0$, we have $x^{1/3} = \sqrt[3]{|x|}$. If $x < 0$, we have from the principal value

$$x^{1/3} = \sqrt[3]{|x|} \angle \left(\frac{1}{3}\pi \right)$$

because $\theta = \pi$ on the negative real axis. In Cartesian form, the preceding is

$$x^{1/3} = \sqrt[3]{|x|}(\cos(\pi/3)+i\,\sin(\pi/3)) = \sqrt[3]{|x|}\left(\frac{1}{2}+i\,\frac{\sqrt{3}}{2} \right)$$

In our preceding graph above, we have the curve $\frac{1}{2}\sqrt[3]{|x|}$ for $x < 0$ because only the real part of $x^{1/3}$ is plotted by MATLAB. This is exactly half the value obtained for the corresponding positive values of x and is confirmed if we study the plot.

To properly graph $x^{1/3}$ over the given interval, we must employ both the real and imaginary parts of this function as in the following code:

```
x=linspace (-2,2,101);
w=x.^(1/3);
plot(x,real(w),'-',x,imag(w),'*')
grid
```

This results in Figure 1.3, to which we have added some labels.

Finally, as we are on the subject of plotting, the reader should be reminded that the command **plot(w)**, where w is a row vector having complex elements, yields a graph in which the imaginary part of w is plotted against the real part. In other words, w is plotted as points in the complex plane. Here is a simple example that yields a circle of radius 10 in the complex plane:

```
>> theta = linspace(0,2*pi,100);
w = 10*exp(i*theta);
plot(w);axis equal
```

The reader should verify this. Note that **axis equal** ensures that a circular plot appears. Without it, MATLAB might use different scales for the two axes, and the circle would be distorted into an ellipse.

We could have also just plotted 100 stars, like this *, for the data points and not obtained a continuous graph if we used plot(w,'*') instead of plot(w). Here is the result. (See Figure 1.4.)

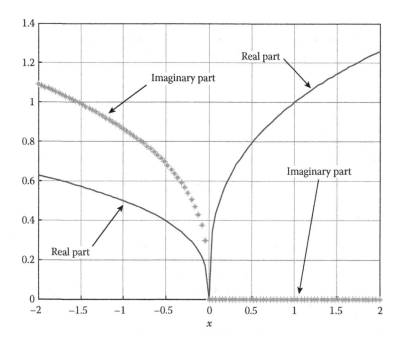

FIGURE 1.3
Plots for $x^{1/3}$.

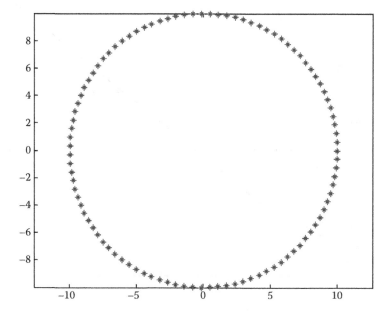

FIGURE 1.4
Values of $10e^{i\theta}$ generated by letting θ go through 100 values between 0 and 2π.

And while we are on the subject of **plot**, here is a further peril involving **plot** when you are using vectors with complex elements together with vectors having real elements. Consider the following:

```
clf
a=[1+i 5+5*i 10+10*i];
b=[1 5 10];
plot(a,'k-','linewidth',2);hold on
text(4,4,'the plot of a')
plot(a,'o')
plot(b,'k-','linewidth',2);
plot(b,'o');
text(2.2,5,'the plot of b');grid
```

We obtain Figure 1.5.

The plot of the vector *a* shows its imaginary elements plotted against their corresponding real parts. They all lie on the line $y = x$, which makes a 45-degree angle with the real axis.

Now the vector *b* has only real elements, which are 1 and 5 and 10. The imaginary part of each element is zero. So, following the logic used by MATLAB in making the plot for *a*, we would expect that these points would occur on the line $y = 0$ at $x = 1$ and 5 and 10. We would expect MATLAB to connect these points with a horizontal line.

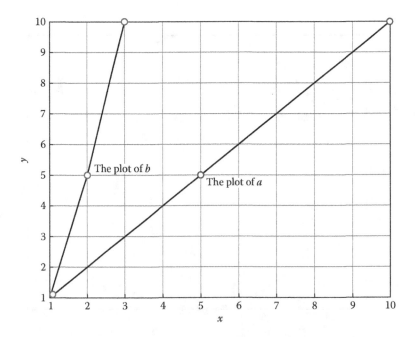

FIGURE 1.5
Perils of using the **plot** function.

Instead, MATLAB, seeing that each element in b is real, plots the value of that element against its *index* in the row vector b. Thus, the first element, one, is plotted against its index, one. The second element has the value 5 and this value is plotted against 2, and finally the third element is 10 and this is plotted at $x = 3$, $y = 10$. The preceding exercise should serve as a warning and an explanation of seemingly strange results.

Exercises

1. Using MATLAB, obtain one numerical value for each of the following by simply asking for a^b.
   a. $(-27i)^{2/3}$
   b. $(1 + i)^{1/15}$
   c. $(-1-i)^{1/9}$

2. Without using MATLAB, but knowing its convention for producing fractional roots, predict whether or not the use of MATLAB will yield these results, and explain your logic:
   a. $(i^{1/2})^2 = i$
   b. $(i^{3/2})^{2/3} = i$
   c. $((-1)^{4/3})^{3/4} = -1$
   d. $((-1)^{3/4})^{4/3} = -1$
   e. $((-1 + i)^{5/3})^{3/5} = -1 + i$
   f. Now confirm your predictions for parts (a) through (e) by using MATLAB.

3. Using **roots**, as discussed, find all values of the following and in each case plot the values using a * to mark their location.
   a. $(1 + i)^{24/5}$
   b. $(3 - 4i)^{5/7}$
   c. $(13 + 7i)^{100/99}$ (be *sure* to use the function **zeros** in this case to create the coefficients for **roots**)

4. Using Equation 1.6, we can sum all the m possible values of $z^{n/m}$. The result is

$$S = \left(\sqrt[m]{r}\right)^n cis\left[\frac{n}{m}\theta\right]\sum_{k=0}^{m-1}cis\left[2k\pi\frac{n}{m}\right]$$

We can show that this summation $S = 0$. In other words, the sum of all possible values of $z^{n/m}$ is zero.

Recall De Moivre's Theorem $(cis\phi)^k = cis(k\phi)$, where k is any integer and ϕ is any number. The result is in all complex variables texts. Thus, we may rewrite the expression for S as

$$S=\left(\sqrt[m]{r}\right)^n cis\left[\frac{n}{m}\theta\right]\sum_{k=0}^{m-1}\left(cis(2\pi n/m)\right)^k$$

We can sum this series using the result given in exercise 3, section 1.1, for the sum of a geometric series that we rewrite here as

$$\sum_{k=0}^{m-1}z^k = \frac{1-z^m}{1-z}\quad z\neq 1$$

Taking $z = cis(2\pi n/m)$ (where n/m is not an integer), we have that

$$\sum_{k=0}^{m-1}\left(cis(2\pi n/m)\right)^k = \frac{1-cis(2\pi nm/m)}{1-cis(2\pi n/m)}=0\quad \text{since } cis(2\pi n) = 1$$

Using MATLAB, find all the values of $(1 + i)^{11/7}$ and sum them together. Note that you will not get exactly zero. This is due to round-off errors in MATLAB. However, the sum should be many orders smaller than the magnitude of any one root.

5. Compare Figures 1 and 2 obtained from this program, explaining differences and similarities:

```
clf; clear
x = linspace (-3,3,301);
w = (x-1).^(2/5);
figure(1)
plot(x,w);grid;
figure(2)
plot(x, real(w),'-',x,imag(w),'*')
grid;
```

6. Modify the program given in the text, which computes $z^{1/m}$ so that when supplied with integers n and a positive m (these numbers have no common factors), it will compute all m values of $z^{n/m}$. Try the problem out by computing $(27i)^{2/3}$ and $(27i)^{-2/3}$.

1.4 Complex Symbolic Algebra

One of the more tedious aspects of the algebra of complex numbers is the manipulation of the symbols in complex expressions, especially the extraction of the real and imaginary parts following a procedure requiring addition, multiplication, and division. Suppose, for example, that a, b, c, d, e, f are all real variables. Consider the equation

$$w = \frac{1}{a + ib} + \frac{c + id}{e + if}$$

if $w = u + iv$, where u and v are real. Our problem is to find u and v in terms of a, b, c, d, e, f.

Problems such as these are solved in MATLAB by using the statements **syms** or **sym**. These two statements are not identical in what they do. The reader might wish to investigate them both in the **Help** window of MATLAB. In general, **syms** is slightly more useful, and we will favor it, but we also use **sym**. Briefly, if a and b are symbols, we write

syms *a*; syms *b*

or more concisely

syms *a b*

We can put any other symbols we might be using in the above statement after a and b. Suppose a is a symbol for a real quantity but b can be complex. Then our code would contain

```
syms a real; syms b
```

However, if both a and b must be real, then we have

```
syms a b real;
```

We could have included other symbols that are real after the a and b.

The following program shows how to obtain the real part of the symbolic expression for u, which is the real part of w given above.

```
syms a b c d e f real
w=1/(a+i*b)+(c+i*d)/(e+i*f);
u=real(w)
  u=expand(u)
  'prettier answer for u'
pretty(u)
```

The output is

u = real((c + d*i)/(e + f*i)) + a/(a^2 + b^2)
u = a/(a^2 + b^2) + (c*e)/(e^2 + f^2) + (d*f)/(e^2 + f^2)

ans = prettier answer for u $\dfrac{a}{a^2 + b^2} + \dfrac{ce}{e^2 + f^2} + \dfrac{df}{e^2 + f^2}$

Notice the following: We did not have to declare w, u, or v as symbols as they are declared equal to a collection of symbolic expressions. We had to use **expand**(u) in our code, otherwise we could not have found an explicit expression that does not involve the **real** operator. The line of code **pretty** is not necessary but does present u as a sum of built-up (as opposed to "shilling") fractions. Had we used **imag** instead of **real**, we would of course have found the imaginary part of our expression u.

Suppose you wished to substitute the numbers 3 and 7 for a and e, respectively, in our expression for u. This is accomplished by adding this line of code to the program just given. You may wish to read up on the MATLAB function **subs** in the **help** window.

```
u = subs(u, [a e], [3 7])
```

This results in the output:

u = (7*c)/(f^2 + 49) + 3/(b^2 + 9) + (d*f)/(f^2 + 49)

where a and e have now been replaced by numbers.

There are some algebraic operations that are sufficiently complicated that MATLAB in its present version will not simplify them or make the real and imaginary parts readily accessible. For example, if $z = x + iy$, then it can be shown (see W section 1.4, Exercises) that

$$z^{1/2} = (x + iy)^{1/2} = \frac{\pm\sqrt{x + \sqrt{x^2 + y^2}}}{\sqrt{2}} \pm i \frac{\sqrt{-x \pm \sqrt{x^2 + y^2}}}{\sqrt{2}}$$

However, we cannot write a program like the one given above that will give this result.

1.4.1 Numbers as Symbols

Sometimes it is useful to have MATLAB treat a number as a symbol and not convert it to a decimal expression. For example, compare the following two MATLAB operations:

```
>> a = sqrt(5)
a = 2.2361
>> a^7
ans = 279.5085
```

```
>> b = sym(sqrt(5))
b = 5^(1/2)
>> b^7
ans = 125*5^(1/2)
```

In the first instance, we raised the square root of five to the seventh power and came up with a decimal answer of around 279.50. In the second instance, we treated the square root of 5 as a symbol. Raising this to the seventh power we get 125 times the square root of 5. Sometimes this is more convenient. If we now want this result as a decimal we add the line

```
>> eval(ans)
```

which again yields

```
ans = 279.5085
```

Now, consider the following where we find $\left(\sqrt{5} + i\sqrt{7}\right)^{11}$:

```
>> (sqrt(5) + i*sqrt(7))^11
ans = −8.5407e + 05 − 1.1650e + 05i
```

We know from the binomial theorem that our final result should be expressible in terms of the square root of 5 and the square root of 7 and integers. But how?

```
>> a = sym(sqrt(5));
>> b = sym(sqrt(7));
>> z = (a + i*b)^11
z = (5^(1/2) + 7^(1/2)*1i)^11
>> expand(z)
ans = −381952*5^(1/2) − 7^(1/2)*44032i
```

This gives the desired result that we can now check against our first answer.

```
>> eval(ans)
ans = −8.5407e + 05 − 1.1650e + 05i
```

which agrees with our first result.

The line of code **who** placed in your program or command window will review for you everything that has been defined as a non-numerical symbol by you or created as a symbol in your program. Thus, using **who** following the above MATLAB code yields the following:

Your variables are

a   ans b   z

For some purposes, it is best to take pi as a symbol. Compare the following:

```
>> sin(1000*pi)
ans = -3.2142e-13
>> sin(1000*sym(pi))
ans = 0
```

The first answer is wrong—it results from round-off error. The second one is correct. By using pi as a symbol, we have MATLAB evaluating $\sin\pi$ and not the sine of a numerical approximation to the irrational number.

Exercises

1. Use MATLAB to extract the real and imaginary parts of the expression

$$w = \frac{1}{x} + \frac{1}{iy} + \frac{i}{\left((x-1)+iy\right)^2}$$

   where x and y are real. Use the pretty command to make your answers as easy to read as possible. Use MATLAB to evaluate your expressions, when $x = 1$ and $y = 1$. Note that for these values the numerical results are easily checked by inspection.

2. Consider the expression

$$z = R + iX = \frac{1}{\dfrac{1}{r_1 + ix_1} + \dfrac{1}{r_2 + ix_2}}$$

   where R, X, r_1, r_2, x_1, x_2 are all real quantities. This notation is common in electrical engineering. Write a MATLAB program that will give explicit expressions for R and X expressed entirely in terms of real quantities. In addition, have your program determine the values of R and X if $r_1 = r_2 = 2$ and $x_1 = x_2 = 4$. Also, consider the case where $x_1 = -x_2 = 4$ and again $r_1 = r_2 = 2$.

3. Using MATLAB, find the real and imaginary parts of

$$\frac{1}{(x+iy)^3}$$

   as explicit functions of x and y, assuming that x and y are real.

4. Let

$$w = u + iv = \left(\frac{\overline{z}}{z} \right)^6$$

where $z = x + iy$. Using MATLAB, obtain explicit real expressions for $u(x, y)$ and $v(x, y)$. Explain why $u^2 + v^2 = 1$, and show using MATLAB that your results satisfy this.

5. Let $w = u + iv = (z + 1/z)^8$, where $z = x + iy$. Using MATLAB, obtain explicit real expressions for $u(x, y)$ and $v(x, y)$. Check your results by substituting, via MATLAB, the values $x = 0, y = 1$ in $u(x, y)$ and $v(x, y)$, and explain how you know that both u and v should be zero.

6. a. Using MATLAB, express the real and imaginary parts of $\left(\sqrt{2} + i\sqrt{3} \right)^{17}$ in terms of integers and $\sqrt{2}$ and $\sqrt{3}$.

   b. Repeat the above but consider

$$\left(\frac{1}{\sqrt{2} - i\sqrt{3}} \right)^{13}$$

2

Loci and Regions in the Complex Plane and Displaying Complex Functions

---

2.1 Meshgrid and Three-Dimensional Plotting

In studying the algebra or calculus of real functions of real variables, you might investigate such equations as $y = e^x$ or $y = \sin x + x^2$ by graphing y as a function of x. The matter of graphing for functions of a complex variable is more complicated.

The following simple example illustrates why. Let $w(z) = z^2$. Since z is complex, we must specify its value with two numbers: the real and imaginary parts. Typically, w assumes complex values, for example, if $z = 1 + 2i$, then $w = (1 + 2i)^2 = -3 + 4i$. Thus, two numerical values are required to specify values of the dependent variable and two for the dependent variable. A four-dimensional space, which we cannot visualize, would be required for a plot.

One approach to our plotting functions of a complex variable $w = f(z) = f(x + iy)$ is to create real functions $u(x,y)$ and $v(x,y)$, the real and imaginary parts of $w(z)$, so that $u(x,y) + iv(x,y) = w = f(z)$. We study the real and imaginary parts of $f(z)$ separately and make three-dimensional plots of $u(x,y)$ and $v(x,y)$. In the case $w = z^2$ mentioned above, we have $u(x,y) = \text{Re}(z^2) = \text{Re}(x^2 - y^2 + i2xy) = x^2 - y^2$ and $v(x,y) = \text{Im}(z^2) = \text{Im}(x^2 - y^2 + i2xy) = 2xy$.

Plots of $u(x,y)$ and $v(x,y)$ for functions of a complex variable are typically performed while x and y are restricted to some region of the complex z plane. We sometimes will also plot $|f(z)| = \sqrt{u^2(x,y) + v^2(x,y)}$ over a region. Recall that a region is a domain in the complex plane perhaps supplemented by some portion of boundary points of the domain. Remember, too, that a domain is an open connected set. The reader may wish to review these terms in a standard text.

A region is composed of an infinite number of points in the x–y plane, and we have no hopes of storing that number of points in our computer in preparation for plotting u and v as functions of x and y. What we do is create a mesh consisting of a rectangular grid of lines in the region of interest. The spacing of the lines is usually but not always uniform. The (x,y) coordinates at the intersections of these lines provide the values for $z = x + iy$ at which we will evaluate the real and imaginary parts of $f(z)$.

To create this mesh, we use the MATLAB® function **meshgrid**. Meshgrid accepts two arguments x and y. These are row vectors whose elements are the x and y coordinates of the points of intersection of the grid that is of interest. For example, suppose you wished to know the values assumed by a function at the points in the x–y plane where the vertical lines $x = 0$, $x = 1$, $x = 2$ intersect the horizontal lines $y = 3, y = 5, y = 6, y = 7$. Note that these lines need not have uniform spacing, and in this case, for the horizontal lines, they do not. One would write in MATLAB

```
>> x=[0 1 2]; y=[3 5 6 7];
```

The output of the function **meshgrid** is then matrices x and y obtained from the following few lines of code:

```
>> x=[0 1 2]; y=[3 5 6 7];
>> [x y]=meshgrid(x,y)
```

$$x = \begin{bmatrix} 0 & 1 & 2 \\ 0 & 1 & 2 \\ 0 & 1 & 2 \\ 0 & 1 & 2 \end{bmatrix}$$

$$y = \begin{bmatrix} 3 & 3 & 3 \\ 5 & 5 & 5 \\ 6 & 6 & 6 \\ 7 & 7 & 7 \end{bmatrix}$$

Note that x is no longer a row vector but a matrix, all of whose rows are identical to one another; these rows are the old row vector $x = [0\ 1\ 2]$. The number of rows is taken to be the number of elements in the old row vector $y = [3\ 5\ 6\ 7]$. One could also say that the number of rows in the new matrix x is equal to the number of rows in the transpose of the old row vector y.

Observe that y is now a matrix, all of whose columns are identical to one another; these columns are the vector that is the transpose of the old row vector y. The number of columns in the new matrix y is taken to be equal to the number of elements in the row vector x; thus, the dimensions of the matrices x and y are identical.

Now consider $z = x + iy$ which in MATLAB is

```
>> format short
>> z=x+i*y
```

$$z = \begin{bmatrix} 0+3.0000i & 1.0000+3.0000i & 2.0000+3.0000i \\ 0+5.0000i & 1.0000+5.0000i & 2.0000+5.0000i \\ 0+6.0000i & 1.0000+6.0000i & 2.0000+6.0000i \\ 0+7.0000i & 1.0000+7.0000i & 2.0000+7.0000i \end{bmatrix}$$

The elements of the preceding matrix consist of the complex number $z = x + iy$ evaluated at the intersections of the grid formed by the lines $x = 0$, $x = 1, x = 2$ with the lines $y = 3, y = 5, y = 6, y = 7$. As we move across the matrix from left to right, we encounter the values of x in the order in which we entered them in our initial row vector, and as we move from top to bottom we encounter the values of y in the same manner.

At this juncture, we are well positioned to evaluate a function $f(z)$ at those values of z contained in the preceding matrix. For example, suppose we want $w = z^2$. We have

```
>> w=z.^2
```

$$w = \begin{bmatrix} -9.0000 & -8.0000+6.0000i & -5.0000+12.0000i \\ -25.0000 & -24.0000+10.0000i & -21.0000+20.0000i \\ -36.0000 & -35.0000+12.0000i & -32.0000+24.0000i \\ -49.0000 & -48.0000+14.0000i & -45.0000+28.0000i \end{bmatrix}$$

Note the necessity for the period (the dot) after the z in the first line. This causes MATLAB to square *each element* in z. Without the period, the matrix z would be squared if that were possible.

We can now make three-dimensional plots of the real and imaginary parts of w as functions of x and y. Calling these $u(x,y)$ and $v(x,y)$, respectively, we can extract their values from the above matrix as follows:

```
>> u=real(w)
```

$$u = \begin{bmatrix} -9 & -8 & -5 \\ -25 & -24 & -21 \\ -36 & -35 & -32 \\ -49 & -48 & -45 \end{bmatrix}$$

```
>> v=imag(w)
```

$$v = \begin{bmatrix} 0 & 6 & 12 \\ 0 & 10 & 20 \\ 0 & 12 & 24 \\ 0 & 14 & 28 \end{bmatrix}$$

MATLAB allows an abundance of choices for three-dimensional plotting. A good place to start is **mesh(x,y,Z)**, where x and y are the matrices describing the rectangular grid, as described above, over which you wish to make the plot in the x–y plane. Z is a real variable that depends on x and y. Do not confuse it with the complex variable z. For example, Z might be the variable v just described. We then have, plotting v,

```
>> mesh(x,y,v);
colormap(gray)
colorbar
xlabel('x');ylabel('y');
title('Imag(z^2)')
```

which results in the gray shaded Figure 2.1.

The appearance of a surface is created by MATLAB's connecting adjacent data points with straight lines. This has been compared to a fishing net with the knots at the data points. We have included on the right a color bar (by including the line of code **colorbar**) that shows how colors (or shades of gray) are assigned to lines used in the plot. Thus, light lines correspond to the most positive values of the function, and the darkest lines correspond to the least positive. You should take a moment to verify that Figure 2.1 is indeed a plot of $\text{Im}(z^2) = 2xy$. For example, where $x = 2$, $y = 7$, the height of the plot is seen to be 28.

Because of my wish to make this an inexpensive book, I have made the preceding plot in shades of gray. This was ensured by using the line of code colormap(gray). Had we not included this line of code, MATLAB would revert to its default of producing a color plot, and the color bar on the right would

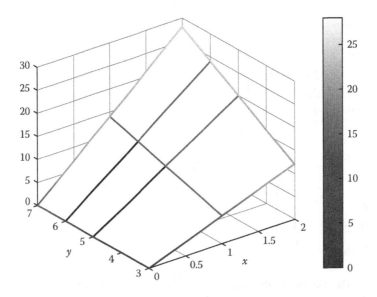

FIGURE 2.1
The imaginary part of z^2.

be in color and would instruct you how to interpret the colors resulting in the "fishnet." You should now try obtaining a color plot on your computer.

The subject of color mapping in MATLAB is rather detailed, and the reader is referred to the topic **colormap** in the help menu.

The plot given in Figure 2.1 is the view of the surface seen by an observer whose azimuth position is at −37.5° and whose elevation is 30°. The meaning of these angles is shown in Figure 2.2, which was downloaded from the MathWorks, Inc., website. These are the default angles that MATLAB uses in providing three-dimensional plots. Sometimes this is not the best angle from which to view a three-dimensional surface. If you attempt a plot of $u = \text{Re}(z^2) = x^2 - y^2$, using the default angles, you will most likely be confused by the picture thus obtained (try this). Instead, after some experimentation, we find that azimuth and elevation of 45 and 60 degrees, respectively, give a better picture. To avoid the default view, enter the line of code **view**(az,el) right after the line of code that generates the plot. You must, of course, assign numerical values to az and el, the desired azimuth and elevation (in *degrees*) to use this code.

This better plot is shown in Figure 2.3, where we also show the code employed. Note that the grid lines used are more numerous than those we employed in generating Figure 2.1. For x we use 10 values uniformly spaced on the interval $0 \le x \le 2$, and for y there are 20 values uniformly spaced on the interval $0 \le y \le 7$. The result is a surface with finer increments in x and y than would have been obtained from the grid formed for Figure 2.1. Note the line of code **view**(45,60) as mentioned.

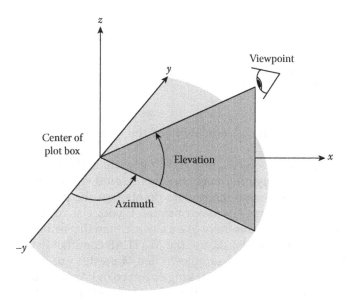

FIGURE 2.2
The default MATLAB viewing angles.

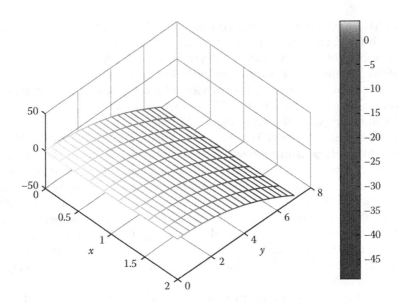

FIGURE 2.3
Real part of z^2.

```
x=linspace(0,2,10);
y=linspace(0,7,20);
[x y]=meshgrid(x,y);
z=x+i*y;
w=z.^2;
u=real(w);
mesh(x,y,u);view(45,60);colormap(gray);colorbar
xlabel('x');ylabel('y')
title ('real part of z^2')
```

Try running the above code without the statement colormap(gray) to obtain a nice plot in color.

If you make the increments in x and y very small, the colored lines forming that plot may merge into a continuous colored surface (the fishnet disappears), one involving so many data points that the plot may print very slowly should you elect to use your printer. If you wish to generate a three-dimensional plot that looks more like a surface than that sometimes given by the function **mesh**, you should use the MATLAB command **surf** in place of mesh. Using **surf**(x,y,u), for example, in lieu of **mesh**(x,y,u) in the preceding code, results in Figure 2.4. (Try making one in color.) Another useful plotting command is **meshz**, which you will explore in problem 1. This places a reference plane under the **mesh** plot of the surface so that you can see how high the surface rises above its lowest point.

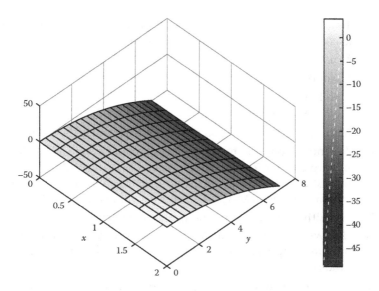

FIGURE 2.4
The real part of z^2.

Exercises

1. Consider the function $f(z) = z + 1/z$. Plot the real and imaginary parts of this function over the region described by $.5 \leq x \leq 2$ and $.5 \leq y \leq 2$. Use 50 uniformly spaced increments in x and y. Use the default values for **view**. Note that you can generate multiple plots from the same program by using such commands as **figure(1)**, **figure(2)**, and so on, placed as code before you call plotting functions such as **mesh**, **meshz, surf**, and so on. Try plots in both color and shades of gray.

   a. Plot each surface using the function **mesh**. Use the default values for the view.

   b. Plot each surface using the MATLAB function **meshz**.

   c. Plot each surface using the MATLAB function **surf**.

   d. Find new values for azimuth and elevation such that you can improve or make more interesting at least one of the above surfaces.

2. The exponential function $f(z) = e^z = e^x \cos y + ie^x \sin y$ is entire (everywhere analytic) and is the most basic transcendental function in complex variable theory. Write a single program that will produce four different plots as described below. The plot should be over the rectangular region $-2 \leq x \leq 2$, $-8 \leq y \leq 8$. Use 40 uniformly spaced data points for x and 80 for y and employ one of the plotting functions

mentioned in problem 1. Work directly with e^z—that is, do not break it into components $u(x,y)$ and $v(x,y)$. Obtain

a. A plot of Re(e^z)

b. A plot of Im(e^z)

c. A plot of $|e^z|$

d. A plot of arg(e^z)

   Suggestion: For part (d) you might not wish to use the default view.

e. We know from any textbook on complex variables that $|e^z| = e^x$ and arg(e^z) = $y + 2k\pi$, k is any integer. Do your results for parts (c) and (d) confirm these facts? Because arg(e^z) is potentially multi-valued, how does MATLAB choose a value for arg(e^z)?

3. We know that the real function of the real variable $f(x) = e^{-x^2}$ is the well-known bell-shaped curve. In the complex plane, $f(z) = e^{-z^2}$ displays more complicated behavior. In the following, make grayscale plots (no color).

a. Using **mesh**, make a three-dimensional plot of $\left|e^{-z^2}\right|$ over the region $-1 \leq x \leq 1$, $-1 \leq y \leq 1$. Use a grid in which there are at least 100 values for x and the same number for y.

b. Similarly, plot the real part of $f(z)$. For variety, use **meshz**.

c. Plot the imaginary part of $f(z)$. For variety, use **surf** for the plot.

d. The function $f(z)$ has the property that $f'(z) = 0$ and $f''(z) \neq 0$ at $z = 0$. A point where the first derivative of an analytic function vanishes but the second does not is called a *saddle point* of the function.[*] In a region containing the saddle point, the three-dimensional plots of the real and imaginary parts of the function should exhibit the shape of a saddle. Is this true of the plots found in parts (b) and (c)? You may wish to adjust the **view** function to verify that saddle shapes have been found.

2.2 Two-Dimensional Plots: The Contour Plot

A useful and popular method for studying the properties of a function of a complex variable $z = x + iy$ is by means of a set of two-dimensional plots called *contour plots*. These are analogous to the plots used by mapmakers to show us the heights of hills via two-dimensional drawings. A contour plot of a mountain might show you a curve indicating where the mountain is 1000 feet high, and another plot drawn on the same paper might show you where it is 3000 feet high.

---

[*] More correctly, this is a first-order saddle point. For more information on saddle points, see any standard text on advanced calculus.

If $f(z) = u(x,y) + iv(x,y)$, we study $f(z)$ by generating a set of curves on which $u(x,y)$ and $v(x,y)$ have prescribed real values. We might also wish to create curves on which $|f(z)|$ assumes some prescribed values. A disadvantage of using the three-dimensional plots described in the previous section is that we must choose our point of view, in the **view** statement, carefully, if we are to appreciate the behavior of the function. An inept choice of the point of view might lead us to overlook some important behavior of the function being studied, or we might obtain a misleading plot. Because contour plots are two dimensional, this problem does not arise.

Contour plots are generated by means of the MATLAB command **contour**. The reader might wish to study this plotting function by typing *help contour* in the MATLAB command window. We will use the contour function with four input arguments even though it can be called with fewer arguments. Thus, we will employ **contour** (x,y,Z,V).

Here x and y are used to describe the rectangular grid over which we seek to evaluate our functions. These are generated from the **meshgrid** command in precisely the same way as was used in such three-dimensional plots as **surf, mesh**, and **meshz** discussed above.

The quantity Z is a real matrix whose elements are specified at the various values of $z = x + iy$ lying at the intersections in the grid. MATLAB assumes that these values in Z are samples taken from a function defined everywhere in a region containing the grid. The symbol V refers to a row vector having real elements. Each element in V specifies a value such that MATLAB will seek to generate, by means of an interpolation process, contour lines on which the presumed continuous function has that value. Thus, if V=[2 4 6], MATLAB will attempt to draw contour lines in which the function studied assumes the values 2 and 4 and 6.

Let us illustrate how to use **contour** with an example. Suppose we wish to see the contour lines on which the imaginary part of the function $f(z) = z^2$ assumes the five integer values between –2 and 2 including the end points (i.e., –2, –1, 0, 1, 2). We will study the function in the region described by $-2.5 \leq x \leq 2.5$, $-2.5 \leq y \leq 2.5$. The following code will do it.

```
x=linspace(-2.5,2.5,100);
y=linspace(-2.5,2.5,100);
n=linspace(-2,2,5);
[x,y]=meshgrid(x,y);
z=x+i*y;
 w=z.^2;
wi=imag(w);
 [c,h]=contour(x,y,wi,n);colormap(gray)
   xlabel('x');ylabel('y');title('Contour Lines for Im(z^2)')
   grid
   clabel(c,h)
```

This generates Figure 2.5.

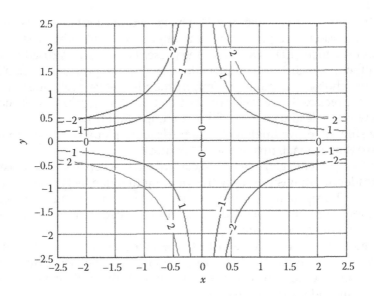

FIGURE 2.5
Contour lines for Im(z²).

The line of code `[c,h]=contour(x,y,wi,n)` creates a matrix `[c]` and a scalar h containing information about the shape of the contours and the numerical values assumed on each contour, in this case, by the function `wi = Im(z²)`. The command `clabel(c,h)` instructs MATLAB to label each contour with the (constant) numerical value that `wi` is assuming on that contour. The numbers in the gap on each contour in Figure 2.5 tell you the value assumed by Im(z²) on that contour. If you delete the line of code colormap(gray), you obtain a set of curves of varying colors. You also want to include the line of code **colorbar** to help you assign a numerical value to those colors.

In general, you are better reading off the numbers placed in the gaps in the curves.

If instead of the line of code `[c,h]=contour(x,y,wi,n);colorbar;` we used instead simply `contour(x,y,wi,n);colorbar;` without creating c and h, and if we deleted `clabel(c,h)`, the numerical values would not appear on the contours. This would simplify the picture, and we could estimate the values assumed by our function on each contour by the color coding.

Recall that $z^2 = x^2 - y^2 + i2xy$ so that Re$(z^2) = x^2 - y^2$ and Im$(z^2) = 2xy$. Thus, the contours on which Im$(z^2) = 2xy$ assumes the values ±2, ±1, 0 are the sets of hyperbolas $2xy = \pm 2$, $2xy = \pm 1$, and $2xy = 0$, which are readily identified in Figure 2.5. The case $2xy = 0$ simply reduces to the x and y axes, as Figure 2.5 again shows.

Suppose we wish to superimpose on Figure 2.5 contours on which $\mathrm{Re}(z^2) = x^2 - y^2$ is constant. We add to the preceding code the lines

```
hold on
wr=real(w);
[d,h] = contour(x,y,wr,n);colormap(gray)
clabel(d,h)
```

which results in Figure 2.6.

By comparing Figure 2.6 with Figure 2.5, we can now see the contour plots on which $\mathrm{Re}(z^2)$ assumes constant values. Notice that these curves have *right angle* intersections with those on which $\mathrm{Im}(z^2)$ assumes constant values. This is a well-known property of an analytic function $f(z)$: the orthogonality, at their intersections, of the curves on which the real and imaginary parts of $f(z)$ assume fixed values.[*] Recall that this orthogonal property need not hold at a point where $f'(z) = 0$, and indeed, the intersections are not orthogonal at the origin in the above. Here $f'(z) = 2z$ is indeed zero.

Suppose you want to make contour plots in order to study some function $u(x,y)$, but you have no idea of what constant values to choose for each contour. You can let MATLAB make the choices for you. In the command **contour** (x,y,Z,V), do not enter a vector for V but instead enter a positive integer (e.g., m), whose value is specified. MATLAB will then provide you with plots

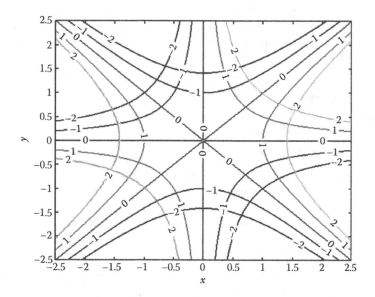

FIGURE 2.6
Contour lines for real and imag (z^2).

---

[*] See, for example, W section 2.5 and S Complex Variables, section 3.12.

on which $u(x,y)$ assumes m different values of its own choosing. The value assumed on each contour by $u(x,y)$ chosen by MATLAB can be read off each plot provided you are using the command `clabel(d,h)`, as described above. Note that it is possible that you might apparently see more than m plots: for example, Im$(z^2) = 2xy = 2$ is satisfied by a curve in the third quadrant and one in the first. Both are branches of the same hyperbola.

Suppose you want the contour on which a function assumes just one numerical value, let us call it v_1. Then in the expression **contour** (x,y,Z,V), enter for V the vector $[v_1 \, v_1]$. The matter is explored in problem 2.

Exercises

1. For the region $-3 \le x \le 3, -3 \le y \le 3$, use MATLAB to plot the contours on which

   a. Re(e^z) has the values 0 and ± 1.

   b. Repeat part (a) but use the imaginary part.

   c. Plot the results of parts (a) and (b) on the same set of coordinates, and verify that the intersections of the curves appear orthogonal.

   d. Repeat part (a) but have MATLAB automatically choose five values on which Re(e^z) is constant. Compare the plot to the one obtained in part (a). Do you see any disadvantages to having MATLAB choose the values?

2. a. For the region $-3 \le x \le 3, -3 \le y \le 3$ use MATLAB to plot the contours on which $\left| e^{z^2} \right|$ has the integer values 2 through 5.

   b. Explain why each of these contours is hyperbolic in shape, and find the equations of these hyperbolas.

   c. Use MATLAB to plot just the contour on which $\left| e^{z^2} \right|$ is 1, and explain why this has degenerated into something that does not look like a hyperbola.

3. a. For the region $-2.5 \le x \le 2.5, -2.5 \le y \le 2.5$, use MATLAB to plot the contours on which $|\cos z - \cosh z|$ assumes the integer values 0 through 4.

   b. You should observe that there is no contour on which $|\cos z - \cosh z| = 0$, although this equation is satisfied at $z = 0$. Explain how one could predict in advance the absence of such a contour.

      Hint: Recall that the zeros of a nonconstant analytic function are isolated (i.e., each zero has a deleted neighborhood throughout which the function is nowhere zero). See, for example, W section 5.7.

4. a. For the region $-2.5 \leq x \leq 2.5$, $-2.5 \leq y \leq 2.5$, use MATLAB to plot the contours on which Re(cos z – cosh z) assumes the integer values 0 through 4. On the same set of axes, plot the contours on which Im(cos z – cosh z) assumes these values.

   b. Observe that the two sets of contours obtained in part (a) intersect at right angles except at one point. What point is this? Explain this in terms of the vanishing of the first derivative of cos z – cosh z at the point in question. See W section 2.5 and S section 3.12 for some help.

2.3 Displaying Regions in the Complex Plane

An inequality such as $|z - i| < 1$ describes a domain in the complex plane. It is the set of points lying inside a disc having radius 1. The center of the disc is at $z = i$. We could, if we wished, make a sketch of this set of points and indicate the bounding circle $|z - i| = 1$ is not to be included in the set; sometimes this is done with a fuzzy or broken line. It is clear in such an elementary problem that MATLAB is of no particular advantage in sketching out the set.

Suppose, however, that matters are more complicated. The reader might first wish to review the meaning of the terms *union* and *intersection* in a mathematics dictionary. For example, imagine we want to sketch the set that is the intersection of the set of points satisfying the set of inequalities $|z - 2| \leq 3$, $|z - 1-i| \leq 2$, Re(z) $\leq$ Im(z). This can be done with MATLAB and is assigned as problem 3. Let us try for starters a problem whose solution is known so that we can confirm that our code works. We look for the intersection of the sets $|z - i| \leq 2$ and $|z + i| \leq 2$.

Although not necessary, it is helpful to convince ourselves that all points satisfying either of these inequalities must be no more than three units from the origin in the z plane. Notice that $|z - i| \geq ||z|-1|$. Thus, if $|z| > 3$, then $|z - i| \leq 2$ cannot be satisfied, and the intersection of the sets $|z - i| \leq 2$ and $|z + i| \leq 2$ cannot exist if $|z| > 3$. Therefore, to study the region that is the intersection of these two sets, with some space left over, we will use **meshgrid** to create a tightly spaced grid satisfying $-4 \leq x \leq 4$ and $-4 \leq y \leq 4$. Note that if you are unsure of how large a region of the complex plane to scrutinize when studying a mathematically prescribed set, you can if you wish simply make an educated guess, erring on the side of generosity. You can later shrink the area of interest to improve your plot. For the present problem, we use the following code:

```
x=linspace(-4,4,1000);
y=linspace(-4,4,1000);
[x,y]=meshgrid(x,y);
z=x+i*y;
```

```
w=abs(z-i);
v=abs(z+i);
ww=(w<=2);
vv=(v<=2);
tt=ww&vv;
meshz(x,y,-tt);colormap(gray)
view(2);
hold on
x=linspace(-4,4,9);
y=x;
[x y]=meshgrid(x,y);
plot(x,y,'r');plot(y,x,'r');axis equal
title('The Intersection of the Sets |z-i|{\leq}2 and |z+i|
  {\leq}2')
xlabel('x');ylabel('y')
```

In the first through third lines, we form a very tightly spaced mesh in the region $-4 \leq x \leq 4$ and $-4 \leq y \leq 4$, where $z = x + iy$ is evaluated at the $1000^2 =$ one million intersections of the mesh. The variable $w = |z - i|$ is found at each of these values of z. Similarly, we find $v = |z + i|$ at these same values. The line of code ww=(w<=2) supplies us with a matrix ww whose elements are equal to 1, where the inequality $|z - i| \leq 2$ is satisfied in the complex plane. Where the inequality is not satisfied, the matrix has the matrix element 0. Similarly, vv=(v<=2) gives us a matrix vv having elements equal to 1 where the inequality $|z + i| \leq 2$ is satisfied, and if the inequality is not satisfied, we get elements of 0. The logical statement tt=ww&vv yields a matrix having ones only where *both* ww and vv have ones. Otherwise, the matrix element is 0. Thus, tt has ones where the inequalities $|z - i| \leq 2$ and $|z + i| \leq$ 2 are both satisfied—that is, at the intersection of the sets specified by these two inequalities. The code meshz(x,y,-tt);colormap(gray) gives a three-dimensional plot of this matrix over the complex z plane. The plot is best viewed from directly above the plane, with a downward view. The MATLAB statement **view(2)** accomplishes this. The surface generated by these closely spaced points through the use of meshz(x,y,-tt) is either 1 unit *below* the x–y plane (where the two inequalities are satisfied) or directly on this plane. The use of the *minus* sign in $-tt$ in the **meshz** argument together with the grayscale **colormap** place a black mark in the complex plane where both $|z - i| \leq 2$ and $|z + i| \leq 2$ are satisfied. Thus, as shown in Figure 2.7, we have a plot of the intersection of the two given sets.

Had we used the code line meshz(x,y,tt); (without the minus sign), we would have generated an image in which the region, which is the union of $|z - i| \leq 2$ and $|z + i| \leq 2$, would have been displayed in white while the remainder of the complex plane would be black; this is contrary to common usage for displaying regions in complex variable texts.

We have not generated the grid of lines that you see in the figure by using the usual command of **grid** following the plot. This command would create a grid based on the data points—one with lines too closely spaced

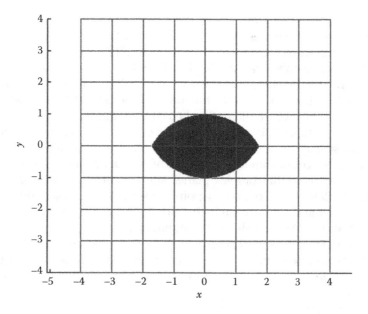

FIGURE 2.7
The intersection of the sets $|z-i| \leq 2$ and $|z+i| \leq 2$.

to be discernible. Lines 13 through 16 of the code, which begins with x=linspace(-4,4,9), instead creates a useful grid of red lines which are uniformly one unit apart. You can adjust this spacing as you see fit. This book is printed without colors, so in the above figures the grid is in black. To generate a black grid on your computer screen, change the "r" to "k" in the two places it appears near the bottom of the program.

Note that the preceding MATLAB code generates a figure based on 10^6 data points, and if you attempt to print this figure you might swamp the buffer in your printer. The easiest method of printing the MATLAB-generated figure, without the risk of overwhelming your printer's buffer with all the data points, is to first display the figure on the computer screen. If you are using the Windows operating system, you should then strike the Print Screen key (while holding down the FN key). This will place a copy of the figure in the clipboard of your computer. You may transfer this image into either a word processing file (like Word) or an image processing file (like Photoshop) from which it is readily printed.

If you are using the Apple operating system, the corresponding technique involves pressing these keys at the same time: Command Shift 3. This places the image on your desktop from which it can be moved into a Word or image processor as described above.

By changing the line of code tt=ww&vv; to tt=ww|vv; we will have replaced an *and* operation with an *or* operation. Now the logical statement tt=ww|vv; yields a matrix tt having ones only where *ww* and *vv* have ones.

The resulting program now yields a plot showing the *union* of the sets $|z - i|$ ≤ 2 and $|z + i|$ ≤ 2. The result is shown in Figure 2.8.

The method presented here has its limitations. If the region being described is unbounded, then we cannot display the entire region as some uniformly colored area of the complex plane; we can only hope to show some portion of that region. For example, the union of the sets $|z| \leq 2$ and $\text{Re}(z) > 0$ is unbounded, a fact we should realize before attempting plots like those in the preceding two figures. We would thus see only a portion of the sought-after set on the screen. Its unboundedness is something we would have to see in advance, or we could convince ourselves of this fact by trying grids covering increasing areas in the complex plane. Another difficulty with the method presented relates to the boundary points of sets. The set described by $|z| \leq 3$ is a closed set and contains all its boundary points, but the set $|z| < 3$ is an open set and contains none of its boundary points. However, were we to apply the MATLAB method described above to displaying both of these sets, we would generate identical figures. So, one must be careful about boundary points, and one should find them by studying the inequalities that one has been given.

We are not limited to codes that find the union or intersection of *two* sets. In fact, the number of sets involved is practically unlimited. Suppose we want to find the intersection of these three sets: $|z - 1| \leq 1$, $|z - i| \leq 1$, and $|z + i| \leq 1$.

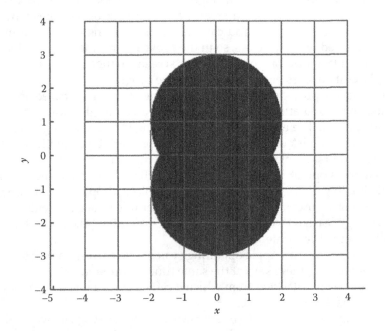

FIGURE 2.8
The union of the sets $|z{-}i| \leq 2$ and $|z{+}i| \leq 2$.

Having generated a matrix z of complex numbers like that used to generate Figure 2.7, we then write code as follows:

```
w=abs(z-1);
v=abs(z-i);
s=abs(z+i);
ww=(w<=1);
vv=(v<=1);
ss=(s<=1);
tt=ww&vv&ss;
```

The remainder of the code would be that used to generate Figure 2.7, beginning with the line meshz(X,Y,-tt);colormap(gray);. Of course, the title of the plot would be different. A similar technique can be used to find the union of these sets.

2.3.1 A Note on pcolor

It may have occurred to the reader that the command **meshz** has been engineered to produce three-dimensional plots and that we are interested in this section in two-dimensional plots in the x–y plane. To ensure this two-dimensional view, we must include the line of code **view(2)** to create a plot in the three-dimensional x–y–z space as seen from directly above as we gaze down on the x–y plane. There is a MATLAB command **pcolor** that will generate the plot in the x–y plane without the additional command of where to view it. The reader might wish to read up on this command in the **help** feature of MATLAB.

Basically, we use **pcolor** (*x,y,m*) where x and y are vectors used to define the elements of a matrix locating the x and y coordinates of points in the Cartesian plane (much as we placed it in meshgrid), and m is a function defined at these points. Thus, by using **pcolor**, we eliminate the line of code **view(2)**. Note that the **colormap** command is used with **pcolor** as it was with **meshz**.

Although we can drop the line **view(2)**, there is an additional line of code that we must often include. When we invoke **pcolor** it creates a checkerboard plot in the x–y plane; the color of each square (or cell) is determined by the numerical value of the function being plotted. But **pcolor** places a narrow border around each square. If there are a great many squares being used (they are densely populated), the borders can obliterate the colors in the squares. To avoid having a border around the squares, one must command them not to exist.

Here is how we rewrite the code at the start of this section to use **pcolor**:

```
x=linspace(-4,4,1000);
y=linspace(-4,4,1000);
[x,y]=meshgrid(x,y);
z=x+i*y;
w=abs(z-i);
v=abs(z+i);
ww=(w<=2);
```

```
vv=(v<=2);
tt=ww&vv;
h=pcolor(x,y,-tt);
set(h,'EdgeColor','none');colormap(gray);grid;axis equal
set(gca,'layer','top'); %required if you hope to see the grid
%sets the grid over the plot
title('The Intersection of the Sets |z-i|{\leq}2 and |z+i|
  {\leq}2')
xlabel('x');ylabel('y')
```

Note that set(h,'EdgeColor','none'); assures that the cells used in the plot will not have outlines. In the section of this book on fractals, The Coda, (i.e., Chapter 7) we favor **pcolor** because we do only plots in the Cartesian plane.

Exercises

1. Write MATLAB code that will display the union of the sets $|z - 1| \leq 1$ and $|z - i| \leq 1/2$.

2. Write MATLAB code that will display the union of the sets $|z - 1| \leq 1$, $|z - i| \leq 1$, $|z + i| \leq 1$.

3. Write MATLAB code that will display the intersection of these sets: $|z - 2| \leq 3$, $|z-1-i| \leq 2$, $\text{Re}(z) \leq \text{Im}(z)$.

4. Write MATLAB code that will display the intersection of these sets: $|\sin z^2| \leq 1$, $|z| \leq 1$.

5. Using the complex plane and the random number generator in MATLAB, you can find π, approximately. The method to be described here is an example of a Monte-Carlo method, and the reader is encouraged to read the history of this subject in the Wikipedia.

   Suppose you draw a square and a circle on a piece of paper. The circle is of a unit radius and the square has sides of length 2, and the circle is inscribed inside the square, which means that it will be tangent to the sides. Suppose you drop some tiny seeds on the square, at random, in such a way that there is uniform probability of a seed landing at any point on the square. Thus, the chance of a seed landing inside the circle is $\pi/4$. This is just the ratio of the area of the circle to the area of the square. We can use this fact to approximately compute π with MATLAB. Place the center of the circle at $z = 1 + i$ in the complex plane. The sides of the square are given by the four lines $x = 0, x = 2, y = 0, y = 2$.

   a. Write a MATLAB code that randomly generates the x coordinates of 1000 points and the y coordinates of 1000 points. Each of these coordinates must be positive and be less than 2.

Hint: Look up the MATLAB documentation for **rand**. Try `x=2*rand(1,1000); y=2*rand(1,1000);` and form the 1000 complex numbers for $z = x + iy$. Plot these as * in the complex plane and show, also, the unit circle centered at $1 + i$.

b. If a point satisfies $|z - 1 - i| < 1$, it will be inside the circle. Using MATLAB, write a line of code that will determine how many points that you have generated satisfy this condition. Use this result to find an approximation to π, and compare this with pi stored by MATLAB.

Hint: The row vector a given by `a=abs(z-1-i)<1;` will produce entries of 1 for any z satisfying the inequality and entries of 0 (zero) where the inequality is not satisfied.

c. Try running the above code repeatedly and notice that your answers differ. Does your estimate of π improve if you average these results? Change your code so as to use 10,000 points. Are your results improved?

2.4 Three-Dimensional and Contour Plots for Functions with Singularities

Most, but not all, of the interesting functions in complex variable theory are ones that are analytic, except possibly at some singular points. Some functions possess isolated singular points; at an isolated singular point the function is analytic throughout a deleted neighborhood (a punctured disc) centered at that singular point. The function is not analytic at the center of the disc. A function with pole singularities exhibits isolated singular points, for example, $f(z) = \dfrac{1}{(z^2 + 9)(z - 1)^2}$ has poles at $z = \pm 3i$ and 1. Here we have a finite number of poles. On the other hand, the function $f(z) = \dfrac{1}{\sin \pi z}$ has an infinite number of poles—there is one at each integer. Another example of an isolated singular point is that of $f(z) = e^{1/z}$ at $z = 0$. Such a function is said to have an isolated essential singularity.* A Laurent expansion of this function about $z = 0$ will be found to have an infinite number of terms with z raised to negative powers.

There are analytic functions whose singularities are not isolated. Then, every neighborhood of a singularity will contain one or more other singularities.

---

* There is also such a thing as a function with a nonisolated essential singular point, for example, $1/\sin(\pi/z)$. Observe that every neighborhood of $z = 0$ will contain another singular point besides $z = 0$ at the origin. You cannot find a neighborhood of $z = 0$ containing just one singular point. We will not treat such functions here.

These singularities most often occur on branch cuts; the branch cut is a means of creating a single-valued branch of what would otherwise be a multivalued function.* An example is $f(z) = z^{1/2}$, a branch of which we might define by means of a branch cut lying along the line $x = 0, y \leq 0$ together with the fact that we choose $f(1) = -1$. Every point on the line $x = 0, y < 0$ is a singular point of $f(z)$, because $f(z)$ changes discontinuously as we cross such a point. Moreover, $f(z)$ fails to be analytic at $z = 0$, which makes $z = 0$ a singular point as well.†

Finally, there are isolated singular points that can be removed through a proper definition of the function at such a point. These singularities are known as *removable*. The most common example is the function $f(z) = \dfrac{\sin z}{z}$. This function is not defined at the origin as you are here dividing 0 by 0. If one asks MATLAB to evaluate such a function at the origin, an error message is generated. However, an application of L'Hôpital's rule shows that the limit of this function as $z \to 0$ is 1. If we define $f(z) = \dfrac{\sin z}{z}$ $z \neq 0$ and $f(0) = 1$, we have a new function that agrees with the old except at $z = 0$. Moreover, the new function is the sum of the convergent power series $1 - \dfrac{z^2}{3!} + \dfrac{z^4}{5!} + \dots$ and is therefore analytic wherever this series converges—in this case, the whole complex plane including the origin. We see that by properly defining a function at an isolated singular point, we can remove its singularity at that point. The original function is said to have a *removable singularity* at this isolated singular point.

2.4.1 Plots When There Are Isolated Singular Points

2.4.1.1 Removable Singular Points

Of the various kinds of singular points described above, the removable singular point presents the least problem when we are making plots, and we treat this case first. After identifying mathematically the location of these points, we take pains in constructing our plots to simply avoid these points in our evaluations. As the function in question is continuous, we can evaluate the function as closely as we wish to the singular point as long as we do not seek to evaluate the function precisely at the singular point. Alternatively, we might simply redefine the function at the singular point so as to remove the singularity and let MATLAB work with the redefined function. The latter course is not usually necessary. Here is one way to obtain a three-dimensional plot of the imaginary part of $f(z) = \dfrac{\sin z}{z}$:

---

* Strictly speaking, a multivalued function is a contradiction in terms because a function has but one value corresponding to each value of its independent value.

† Note that we might have taken $f(1) = 1$, which would just mean we are using a different branch of $z^{1/2}$.

```
x=linspace(-5,5,101);
y=linspace(-1,1,51);
[x,y]=meshgrid(x,y);
z=x+i*y;
w=sin(z+eps)./(z+eps);
wi=imag(w);
   meshz(x,y,wi);colormap gray
   grid on
   title('The Imaginary Part of sin(z)/z')
 xlabel('x');ylabel('y')
```

The result is presented in Figure 2.9.

Notice that the choices x=linspace(-5,5,101) and y=linspace(-1,1,51) mean that one element in the row vector for x is at 0, and one element in the row vector for y is at 0. This causes the complex variable z in our calculation to assume, among other values, the value 0. The function $w = \dfrac{\sin z}{z}$ is undefined at $z = 0$. Had we used the line of code w=sin(z)./z and run the program, MATLAB would have generated a warning message that we have divided by zero, although it is likely that the program will still execute. To sidestep this problem, we use instead the code w=sin(z+eps)./(z+eps);. The difference between the functions $\sin z/z$ and $\sin(z + eps)/(z + eps)$ is numerically insignificant, and certainly no difference could be perceived in the graphs of the two functions.

There are other ways to avoid the division by zero. We could take w=sin(z)./z in the above code but replace z=x+i*y; with z=x+i*y+eps; In this way, z is never zero in our calculations, although the values used

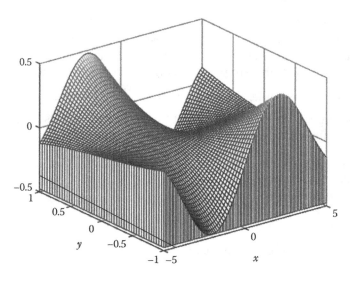

FIGURE 2.9

The imaginary part of $\dfrac{\sin(z)}{z}$.

for z would be effectively indistinguishable from those used in the above program as they differ only by the number eps, which is of the order of 10^{-16}.

A yet more sophisticated ploy is to again take w=sin(z)./z. Returning to the program that generates Figure 2.9 given above, after z=x+i*y, we place the additional line z=z+(z==0)*eps. The logical operator (z==0) is zero *except* if z is zero. The operator then assumes the value 1. This means that should the value of z on the right assume the value zero, it will be replaced by the number *eps*. Other values of z are unaffected.

Finally, the least sophisticated way to deal with the removable singularity is to replace the line of code x=linspace(-5,5,101);with x=linspace(-5,5,100);. In this way, there is no element in the row vector for x that is equal to zero, which means that no value of z that is zero is in the preceding program.* Division by zero is avoided. The drawback here is that we do not evaluate our function as close to the removable singular point as we do in the preceding methods.

2.4.1.2 Pole Singularities

Recall that if a function $f(z)$ has a pole of order N at the point z_0, then this function has a Laurent expansion valid in a deleted neighborhood of that point, and the following must be true: $\lim_{z \to z_0} (z - z_0)^N f(z) = c_{-N}$, where c_{-N} is neither zero nor infinity. The existence of the Laurent expansion means that in a neighborhood of z_0, $f(z)$ has the representation $f(z) = \frac{1}{(z - z_0)^N} \phi(z)$, where $\phi(z)$ is analytic and nonzero at z_0, and $\phi(z_0) = c_{-N}$. Thus, near z_0, the function $f(z)$ varies as, and "blows up" as, $\frac{c_{-N}}{(z - z_0)^N}$. If we are to plot the real or imaginary parts of $f(z)$ or $|f(z)|$, we must not include z_0 as a data point lest we divide by zero. However, the situation is a bit more complicated than that; we may run into trouble simply by trying to plot $f(z)$ "close" to z_0.

Consider, for example, the function $f(z) = \frac{\cos z}{z}$, which has a simple pole at the origin. Suppose we wish to plot the magnitude of this function in the complex plane near the pole. We use the following code:

```
x=linspace(-1,1,200);
y=linspace(-1,1,200);
[x,y]=meshgrid(x,y);
z=x+i*y;
w=cos(z)./(z);
wm=abs(w);
  meshz(x,y,wm);colormap gray
  grid on
```

---

* Recall that in the statement x=linspace(a,b,n) one element in the row vector for x is exactly halfway between a and b if n is odd.

```
title('The Magnitude of cos(z)/z')
xlabel('x');ylabel('y')
```

which results in Figure 2.10.

This result is not very useful if we wish to study $|f(z)|$ near the origin. The problem is that MATLAB finds the largest value of $|f(z)|$ generated by the code and creates a scale such that this value is the highest shown in the plot. In the present problem, this maximum occurs where $|z|$ is smallest, at approximately $0.00502 + i\ 0.00502$, and here $|f(z)|$ is approximately 100. To accommodate this value, the graph will not show much detailed behavior of the function except very close to the pole.

If we used x=linspace(-1,1,50);y=linspace(-1,1,50); in place of x=linspace(-1,1,200);y=linspace(-1,1,200); we would not be evaluating $|f(z)|$ so close to the pole, and we would obtain the plot shown in Figure 2.11.

Figure 2.11 still makes the presence of the pole apparent while showing more about the behavior of the function near $z = 0$.

The order of a pole of a function tells you how rapidly the magnitude of the function rises to infinity as the pole is approached. Thus, a function will "blow up" more rapidly near a pole of order 2 than near a pole of order 1. Let us investigate the behavior of $f(z) = \dfrac{1}{z(z-2)^2}$ near the two poles. The following code will generate a plot of $|f(z)|$ over a region:

```
x=linspace(-1,4,150);
y=linspace(-1,1,150);
[x,y]=meshgrid(x,y);
z=x+i*y;
f=z.*(z-2).^2;
```

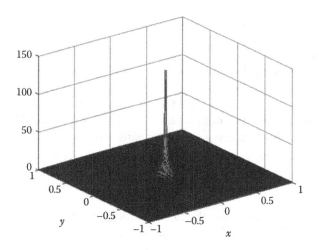

FIGURE 2.10

The magnitude of $\dfrac{\cos(z)}{z}$.

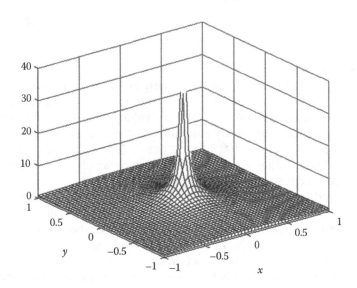

FIGURE 2.11

A better plot of the magnitude of $\dfrac{\cos(z)}{z}$.

```
f=1./f;
f=abs(f);meshz(x,y,f);
axis([-.5 2.5 -1 1 0 30]);colormap gray
title('The Magnitude of 1/(z(z-2).^2)')
xlabel('x');ylabel('y')
view(10,15)
```

The resulting plot is as shown in Figure 2.12.

The values chosen for the elements of x and y in the code ensure that we never seek to evaluate $f(z)$ at the poles $z = 0$ and $z = 2$. There is no division by zero.

Note especially the line of code: `axis([-.5 2.5 -1 1 0 30])`.

This establishes the limits for the three axes. Thus, we consider only values $-.5 \le x \le 2.5$, $-1 \le y \le 1$, and $0 \le z \le 30$. The last of these is especially impor-tant. Note that we have used a bold **z** to distinguish this axis from the com-plex variable z. If we had not placed a bound on **z**, the MATLAB program would have established the **z** scale by using the very large value assumed by $\left|\dfrac{1}{z(z-2)^2}\right|$ near the pole of second order at $z = 2$. This value is sufficiently large that the remainder of the plot, away from the pole, would look like a flat surface.

We have elected not to use the default values for the **view** statement built into MATLAB. This would produce an azimuth of −37.5 and an elevation of 30 degrees. The values that we have selected of 10 and 15 degrees are some-what more satisfactory. The reader might wish to experiment with different values after studying the view command in the MATLAB help folder.

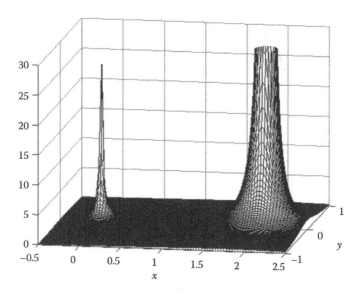

FIGURE 2.12
The magnitude of $\dfrac{1}{z(z-2)^2}$.

The proper choice of the arguments in **view** may be critical when we study functions with pole singularities. If a function has more than one pole singularity, it is possible that we might not see all of the poles when we make a three-dimensional plot of the magnitude of this function if our perspective in viewing the plot is poorly chosen. For example, if the line of code in the preceding program is changed from view(10,15) to view(90,15), the resulting plot will show only the second-order pole at $z = 2$ because the simple pole at the origin will be obscured. To avoid these situations, we can make contour plots of the magnitude of the function under study.

If a function $f(z)$ has a pole of order N at z_0, we have in a neighborhood of z_0 that $|f(z)| = \left|\dfrac{\phi(z)}{(z-z_0)^N}\right| \approx \left|\dfrac{C_{-N}}{(z-z_0)^N}\right|$. The approximation becomes increasingly valid as z approaches z_0. Here $\phi(z)$ is a function that is analytic in a neighborhood of z_0, while $C_{-N} = \phi(z_0) \neq 0$.

Thus, near a pole, approximately, the magnitude of $f(z)$ is constant on contours on which $|z - z_0|$ is constant (i.e., on circles centered at z_0). The magnitude grows larger as we consider circles of diminishing radius. Contour plots of the magnitude of a function near a pole are thus circles concentric with the pole. At some point, sufficiently far from the pole, these plots may cease to be circular, because we are beyond the neighborhood in which $|f(z)| \approx \left|\dfrac{C_{-N}}{(z-z_0)^N}\right|$ is a fair approximation. The preceding discussion should be illuminated by the following example where we plot the contour

lines on which the magnitude of $f(z) = \dfrac{1}{z(z-2)^2}$ assumes certain prescribed
values. Note this was the function just studied with a three-dimensional
plot of its magnitude.

```
clf;clear
x=linspace(-1,3,100);
y=linspace(-1,1,100);
 n=[.5 1 2];
[x,y]=meshgrid(x,y);
z=x+i*y;
w=1./(z.*(z-2).^2);
wm=abs(w);
   [c,h]=contour(x,y,wm,n);
axis equal
 grid
    hold on
    clabel(c,h);
    title('Contour Lines Showing |1/(z(z-2)^2)|')
```

We have plotted the contour lines on which $\left| f(z) \right| = \left| \dfrac{1}{z(z-2)^2} \right|$ assumes the
values 0.5, 1, and 2. There are lines that are nearly circular around the poles at
$z = 0$ and $z = 2$. On these, $\left| f(z) \right|$ is 1 and 2, with, as you would expect, the higher
value resulting in contours that are closer to the poles. The value $\left| f(z) \right| = .5$
does not occur in close proximity to the poles and in fact results in a contour enclosing both poles. Figure 2.13 should be compared with Figure 2.12,
which should make clear why the contours in Figure 2.13 for values 1 and 2
lie closer to the pole at $z = 0$ than they do for the pole at $z = 2$.

Note the line of code **axis equal**. This ensures that equal units of measure-
ment along the x and y axes are represented by equal intervals on the figure
generated by MATLAB.

In this way, a contour will appear as a circle on your computer screen
where appropriate.

2.4.1.3 Essential Singularities

Recall that if a function has an isolated essential singularity at z_0, then the
function is analytic in a deleted neighborhood of z_0 and has in this deleted
neighborhood a Laurent series representation of the form $f(z) = \displaystyle\sum_{n=-\infty}^{\infty} c_n(z-z_0)^n$.

The portion of the series containing the negative exponents $\displaystyle\sum_{n=-\infty}^{-1} c_n(z-z_0)^n$ is
known as the *principal part* and has an infinite number of terms; in other
words, we cannot identify any exponent as being the most negative one in
the Laurent series.

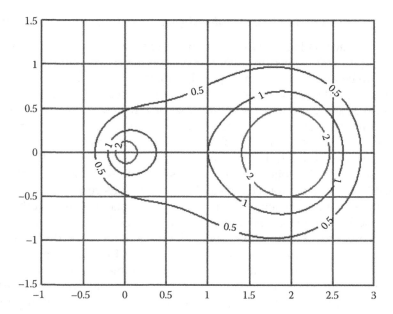

FIGURE 2.13

Contour lines showing $\left|\dfrac{1}{(z(z-2)^2)}\right|$.

The behavior of a function near an essential singularity is remarkably strange. Picard's theorem, which is proved in advanced courses in complex variable theory, asserts that in *every* neighborhood of an essential singular point, a function assumes every possible (finite) complex value, with possibly one exception, an *infinite number* of times.[*]

To appreciate how strange such a singularity is, consider the function $f(z) = e^{1/z} = \ldots \dfrac{1}{z^3 3!} + \dfrac{1}{z^2 2!} + \dfrac{1}{z} + 1$, $z \neq 0$, which has an essential singularity at $z = 0$. Suppose we approach the origin along the real axis and study the function. We have $f(z) = f(x) = e^{1/x}$. Approaching the origin from the right, through positive values of x, we have $\lim\limits_{x \to 0+} e^{1/x} = \infty$. Approaching from the left, through negative values of x, we have $\lim\limits_{x \to 0-} e^{1/x} = 0$. If we approach the origin from the positive imaginary axis, we have $\lim\limits_{y \to 0+} e^{-i/y} = \lim\limits_{y \to 0+} \cos y - i \sin y$, which has no limit, although the magnitude of this expression is fixed at the value one as y shrinks to zero. Approaching the origin from the negative y axis similarly has no limit, although we are looking at an expression whose magnitude is also one. Because of the difficulties involved in making three-dimensional plots of functions near their essential singularities, we will avoid such ventures.

---

[*] See, for example, Conway, John, *Functions of One Complex Variable I* (2nd ed.). Springer, 1978.

2.4.1.4 Branch Cut and Branch Point Singularities: Three-Dimensional Plots

Functions such as $\log z$, z^β (β is not an integer), and $\sin^{-1} z$ are not, strictly speaking, functions, because without our having further information they appear to be multivalued. The log of 1, for example, might be 0 or $i2\pi$ or in fact $i2k\pi$ where k is any integer, while $z^{1/2}$ evaluated at -1 could be $\pm i$. We establish analytic branches of these functions, which are single valued, by performing two tasks: we choose a branch cut to describe the function and assign at some point in the complex plane, not on the cut, a specific numerical value for our branch. This value cannot be chosen haphazardly but must be chosen from the collection of permissible values of the original multi-valued function. The branch cut can be thought of as cutting, or removing, all the points along its length from the complex plane. This, together with the numerical value just mentioned, creates a function that is analytic in a domain, which is the complex plane with points removed by the branch cut. Some branches require more than one branch cut for their complete speci-fication. Not every multivalued function of a complex variable can be used to create an analytic single-valued branch; for example, one could create a single-valued function from $\log \bar{z}$, but it is nowhere analytic. If all possible branch cuts used to create an analytic function have a point in common, we refer to that as a *branch point* of the analytic functions being created. For example, $\log z$ has a branch point at $z = 0$ since all branch cuts for all pos-sible branches of the log must go through the origin. Similarly, all branches of $(z - 1)^{1/2}$ pass through $z = 1$, which means that $z = 1$ is a branch point of this function.

An alternative interpretation of a branch cut is that it is a line, which when crossed, causes the function defined by the cut to change discontinuously. The function in question is analytic but not on the cut.

If we ask MATLAB for the numerical value of a function such as $\log z$ or $\tan^{-1} z$, which, as they stand, are multivalued, how does it decide which value to produce? Moreover, if we were to ask MATLAB for the three-dimensional plots of the real or the imaginary parts or the magnitude of such a function, what branch would it choose to plot? Unfortunately, the Help command in MATLAB does not provide much help here. As an illustration, try typing *Help sqrt*. These questions are best resolved through experimentation with MATLAB. Let us tackle the matter of **sqrt**(z). We use the following code to get the imaginary part of this function:

```
x=linspace(-1,1,50);
y=linspace(-1,1,50);
[X,Y]=meshgrid(x,y);
z=X+i*Y;
w=sqrt(z);
wi=imag(w);
meshz(X,Y,wi);colormap gray
grid on;view(45,30)
```

```
title('The Imaginary Part of the Square Root of z')
xlabel('x');ylabel('y')
```

which results in Figure 2.14.

Notice that the function plotted changes discontinuously along the line $y = 0$, $x \leq 0$. This line must be the branch cut for the function **sqrt(z)**. Had we asked MATLAB for a plot of the real part of $z^{1/2}$, we would have obtained a plot with no discontinuity. In hunting for discontinuities associated with branch cuts, we should make plots for both the real and imaginary portions of the function in question. A discontinuity in *one* or *both* is an indicator of a branch cut.

If we ask MATLAB for the numerical value of say sqrt(1), we get 1 as a result. Thus, it would appear that $z^{1/2}$ is computed in MATLAB by means of a branch cut along the negative real axis and that the numerical values obtained are consistent with the choice $z^{1/2} = \left(e^{Logz}\right)^{1/2} = e^{\frac{1}{2}Logz}$, where $Logz$ is the principal branch of the logarithm, which is defined by a branch cut along the negative real axis and includes $z = 0$ as shown in Figure 2.15.

For the principal branch of the log, we have $Log z = Log|z| + i \arg z$, where $-\pi < \arg z < \pi$. Actually, MATLAB uses not the principal *branch* of the log but the principal *value* defined by $Log z = Log|z| + i \arg z$, where $-\pi < \arg z \leq \pi$. There is a subtle distinction between the principal branch and the principal value. Notice that $<\pi$ (in the principal branch) has been replaced by $\leq\pi$ (in the principal value). In this way, we can ask MATLAB to compute, for example, square roots of negative real numbers; such numbers of course have arguments (angles) of π in the expression $z^{1/2} = \left(e^{Logz}\right)^{1/2} = e^{\frac{1}{2}Log z} = e^{\frac{1}{2}Log|z|+\frac{i}{2}\arg z}$.

We cannot seek the square root of negative real numbers if we are using the

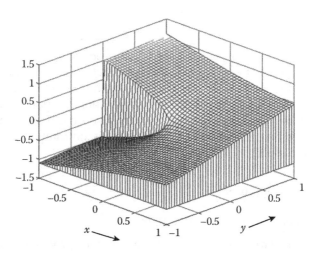

FIGURE 2.14
The imaginary part of the square root of z.

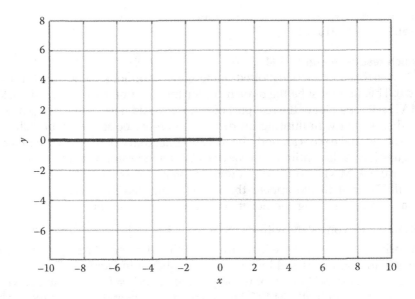

FIGURE 2.15
Branch cut for Log(z) is $y = 0$, $-\infty \leq x = 0$.

principal branch, because these numbers have been removed (by a branch cut) from the plane in which the principal branch of $z^{1/2}$ is defined. The function $z^{1/2}$ is analytic in that cut plane.

Using the principal value from MATLAB, we have

```
>> format short
>> sqrt(-1)
ans = 0 + 1.0000i
```

since $Log(-1) = i\pi$ and $e^{\frac{1}{2}Log(-1)} = e^{\frac{1}{2}Log|(-1)|+\frac{i}{2}arg(-1)} = e^{i\frac{\pi}{2}} = i.$

MATLAB will also tell you that the square root of 0 is zero by using the limit of the real part of Log z as $z \to 0+$. The function produced by MATLAB for $z^{1/2}$ is actually not analytic anywhere on the line $y = 0$, $x \leq 0$, even though MATLAB yields numerical values for the function for points on this line. The difficulty is that if you move from a point on the line downward to a point below the line, the value of $z^{1/2}$ changes discontinuously. Consider the following two results:

```
>> sqrt(-3)
ans = 0 + 1.732050807756888i
>> sqrt(-3-i*eps)
ans = 0.00000000000000 - 1.732050807756888i
```

Moving from -3 to its *very* near neighbor $-3-i*$eps produces results for the square root that are so different that you feel that you are falling off a cliff. Indeed, such a cliff is shown in Figure 2.14.

If we ask MATLAB to compute the logarithm of $z \neq 0$, it will, as expected, produce the principal value. Asking for the log of zero results in a warning and the result of minus infinity. In general, MATLAB computes z^β, where β is not an integer, by means of $e^{\beta Log\, z}$, where the principal value is used as the capital L indicates. Error messages may arise if $z = 0$. If you ask MATLAB for $0^{\wedge}(1/2)$, it will produce the value zero, but asking for $0^{\wedge}(1/2 + i)$ will inconsistently give you the result NaN meaning "not a number." An expression like $(z - z_0)^\beta$ is computed as $e^{\beta Log\,(z-z_0)}$. The function is discontinuous if we cross the branch cut for $Log\,(z - z_0)$—that is, a semi-infinite line beginning at $z = z_0$ parallel to the real axis and extending to the left.

If we ask MATLAB to evaluate an expression such as $f(z) = (az + b)^\beta$, where β is any number that is not an integer, then MATLAB uses $f(z) = e^{\beta Log(az + b)}$ provided $(az + b) \neq 0$. Because the principal value of the log used here has a branch cut where $(az + b)$ is negative real or zero, $f(z)$, as computed by MATLAB, will exhibit such a cut, and a three-dimensional plot of either or both the real and imaginary parts of the function will exhibit a discontinuity there. The matter is explored in problems 9 and 12.

It is more challenging to consider how even more complicated functions such as $f(z) = (z^2 - 1)^{1/2} = (z - 1)^{1/2} (z + 1)^{1/2}$ and $f(z) = \sin^{-1} z = -i\, \log(iz + (1 - z^2)^{1/2})$ are treated by MATLAB. We could try to figure out how MATLAB chooses cuts for such functions by making three-dimensional plots of the real and imaginary parts. Let us study $f(z) = \sin^{-1} z = -i\, \log(iz + (1-z^2)^{1/2})$. The multivalued function $(1 - z^2)^{1/2}$ can be defined by a variety of branch cuts.[*] Some experimentation involving MATLAB computation of numerical values of this function at selected values of z, or three-dimensional plots of the real and imaginary parts of the function (see problem 10), shows that the branch used by MATLAB is defined by cuts along the lines $y = 0$, $x \geq 1$ and $y = 0$, $x \leq -1$. The function is chosen to be 1 when $z = 0$. MATLAB evaluates the logarithm using the principal value. However, no branch cut is needed for this log. The function $Log\, z$ is discontinuous along the line where $Im(z) = 0$, $Re(z) \leq 0$. Therefore, using the function $-i\, \log(iz + (1-z^2)^{1/2})$, we must look to see where the argument $iz + (1 - z^2)^{1/2}$ is zero or negative real. One can show almost immediately that it can never be zero with the branch of the square root we are using, and a slightly more complicated argument shows that it can never be negative real. Thus, the branch of $(1 - z^2)^{1/2}$ that MATLAB chooses, and the decision to use the principal value of the log, is sufficient to define the branch of the inverse sine produced by MATLAB.

A plot of the imaginary part of $\sin^{-1} z$ is given in Figure 2.16 for the region $-2 \leq x \leq 2$, $-2 \leq y \leq 2$. Note the discontinuities as we cross the branch cuts $y = 0$, $x \geq 1$ and $y = 0$, $x \leq 1$. The real part exhibits no such discontinuity. Note, however, that in problem 11 where you study asinh(z), it is the real part of the function that is discontinuous across the branch cut.

---

[*] The reader may wish to consult section 3.8 of W for a discussion of branch cuts of some similar functions.

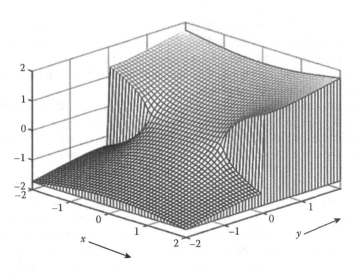

FIGURE 2.16
The imaginary part of arcsin z.

Exercises

1. a. Show that the function $f(z) = \dfrac{e^z - 1}{z}$ has a removable singularity
   at $z = 0$ through the use of either L'Hôpital's rule or a Maclaurin
   series expansion of the numerator. Be sure to state how $f(0)$
   should be defined in order to remove the singularity.

   b. Make a plot of the surface describing $|f(z)|$ over the region
   $-2 \le x \le 2$, $-2 \le y \le 2$; use at least 1000 data points. Your code
   should be written in such a way that you never divide by zero
   (i.e., MATLAB should not give you a warning message).

   c. Repeat part (b) but use the real part of the function $f(z)$.

   d. Repeat part (b) but use the imaginary part of $f(z)$.

2. a. Show that the function $f(z) = \dfrac{\sin(z - e)}{\text{Log}(z) - 1}$ has a removable singularity at $z = e$ through the use of L'Hôpital's rule. Be sure to state
   how $f(e)$ should be defined in order to remove the singularity.

   b. Make a plot of the surface describing $|f(z)|$ over the region $.5 \le x$
   ≤ 4, $-2 \le y \le 2$; use at least 1000 data points. Your code should be
   written in such a way that you never divide by zero (i.e., MATLAB
   should not give you a warning message). You should provide a
   view statement in your code such that you can view the height of
   your surface at the removable singular point.

    c. Repeat part (b) but use the real part of the function $f(z)$.

    d. Repeat part (b) but use the imaginary part of $f(z)$.

3. Make a plot of the magnitude of $f(z) = \dfrac{1}{z(z-i)^2}$. Choose the limits on your axes and the arguments in your **view** statement such that you can clearly see the behavior near both poles. Your plot should demonstrate that the surface rises more steeply near the pole of order 2 than near the pole of order 1.

4. a. Where are the two poles of $f(z) = \dfrac{z}{\sin \pi z}$ that are nearest the origin, and what is their order? Where is the removable singularity? Prove that it is removable.

    b. Make a plot of the magnitude of this function that clearly shows the behavior near these poles, but do not include any other poles in your plot. Make certain that your program does not seek to divide by zero.

*In the following three problems, be sure to use axes of equal scales (the **axis equal** command in MATLAB does this) so that contour plots near the poles will appear nearly as circles if you are sufficiently close to the poles.*

5. Make a plot of the contours on which the magnitude of $f(z) = \dfrac{1}{z(z^2+1)}$ assumes the values 0.5, 1, and 5.

    Suggestion: Use the region described by $|x| \le 2$, $|y| \le 2$. Do the contours encircle the individual poles? Explain.

6. Make a plot of the contours on which the magnitude of $f(z) = \dfrac{1}{\sin(\pi z)}$ assumes the values 0.5, 1, and 3. Use a region that includes only the poles at 0 and ± 1. Which contours encircle the poles?

7. a. Identify the location of the pole of $\dfrac{1}{(\text{Log} \, z - 1)^2}$, and state its order.

    b. Make contour plots on which the magnitude of this function assumes the values 1, 2, and 10. A suggested region to use satisfies $0 \le x \le 10$, $-5 \le y \le 5$. Place an asterisk * at the location of the pole, on your plot. Comment on how well these contours appear to be circles surrounding and concentric with the pole.

8. Make three-dimensional plots of both the real and imaginary parts of $f(z) = \text{Log} \, z$ for the region $-2 \le x \le 2$, $-2 \le y \le 2$. The plot of the real part should clearly show the discontinuous behavior of $\text{Log} \, |z|$ at $z = 0$. The plot of the imaginary part should show the discontinuous nature of $\text{Log} \, z$ at the branch cut for this function.

9. Consider the function $f(z) = (iz - 1)^{1/3}$. Where do you expect the branch cut for this function, as created by MATLAB, to lie? Verify

the branch cut that MATLAB uses in this case by making three-dimensional plots of the real and imaginary plots of this function over the region $-2 \le x \le 2$, $-2 \le y \le 2$.

10. a. Consider the function $f(z) = (1 - z^2)^{1/2}$. Establish the branch cut that MATLAB uses in evaluating this function by making three-dimensional plots of the real and imaginary plots of this function over the region $-2 \le x \le 2$, $-.25 \le y \le .25$. For **view** a good choice is **view**(–30,70). What value does MATLAB produce at $z = 0$?

   b. Repeat part (a) but choose $f(z) = (z^2 - 1)^{1/2}$. Notice that the branch cuts are now different from those in part (a). Where are they? What value does MATLAB choose for $f(0)$?

11. Consider the function $f(z) = \sinh^{-1}(z) = \log(z + (z^2 + 1)^{1/2})$, the inverse hyperbolic sine of z. Establish the branch cut that MATLAB uses in evaluating this function by making three-dimensional plots of the real and imaginary plots of the function $\sinh^{-1} z$ over the region $-2 \le x \le 2$, $-2 \le y \le 2$. Explain why no branch cut arises from the logarithm, and explain what branch of the log is used by MATLAB.

12. Consider the function $f(z) = z^i$. Establish the branch cut that MATLAB uses in evaluating this function by making three-dimensional plots of the real and imaginary plots of this function over the region $-2 \le x \le 2$, $-2 \le y \le 2$. What relationship does this have to the branch cut for Log z?

   Suggestion: Try **view**(–125,30) to view your plot. What numerical values does MATLAB assign to 1^i and i^i?

2.5 Contour Plots Affected by Branch Cuts

This section is in some sense a continuation of the previous one, but here we focus on how to create contour plots when the function being studied exhibits a branch cut in the domain in which we are seeking the plot. To see how this might arise, suppose we seek a contour plot of the imaginary part of Log z. We proceed with this code:

```
x=linspace(-5,5,1000);
y=x;
[x y]=meshgrid(x,y);
z=x+i*y;
w=log(z);
wm=imag(w);
[g h]=contour(x,y,wm);
clabel(g,h);grid
```

whose output is as shown in the plot presented in Figure 2.17.

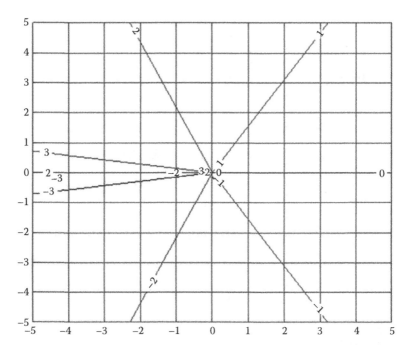

FIGURE 2.17
Contour lines for the imaginary part of Log z.

Notice the jumble of contour lines and numbers near the negative real axis. The imaginary part of Log z is the principal value of the angle of z (i.e., the principal value of arg(z)). The curves on which this angle assumes constant values are simply rays extending out from the origin. The command **contour** should produce rays spaced at equal angles. This is not the case in the above. The difficulty arises from the branch cut for Log z, which extends out from the origin along the negative real axis. Just above the cut we have arg $z = \pi$, and just below the cut arg $z = -\pi$. In constructing the contour lines, MATLAB erroneously assumes that Im(Log z) assumes all possible real values lying between $-\pi$ and π in any domain containing the negative real axis. It does not allow for this function having a step-type discontinuity. It is possible that later versions of MATLAB will sense the discontinuity and that what we will describe here is not necessary. As noted, we are using here MATLAB release 2015a.

To prevent what just happened, before we invoke the **contour** command, we must insert a line of code asserting that if z lies in a narrow band containing the negative real axis, then z "is not a number." This is done below and on the next page:

```
x=linspace(-5,5,1000);
y=x;
[x y]=meshgrid(x,y);
z=x+i*y;
```

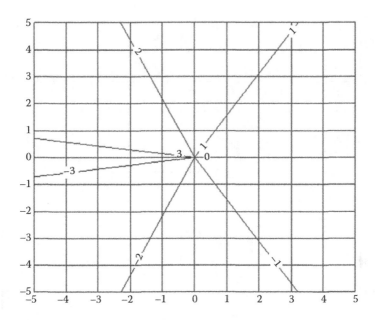

FIGURE 2.18
An improved version of the preceding plot.

```
w=log(z);
wm=imag(w);
del=.01;
wm(abs(y)<del&x<=0)=nan;
[g h]=contour(x,y,wm);
clabel(g,h);grid
```

Notice the lines of code **del=.01;** and **wm(abs(y)<del&x<=0)=nan;**

This indicates that if z lies in a strip of width 2* del straddling the negative real axis, that imag(w) is not to be regarded as a number.

The output of the program, which displays the desired result, is as shown in Figure 2.18.

Some experimentation can be required in choosing del. If del is chosen too small (e.g., del = .001), then the same problem arises as when we did not exclude the real axis from our domain of consideration. You will once again get Figure 2.17. However, if you choose del too large (e.g., del = .1), you will find that some portion of the desired contours are missing. You should try both these values to see the results.

Example 2.1

Using the ideas presented above, obtain a satisfactory plot of the contour lines of the imaginary part of $w = \text{Log}\dfrac{(z-1)}{(z+1)}$. Let MATLAB choose the lines, and then change the program so that it plots exactly six lines.

Solution:

This function has a branch cut where $(z + 1)/(z - 1)$ is zero or negative real. Some study (which you should do now) reveals that this is only where $y = 0$, $-1 \leq x \leq 1$. We remove from consideration a strip containing points along abs(y)<del, $-1 \leq x \leq 1$. We try del =.01. The following code is used:

```
clf
x=linspace(-5,5,1000);
y=x;
[x y]=meshgrid(x,y);
figure(1)
z=x+i*y;
w=log((z-1)./(z+1));
wm=imag(w);
del=.01;
wm(abs(y)<del&-1<=x<=1)=nan;
[g h]=contour (x,y,wm,6);
clabel(g,h);grid
title('contour lines for Im log((z-1)/(z+1))')
xlabel('x');ylabel('y');
```

the output is as shown in Figure 2.19.

We have elected to use six contours as is indicated by the line
`[g h]=contour(x,y,wm,6);`

The reader is invited to play with the previous program to see the effects of changing del, the number of contours plotted, as well as the parameters in linspace.

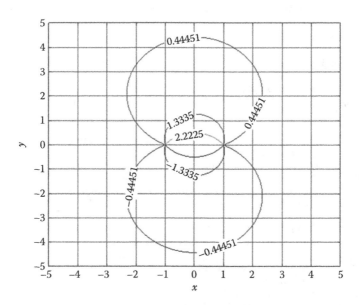

FIGURE 2.19

Contour lines for $\mathrm{Im}\ \log\left(\dfrac{(z-1)}{z+1}\right)$.

One should not think that the branch cut can create problems only in the imaginary part of an analytic function. Here is an instance where the branch cut appears in both real and imaginary parts.

Example 2.2

Obtain the contour lines for both the real and imaginary parts of the function $f(z) = u(x,y) + iv(x,y) = z^i$, where the principal branch of the function is used.

Solution:

$$u(x,y) + iv(x,y) = z^i = e^{i(\text{Log}\,z)} = e^{i(\text{Log}|z| + i\arg(z))}$$

$$= e^{-\arg(z)}\cos\left(\text{Log}|z|\right) + ie^{-\arg(z)}\sin\left(\text{Log}|z|\right)$$

Both the real and imaginary parts of the above are discontinuous as we cross the negative real axis because of the discontinuity in the principal value of arg(z). Both real and imaginary parts are discontinuous as $|z| \to 0$ because of the singularity in $\text{Log}|z|$. Thus, in computing the contour lines for both the real and imaginary parts of this function, we must again exclude points from a strip that extends outward from the origin along the negative real axis. We use the same strip as we used in creating Figure 2.18. We seek the contours on which the levels of 0 through 5 (integer values) are achieved. The following code will do the job. If you run this program, the plot will be in color.

```
%example 2 sec 2_5
clf
x=linspace(-5,5,1000);
y=x;
[x y]=meshgrid(x,y);
z=x+i*y;
w=z.^i;
del=.01
levels=(0:5);%the values 0,1...5 for the contours
wr=real(w);wm=imag(w);
wr(abs(y)<del&-1<=x<=1)=nan;%accounts for branch cut
figure(1)
[g, h]=contour(x,y,wr,levels);
clabel(g,h);grid
title('contour lines for Re z^i')
xlabel('x');ylabel('y');
wm(abs(y)<del&-1<=x<=1)=nan;%accounts for branch cut
figure(2)
[G, H]=contour(x,y,wm,levels);
clabel(G,H);grid
xlabel('x');ylabel('y');
title('contour lines for Im z^i')
```

whose output is as shown in Figures 2.20 and 2.21.

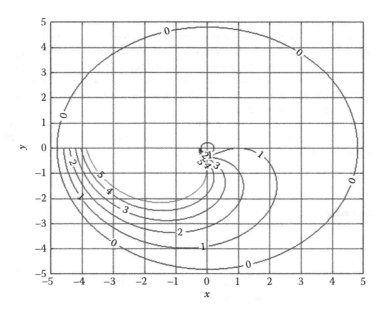

FIGURE 2.20
Contour lines for Re z^i.

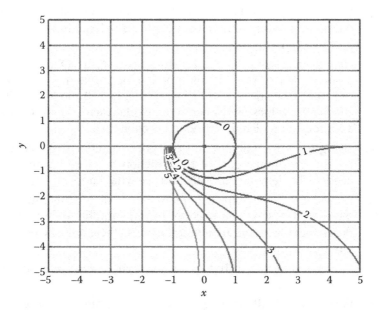

FIGURE 2.21
Contour lines for Im z^i.

If you were to make both sets of plots on the same set of axes, you would find that the sets of curves intersect at right angles. This is a well-known property of analytic functions (see W section 2.5 and S section 3.12). If you were to delete the lines of code `wr(abs(y)<del&-1<=x<=1)=nan; wm(abs(y)<del&-1<=x<=1)=nan`, you would find that the curves form a jumble along the negative real axis due to the discontinuity caused by the branch cut described above.

The near absence of contour lines in the upper half plane in the two figures can be remedied if we seek some contours having numerical values between 0 and 1.

Exercises

1. a. Consider $f(z) = \text{Log}(z^2 + 1)$. Obtain contour line plots for the real and imaginary parts of this function. Before doing so, determine the location of the branch cut that makes a domain of analyticity for this function, and in your code eliminate points on this cut through the use of NaN. Do this for both plots. Use a grid like that in Example 2.1.

   b. Try out your program but do not include the use of NaN, and see what happens to your plots. Are both affected?

2. a. Consider $f(z) = \sin(\text{Log}\,z)$. This function is obviously discontinuous as we cross the branch cut $y = 0$, $x \le 0$. Explain why the real part is *not* discontinuous.

   Hint: Recall $\sin(a + ib) = \sin a \cosh b + i \cos a \sinh b$. Obtain contour plots of the real part of this function, noting that you need not use NaN. Consider a grid filling $|x| \le 2$, $|y| \le 2$ and let MATLAB choose the values on the contours. Explain why the unit circle is a contour on which this function is equal to 0.

   b. Using the same domain and procedure as above, plot contours for the imaginary part of the function but insert NaN where needed. Try leaving out NaN and see what happens.

3. a. Consider $f(z) = \text{Log}(\sin(z))$. Explain why to define this function in the complex plane you need an infinite number of branch cuts. Prove that these cuts are along the lines where any one of the following is true:

   $y = 0$, $\sin x \le 0$, or the set of lines where $\cos x = 0$, $\sin x < 0$, $-\infty < y < \infty$.

   b. Write a MATLAB plot to generate two separate contour program of the real and imaginary parts of $f(z)$.

Suggestion: Consider the space $-6 < x < 6$, and $-2 < y < 2$, and let the **contour** function choose the values on the contours. You might also wish to make both plots on the same set of axes to verify that they intersect at right angles. Be sure to use NaNs to avoid evaluating the imaginary part of $f(z)$ at or near the branch cuts. Explain why this is not necessary for the contour lines for the real part.

4. a. Consider $f(z) = z^z$, where we use the principal branch. Explain why both the real and imaginary parts of this function are discontinuous in any domain containing any portion of $y = 0$, $x \leq 0$.

b. Obtain a set of contour plots for both the real and imaginary parts of this function. A suggested range for both x and y is that their absolute value be less than or equal to 2. Use NaNs where required. Let the contour function choose the values assumed by the function.

c. Obtain a contour plot showing only where the imaginary part of this function assumes the value 1. A suggested range for both x and y is that their absolute value be less than or equal to 5. Notice that one of the contours appears coincident with a portion of the positive real axis. This would seem to be wrong, because if $z = x$ is positive real, then $z^z = x^x$ must be real and cannot have an imaginary part equal to one. Can you explain this? Think in terms of asymptotic behavior.

3

Sequences, Series, Limits, and Integrals

Complex variable theory is mostly about the properties of analytic functions and the integration of these functions. Among the most salient features of analytic functions is the fact that they are the sums of convergent infinite series. Thus, we must have some understanding of infinite series to comprehend analytic functions. We must also know what it means to say that a series is convergent, and here we rely on the fundamental concepts of sequences and their limits. Integrals in the complex plane, like the real integrals of elementary calculus, are the limits of sums, and here also we are concerned with limits.

MATLAB® can be a useful tool in determining how rapidly a sequence of functions approaches its limit—if indeed it has one. Additionally, a finite portion of an infinite series—say the first N terms—can be a useful device for approximating not only an analytic function but also an integral, and a numerical comparison employing MATLAB can be useful to investigate the validity of approximations. Infinite series, as a means of approximating functions, do have their limitations, and sometimes infinite products are preferable; we do not have the space to explore that subject here, and the reader is referred to Chapter 9 of W.

3.1 Sequences

In complex variable theory, we deal with both infinite and finite sequences, the former being more common. A finite sequence of complex numbers is simply a list $z_1, z_2, z_3, ..., z_n$ of n numbers, where n is any positive integer. Sometimes we begin the list with z_0 instead of z_1 so that the list has $n + 1$ elements instead of the n just shown. If n is allowed to grow without bound, we have an infinite sequence. An example of a finite sequence might be i, i^2, i^3, i^4, i^5. Thus, the nth member or element in the sequence is i^n. If we were to let n here run beyond the number 5 and increase all the way to infinity, we would have an infinite sequence.

Sometimes we have sequences of functions. Each element in the sequence is defined by a function whose value typically depends on the index that locates that function in the list defining the sequence. Here is an infinite sequence of functions: $\sin z, \sin 2z, \sin 3z, ...$, whose nth element is $\sin nz$. Here n is allowed to grow without bound. Restricting the size of n would result in a finite sequence.

The most useful infinite sequences have a *limit*. This has a precise mathematical meaning. Suppose the sequence is of the form

$$p_1(z), p_2(z), p_3(z), \dots p_n(z), p_{n+1}(z), \dots \tag{3.1}$$

where the terms in the sequence might be functions of z, like $\sin nz$, $n = 1$, $2, \dots$, or simply constants like i^n.

The sequence has a limit (or "converges to the limit") $P(z)$ if the following is true.

Given any positive number ε, there exists a positive integer N such that

$$|P(z) - p_n(z)| < \varepsilon \text{ for all } n > N \tag{3.2}$$

The preceding says simply that if we go to the Nth term in the sequence, then the magnitude of the difference between all terms beyond the Nth (to the right of the Nth) and $P(z)$ will be less than any preassigned positive number ε that you choose, no matter how small. N almost always depends on the value chosen for ε, and the smaller you make ε the larger must be N. If the terms $p_n(z)$ are not constants but are functions of z, then almost invariably N will depend on z as well as ε. Loosely speaking, if a sequence has a limit $P(z)$, then its terms must cluster closer and closer to $P(z)$, in the complex plane, as you move to the right in Equation 3.1. A sequence might have a limit for some values of z but not others. For the latter, it would be futile to try to find $P(z)$. The reader might wish to see W section 5.2 and S sections 2.15 through 2.17 for more information on sequences.

Example 3.1

Let us consider the simple sequence

$$1 + 1/(1 + i), 1 + 1/(1 + i)^2, 1 + 1/(1 + i)^3, \dots$$

whose nth term is $p_n(z) = 1 + \dfrac{1}{(1+i)^n}$. We use MATLAB to plot the elements of this sequence.

Solution:

It should be obvious that the limit of this sequence is $P = 1$, since

$$\lim_{n\to\infty}\left(1 + \frac{1}{(1+i)^n}\right) = 1 + \lim_{n\to\infty}\frac{1}{(1+i)^n} = 1$$

This follows from the well-known fact that $\lim_{n\to\infty}\dfrac{1}{w^n} = 0$ if $|w|>1$. We can take $w = 1 + i$.

The reader should verify that we can take N in Equation 3.2 as any integer N satisfying $N > \dfrac{\text{Log}\dfrac{1}{\varepsilon}}{|\text{Log}(1+i)|}$. Thus, if we use MATLAB to plot the elements $p_n(z)$ of this sequence in the complex plane, they should cluster more and more closely around the point $1 + i0$. We prove this with the following code and Figure 3.1:

```
N=15;
p=ones(1,N);rp=real(p);ip=imag(p);
    for n=1:N
p(n)=1+1/(1+i)^n;
rp(n)=real(p(n));
ip(n)=imag(p(n));
b=num2str(n)
plot(rp(n),ip(n));
    text(rp(n),ip(n),b,'FontSize',6);hold on
    pause(.25)
n
p(n)
    end
grid
    plot(p,'-')
    title('The First 15 Terms in the Sequence 1+1/(1+i)^n')
    hold off
```

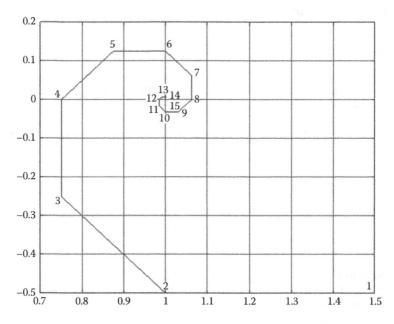

FIGURE 3.1
The first 15 terms in the sequence $\dfrac{1+1}{(1+i)^n}$.

The line of code

```
p=ones(1,N);rp=real(p);ip=imag(p);
```

is not, strictly speaking, necessary. It tells the MATLAB compiler to set aside space for vectors, *p*, *rp*, and *ip* having *N* elements. By doing this before the FOR loop is executed, we save computation time. Although time is not an issue in this program, it is a good habit to create space in the beginning for any array that is to be generated.

Note in the code that the statement **pause(.25)** places a pause of one-quarter second between the plotting of each point on the figure. In this way one can follow visually the generation of each point. It should be evident that the points are clustering increasingly close to the limit of the sequence at $z = 1$.

The statement b=num2str(n) converts the number *n* to a character *b*. This number can be placed at the point in the complex plane corresponding to *n*th element in the sequence. We have arranged to do this with a small-sized font. This is accomplished with

```
text(pr(n),pi(n), b,'FontSize',6);hold on
```

Here the number 6 is the font size. Using a larger integer than 6 here will allow the values of *n* to be printed larger, but there will be increased crowding of these numbers near the limit of the sequence at 1.

Some sequences are generated through an iterative scheme. Each new element is obtained by putting the previous element or elements through a defined process, as presented in Example 3.2.

Example 3.2

Investigate the sequence $i, i^i, i^{\left(i^i\right)}, i^{\left(i^{\left(i^i\right)}\right)}, \dots$ Thus, the *n*th element p_n is obtained by taking the previous element p_{n-1} and using it for the exponent of *i* so that $p_n = i^{p_{n-1}}$, where $n \geq 2$. We will agree to take $p_1 = i$. We use the principal value in each of these calculations, so that, for example,

$$i^i = \left(e^{i\text{Log}(i)}\right) = e^{ii\pi/2} = e^{-\pi/2}$$

while,

$$i^{\left(i^i\right)} = i^{\left(e^{-\pi/2}\right)} = e^{\left(e^{-\pi/2}\right)\text{Log}(i)} = e^{\left(e^{-\pi/2}\right)(i\pi/2)}$$

Solution:
Using MATLAB we can make a plot of the elements of this sequence in the complex plane. Here is the code and the resulting Figure 3.2 when we use the first 50 terms:

```
clear
clf
```

```
N=50;
p=ones(1,N);rp=real(p); ip=imag(p);
p(1)=i;
  for n=2:N
p(n)=i^p(n-1);
rp(n)=real(p(n));
ip(n)=imag(p(n));
b=num2str(n)
  plot(rp(n),ip(n),'*');
pause(.3);% gives a .3 sec delay in plotting
hold on
  % text ( rp(n) , ip(n) , b , 'FontSize' , 6 );hold on
n
p(n)
  end
  plot(p(1),'*');
  grid
  title('Sequence from p_n=i^{p_{n-1}}')
hold off
```

Note that if you remove the % sign at the start of

```
% text( rp(n),ip(n), b,'FontSize', 6 );hold on
```

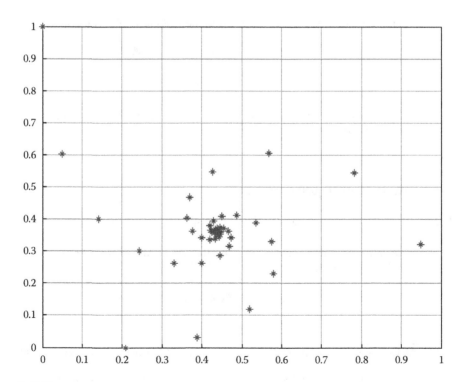

FIGURE 3.2
Sequence from $p_n = i_{n-1}^p$.

you will place at each point on the plot the index n corresponding to each point in the plot. If you do this, you should eliminate the line `plot(rp(n),ip(n),'*');`. The numbers for n will be hard to read on the plot near the limit of the sequence, where they become crowded.

Why are there three spiral arms? Does the sequence in question have a limit, and what it is it? This problem has been studied in two papers by Greg Packer, Steve Abbott, and Steve Roberts.[*] The limit is close to 0.4383 + i0.3606. We can obtain the limit by means of MATLAB.

Notice that $\lim_{n\to\infty} p_{n-1} = \lim_{n\to\infty} p_n = p$, assuming the limit p exists. Now observe that $\lim_{n\to\infty} p_n = \lim_{n\to\infty} i^{p_{n-1}}$. The left side of the preceding is this p. Let us assume that we can swap the limit and exponentiation process of the right side.[†] We thus have the equation

$$p = i^p \tag{3.3}$$

which must be satisfied by the limit p. We can solve this equation in the complex plane using MATLAB. Let $p = x + iy$. Using this in Equation 3.3, we have

$$x + iy = i^{x+iy} = e^{(i\pi/2)(x+iy)} = e^{-\pi y/2} e^{(i\pi x/2)}$$

Let us equate corresponding parts from both sides of this equation:

$$x = e^{-y\pi/2} \cos(\pi x/2) \tag{3.4}$$

$$y = e^{-y\pi/2} \sin(\pi x/2) \tag{3.5}$$

Dividing the second of these by the first and making a little rearrangement, we have

$$y = x \tan(\pi x/2) \tag{3.6}$$

We use this in Equation 3.4 to get

$$x = e^{-(\pi x/2)(\tan \pi x/2)} \cos(\pi x/2)$$

We can solve this equation by using the MATLAB function **fzero** to find the solutions (if any) of

$$x - e^{-(\pi x/2) \tan \pi x/2} \cos(\pi x/2) = 0 \tag{3.7}$$

---

[*] Greg Packer and Steve Abbott, "Complex Power Iteration," *The Mathematical Gazette* 81 (492, Nov. 1997): 431–434.

[†] The justification for this swap lies with the "Continuous Function Theorem for Sequences." See, for example, George B. Thomas, Ross L. Finney, Maurice D. Weir, and Frank R. Giordano, *Thomas' Calculus*, 10th ed. (Reading, MA: Addison-Wesley, 2000), 614.

The code follows:

```
% considers i^p=p solution
format long
myfunc='exp(-pi/2*x.*tan(pi*x./2)).*cos(pi*x./2)-x'
[x,value]=fzero(myfunc,[0 10])
y=x*tan(pi*x/2);
z=x+i*y
check=i^(z)
```

The reader may wish to review the function **fzero** in the MATLAB help file. Notice that the argument [0 2] that we have used means that we are seeking a solution of Equation 3.7 in the interval $0 < x < 2$. We were guided by the approximate location of the limit shown in Figure 3.2. The output for x is 0.438282936727032. The line of code y=x*tan(pi*x/2) uses Equation 3.6 and yields $y = 0.360592471871385$. Observe that these values of x and y are consistent with the coordinates of the limit point that we can approximately visualize from Figure 3.2. Our final line of code computes check= i^(z). According to Equation 3.3, if z is the limit of our sequence, then the value of *check* should be the same as z. The agreement to within the 16 digits displayed by MATLAB is perfect.

A relatively recent use of sequences having complex elements, and MATLAB, is in the generation of *fractal sets*. These are sets of points in the complex plane describing a shape that is not one dimensional (like a straight line) or two dimensional (like a disc). These shapes have fractional dimensionality. At the end of this book you will find a Coda devoted to fractals and there find references that will explain what is meant by a fractional dimension, a topic not treated in this book.

In preparation for dealing with fractals, and one particular kind, the *Mandelbrot Set*, we introduce here the concept of *divergence to infinity*. Suppose we are given the sequence in Equation 3.1. The sequence diverges to infinity if the following is satisfied: if you are given any positive number ρ there exists a number N such that

$$|p_n(z)| > \rho \text{ for all } n > N \tag{3.8}$$

where typically N depends on ρ and z.

Thus, if we consider terms in the sequence to the right of the Nth term, we find that the magnitude of all of them will exceed any *a priori* number chosen in advance. Here is a sequence that diverges to infinity $e^z, e^{2z}, e^{3z}, ..., e^{nz}, e^{(n+1)z}, ...$, where $z = x + iy$ and we take Re$(z) > 0$.

Recall that $|e^{nz}| = e^{nx}$ and that this expression grows with increasing n. To satisfy Equation 3.8, we require that $e^{nx} > \rho$ for $n > N$. Hence, by taking logs, we see that if $x > 0$, we can choose N as any integer greater than $\frac{\log\rho}{x}$.

A sequence might diverge but not to infinity—an example being the sequence whose nth element is i^n. A more interesting example is the sequence $p(n) = \sin(n\pi/10)z^n$, where $n = 1, 2, 3, ...$ If $|z| \geq 1$, this sequence diverges, but not to infinity. The matter is considered in problem 3.

Exercises

1. All textbooks on calculus have the formula $e^x = \lim_{n \to \infty}\left(1+\dfrac{x}{n}\right)^n$, where x is any real number. By studying the derivation of this formula, we find that it applies equally well to the case where z is any number, not just reals.* Thus, we can say that e^z is the limit of the sequence $(1 + z)$, $(1 + z/2)^2$, $(1 + z/3)^3$, ..., $(1 + z/n)^n$, ... Take $z = (1 + i)$ and make a plot similar to that in Figure 3.2 in which you display the first 100 terms of this sequence. Also display in this plot the value of e^{1+i}.

2. a. Consider the sequence in which each term is obtained by cubing the preceding term and squaring this same term, summing these and adding them to $i/2$. Thus, $p(n) = [p(n-1)]^3 + [p(n-1)]^2 + i/2$. Taking $p(1) = 0$, obtain a plot like Figure 3.1, which generates a line connecting the first 1,000 terms in the sequence.

      Suggestion: Use the **xlim** and/or **ylim** functions of MATLAB to show the portion of the x–y plane near what is the apparent limit of this sequence. From your plot or from the value of $p(1000)$, estimate the limit of this sequence.

   b. Repeat part (a) but take $p(1) = .9*i$. Compare $p(1000)$ arising from each case. Do you suspect that both sequences are heading for the same limit?

   c. Investigate what happens if you take $p(1) = i$. Leave off the xlim and ylim. Does the sequence appear to have a limit?

   d. No matter what you take for $p(1)$, if this sequence has a limit, let us call it p, show that it must satisfy the equation $p^3 + p^2 - p + i/2 = 0$. Find the roots of this equation by using the MATLAB function **roots**. Which of the three roots is close to the limit you estimated in parts (a) and (b)? How good is the agreement between the result obtained here for the limit and the result for parts (a), (b), and (c)?

3. Consider the sequence $p(n) = \sin(n\pi/10)z^n$, where $n = 1, 2, ...$

   a. Let $z = 1 + i/10$. Obtain a plot comparable to Figure 3.2 showing the first 100 terms of this sequence.

      Suggestion: Put a pause statement in your code giving a pause of around 0.5 second as each point is plotted so that you can see how the points are generated. What terms in your sequence have the smallest absolute magnitude? Prove mathematically that this sequence cannot diverge to infinity.

---

* See, for example, K. Knopp, *Infinite Sequences and Series*, English edition (Mineola, NY: Dover, 1956): 153–154.

b. Let $p(n) = \sin(0.1 + n\pi/10)z^n$. Using the same value of z, obtain a plot like the one in part (a). Notice the similarity in the two plots. Prove mathematically that the sequence in this case does diverge to infinity. Write your MATLAB program in such a way that it will tell you the term with the smallest magnitude and the value of n used in generating that term.

Hint: See the MATLAB function $[Y,I] = \mathbf{min}(p)$.

4. a. In Example 3.2 we considered the sequence in which $p(n) = i^{p(n-1)}$, where $n \geq 2$ and $p(1) = i$. Let us consider here something of the reverse: $p(n) = [p(n-1)]^i$, where $n \geq 2$ and $p(1) = i$. Compute and plot the first million terms of this sequence, and connect the points with a solid line in the order in which they are generated. Do not use the pause statement employed in Example 3.2 for obvious reasons. It is not necessary to separate the real and imaginary parts of p. Just use $\mathbf{plot}(p)$ for your plot. Using the **tic** and **toc** functions figure out how much time MATLAB took to solve this. On the author's laptop computer, purchased in 2014, this required about 0.2 seconds.

b. Explain why the plot is so simple (compare it to Example 3.2) and why the sequence diverges. Does it diverge to infinity?

3.2 Infinite Series and Their Convergence

The topic of infinite series is central to complex variable theory. Infinite series are not just a useful tool in numerical computation—they have a close connection to the functions we use in complex variables: analytic functions. An analytic function can be represented in a domain by the infinite series called a Taylor series: a power series convergent in a domain has a sum that is an analytic function. Laurent series play a special role in the evaluation of contour integrals, and we study them here, too.

For our purposes, an infinite series is an expression of the form $\sum_{n=0}^{\infty} u_n(z)$, or with different indexing, $\sum_{n=1}^{\infty} u_n(z)$. Here the $u_n(z)$, are usually, but not always, functions of z and almost always depend on the index n. With the second indexing, examples of such series are $\sum_{n=0}^{\infty} \dfrac{e^{inz}}{n+1}$ and $\sum_{n=1}^{\infty} n^2(z+i)^n$, while a series with no z such as $\sum_{n=1}^{\infty} \dfrac{1}{n^2}$ also qualifies as infinite. The most useful series *converge*. Loosely speaking, it means that if you look at the first N terms in the summation as in $\sum_{n=1}^{N} u_n(z) = u_1(z) + u_2(z) + ... + u_N(z)$, we get closer and closer to some value (which depends on z) as N gets bigger and bigger. Convergence has a precise mathematical meaning: the series $\sum\limits_{n=1}^{\infty} u_n(z)$ converges to the function $S(z)$, which is written $\sum\limits_{n=1}^{\infty} u_n(z) = S(z)$, if the sequence formed by taking the Nth partial sums (where $N = 1, 2, \ldots$), namely, $u_1(z)$, $u_1(z) + u_2(z)$, $u_1(z) + u_2(z) + \ldots + u_N(z)$ converges to $S(z)$.

A Taylor series, which has the form $\sum\limits_{n=1}^{\infty} c_n(z - z_0)^{n-1}$, typically converges to a sum $S(z)$ when z is limited to some neighborhood of the point z_0.

Replacing $S(z)$ by the more familiar function $f(z)$, we know from complex variable theory that if $f(z)$ is analytic at the point z_0, we can expand this function in a Taylor series—that is, a series that converges to $f(z)$ in a circle described by $|z_0 - z_0| < r$, where r is the distance from z_0 to the singularity of $f(z)$ lying closest to z_0. We call z_0 the center of expansion. Thus, with a slight reindexing, we write

$$f(z) = \sum_{n=0}^{\infty} c_n(z - z_0)^n \tag{3.9}$$

where

$$c_n = \frac{f^n(z_0)}{n!} \tag{3.10}$$

The above two statements give us Taylor's theorem.

Taylor's theorem with Remainder is as follows:

$$f(z) = \sum_{n=0}^{N-1} c_n(z - z_0)^n + R_N \tag{3.11}$$

where the same values for c_n are used as in Equation (3.9). R_N is known as the remainder in the series and is given by this integral:

$$R_N = \frac{(z - z_0)^N}{2\pi i} \oint_C \frac{f(z')}{(z' - z_0)^N (z' - z)} dz'$$

where the integration is done in the complex plane, using the variable z', around the circle $|z' - z_0| = b$, where b is a positive real number less than r given above. Thus, the circle excludes any singular points of $f(z')$. It is not hard to show that as $n \to \infty$, the remainder vanishes. If we delete the remainder in Equation 3.11, we may approximate $f(z)$ with a series containing N terms as follows:

$$f(z) \approx \sum_{n=0}^{N-1} c_n(z - z_0)^n = c_0(z - z_0)^0 + c_1(z - z_0)^1 + \ldots + c_{N-1}(z - z_0)^{N-1} \tag{3.12}$$

MATLAB provides a convenient way of comparing an analytic function with its Taylor series approximation.

Example 3.3

Let us look at the series $e^z = \sum_{n=0}^{\infty} c_n z^n$, the Taylor series expansion about the origin. Since e^z is an entire function, this expansion is valid within a circle of infinite radius centered at the origin. Here $c_n = \dfrac{1}{n!}$. Using the approximation in Equation 3.12, we have

$$e^z \approx 1 + z + \frac{z^2}{2!} + \frac{z^3}{3!} + \dots + \frac{z^{N-1}}{(N-1)!} \tag{3.13}$$

Setting $z = 2 + 2i$, we will use MATLAB to compute $e^{(2+2i)}$ and compare this with the preceding series approximation where we consider $N = 1, 2, \dots, 11$ terms. Thus, we are ultimately using the 10th power in our final approximation. The code to do this is

```
Nmax=11;
a=0:Nmax-1;
q=1./factorial(a);
b=(2+2*i).^a;
SN=q.*b;
format long
S=(cumsum(SN)).';% this sums all the elements in the row
%vector SN
N=[1:Nmax]';
disp('number of terms series approximation')
[N S]
 disp('MATLAB computes exp(2+2i)')
exp(2+2*i)
```

The output is

```
number of terms                           series approximation

ans =
 1.000000000000000 + 0.000000000000000i 1.000000000000000 +
   0.000000000000000i
 2.000000000000000 + 0.000000000000000i 3.000000000000000 +
   2.000000000000000i
 3.000000000000000 + 0.000000000000000i 3.000000000000000 +
   6.000000000000000i
 4.000000000000000 + 0.000000000000000i 0.333333333333333 +
   8.666666666666666i
 5.000000000000000 + 0.000000000000000i -2.333333333333333 +
   8.666666666666666i
 6.000000000000000 + 0.000000000000000i -3.400000000000000 +
   7.600000000000000i
 7.000000000000000 + 0.000000000000000i -3.400000000000000 +
   6.888888888888888i
 8.000000000000000 + 0.000000000000000i -3.196825396825396 +
   6.685714285714285i
 9.000000000000000 + 0.000000000000000i -3.095238095238095 +
   6.685714285714285i
```

```
10.000000000000000 + 0.000000000000000i -3.072663139329805 +
     6.7082892416225574i
11.000000000000000 + 0.000000000000000i -3.072663139329805 +
     6.7173192239858890i

MATLAB computes exp(2+2i)

ans = -3.074932320639359 + 6.718849697428250i
```

The left-hand column shows the number of terms we have elected to use in our series approximation (Equation 3.11), while the adjacent columns give the real and imaginary parts of the resulting sum. At the bottom of the output, we see what value MATLAB gives for e^{2+2i}. By the time we use a series with 11 terms, the disparity between the approximate and "exact" value is tiny.

The command $S=(cumsum(SN))$.'; in the above code may be unfamiliar. If the argument of **cumsum** is a vector, either row or column, the resulting vector is a vector that has elements that are the cumulative sum of the elements in the argument. For example, from MATLAB:

cumsum([2 7 9])

ans = 2 9 18

Thus, $S=(cumsum(SN))$.'; will sum the elements in the row vector SN;, next the ' (prime) converts this to a column vector, and the . (dot) before the prime prevents this operation from taking the conjugate of these complex elements.

The representation of e^z by a polynomial with N terms as in Equation 3.13 may seem puzzling. According to the fundamental theorem of algebra, such a polynomial has $N-1$ roots (i.e., vanishes at $N-1$ locations in the complex plane). Thus, the more terms used in the series, the greater is the number of these points. However, e^z is known to be nonzero throughout the complex plane. (See W section 5.7 and S problem 2.59.) Thus, the series might appear to become increasingly not valid as N increases. This paradox is studied in problem 3.

Example 3.4

Here we look at a series that is valid only within a domain having a finite radius.

Consider $\dfrac{1}{\cosh z} = \displaystyle\sum_{n=0}^{\infty} c_n z^n$, where we are using a Maclaurin expansion (i.e., a Taylor series valid in a neighborhood containing the origin). The singularities of the function on the left side lying closest to $z_0 = 0$ are those where $\cosh z$ vanishes—that is, at $z = \pm i\dfrac{\pi}{2}$. Thus, the series is not valid outside the domain $|z| < \dfrac{\pi}{2}$. Let us obtain the first three nonzero coefficients in the Taylor expansion. We could use Equation 3.10 where $f(z) = \dfrac{1}{\cosh z}$, but the differentiation quickly becomes tedious. Instead one might do a long division as follows:

$$\frac{1}{\cosh z} = \frac{1}{1 + \dfrac{z^2}{2!} + \dfrac{z^4}{4!} + \ldots} \approx 1 - \frac{z^2}{2} + \frac{5z^4}{24} \tag{3.14}$$

The reader should fill in the details of this calculation. Note that on the above right,

$$\frac{1}{\cosh z}\bigg|_{z=0} = 1, \quad \frac{1}{2!}\frac{d^2}{dz^2}\frac{1}{\cosh z}\bigg|_{z=0} = \frac{-1}{2}, \quad \frac{1}{4!}\frac{d^4}{dz^4}\frac{1}{\cosh z}\bigg|_{z=0} = \frac{5}{24}$$

The derivatives of odd order of $\dfrac{1}{\cosh z}$ all vanish, $z = 0$. Although the business of obtaining the coefficients in this Taylor expansion can be tedious, we will see in a moment that MATLAB makes it simpler.

The approximation sign is used in Equation 3.14 because we have used only the first three terms in the Maclaurin series. If we had used an infinite number of terms, we would still require the *caveat* $|z| < \pi/2$. Let us compare the left side of the Equation 3.14 with the right side when z is real and confined to the real axis (where both sides are real). We make the MATLAB plot as shown in Figure 3.3.

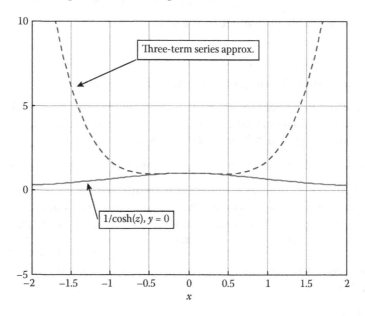

FIGURE 3.3

$\dfrac{1}{\cosh(z)}$ and its series approximation.

This plot arises from this code:

```
clf
x=linspace(-2,2,100);
z=x+i*0;
w=1./cosh(z);
u=real(w);
v=imag(w);
plot(x,u)
grid
hold on
ser=1-z.^2/2+5*z.^4/4;
plot(x,ser,'--r','LineWidth',2)
ylim([-5,10]);
xlabel('x');ylabel('1/cosh(z) and its series approximation')
```

The series we use is a three-term approximation to an infinite series that is valid only for $|z| < \frac{\pi}{2}$. If we confine z to the real axis, this restriction becomes $-\frac{\pi}{2} < x < \frac{\pi}{2}$, $y = 0$ even though 1/cosh(z) is analytic at the endpoints of this interval $z = \pm\frac{\pi}{2}$. Notice from our figure that the three-term series is a fair approximation to $\frac{1}{\cosh z}$ in the interval $-.5 \le x \le .5$, $y = 0$. Had we used more terms, we could have broadened this interval. Notice, too, that for $x < \frac{-\pi}{2}$ or $x > \frac{\pi}{2}$, $y = 0$ the series in no way approximates $\frac{1}{\cosh z}$, and the situation could not be improved by increasing the number of terms.

3.2.1 Using MATLAB to Obtain the Series Coefficients

Had we attempted to obtain the first 10 coefficients in the above series by the process of long division that we just used, we would face a tedious problem. Just as forbidding would be our getting these coefficients via Equation 3.10 and successive differentiation of $\frac{1}{\cosh z}$. Here MATLAB comes to our rescue.

MATLAB can be used to differentiate any differentiable function as many times as is practically necessary. We could thus obtain any derivative of 1/cosh z through the MATLAB function **diff**. Here is an example of our obtaining the fourth derivative of 1/cosh z. We are working in the Command window of MATLAB. You must have the Symbolic Math Toolbox installed to accomplish this. You might wish to study the commands **syms** and **diff** in MATLAB help.

```
>> syms z a
a=diff(1/cosh(z),4)
a = 5/cosh(z) - (28*sinh(z)^2)/cosh(z)^3 + (24*sinh(z)^4)/cosh(z)^5
```

Had we wanted say the 20th derivative, we would have put a 20 instead of a 4 in the diff expression. Note that the line of code **syms** z is required to establish z as a symbolic variable.

Suppose we wished to evaluate the preceding answer at $z = 0$ and divide that result by 4! in order to obtain the coefficient of z^4 in the Taylor expansion of $1/\cosh(z)$ about $z = 0$. To do this, we create a *function handle* and an *anonymous function* in MATLAB, topics the reader might wish to review.

```
syms z b
b(z)=@(z)diff(1/cosh(z),4)
the_coeff=b(0)/factorial(4)
```

The output is identical to the $5/24 = \dfrac{1}{4!} \dfrac{d^4}{dz^4} \dfrac{1}{\cosh z}\Big|_{z=0}$ we used in our series above.

There is an even easier way to obtain the Taylor series of an analytic function by means of MATLAB. Suppose we want the first three terms in the Taylor expansion of $\dfrac{1}{\cosh z}$ about $z = i$. We use the MATLAB function **taylor**, in the Symbolic Math Toolbox, from the Command window as follows:

```
>> syms z T
>> T=taylor(1/cosh(z),z,i,'order',3)
```

The output is

```
T = 1/cosh(i) - (sin(1)*(z - i)*i)/cosh(i)^2 - ((sin(1)^2/
    cosh(i)^2 + 1/2)*(z - i)^2)/cosh(i)
```

The series contains the first three terms: $(z - i)^0$, $(z - i)^1$, $(z - i)^2$.

The reader should investigate the function **taylor** in MATLAB help. Notice that had we wanted a different number of terms, we would replace 3 above with that number. The reader should convince himself or herself that the Taylor expansion of $\dfrac{1}{\cosh z}$ about i is valid within a circle in the z plane centered at i and having a radius $\dfrac{\pi}{2} - 1$.

Suppose we want to have the above series as a function that depends on z. And we want to get the numerical value of this three-term series at $.1+.1*i$. The following code will accomplish this:

```
syms z b
b(z)=@(z)taylor(1/cosh(z),z,i,'order',3)
b(.1+.1*i)
eval(b(.1+.1*i))
```

The output is

```
b(z) = 1/cosh(i) - (sin(1)*(z - i)*i)/cosh(i)^2 - ((sin(1)^2/
       cosh(i)^2 + 1/2)* (z - i)^2)/cosh(i)
ans = ((sin(1)^2/cosh(i)^2 + 1/2)*(4/5 + (9*i)/50))/cosh(i) +
      (sin(1)*(-9/10 - i/10))/cosh(i)^2 + 1/cosh(i)
ans = 3.5883 + 0.6864i
```

3.2.2 Laurent Series

If a function $f(z)$ is analytic in a ring-shaped domain described by $r1 < |z-z_0| < r_2$, where z_0 is the center of the ring and r_1 and r_2 are the inner and outer radii, then $f(z)$ can be written as the sum of a Laurent series when z is in this ring, and we have $f(z) = \sum_{n=-\infty}^{\infty} c_n(z-z_0)^n$. The series on the right is said to be the Laurent series expansion of the function on the left. If a function is analytic in a deleted neighborhood of z_0 described by $0 < |z-z_0| < r_2$, then a Laurent expansion of this function is also possible, and it is valid in this "punctured disc." The coefficients of the series in either case can be expressed as contour integrations around closed contours in either the ring or the punctured disc (deleted neighborhood) centered at z_0. The reader might wish to consult a book on complex variables to recall what these integrals are (e.g., W section 5.6 and S section 6.9).

Ordinarily we do not obtain the coefficients by means of these integrals but instead use the method of partial fractions or changes of variable within a known Taylor series. For example, e^{1/z^2} is analytic in a deleted neighborhood of $z = 0$. Recall that e^w has the Taylor expansion $1 + w + \dfrac{w^2}{2!} + \dfrac{w^3}{3!} + ...$ valid for all finite w. We can replace w with $1/z^2$ in the preceding (as long as $z \neq 0$) and obtain the Laurent expansion $e^{1/z^2} = 1 + \dfrac{1}{z^2} + \dfrac{1}{2!z^4} + \dfrac{1}{3!z^6} + ...,$ which is valid for all $z \neq 0$.

Can we use MATLAB to obtain the coefficients in a Laurent series expansion much as we used it to obtain the coefficients in a Taylor series? The answer is no... and yes—there is no function **laurent** like **taylor** in the release 2015a. However, there is a quirk in the **taylor** function that allows it to yield Laurent expansions of functions.

Example 3.5

Use MATLAB to obtain the first few terms in the Laurent expansion of $f(z) = \dfrac{1}{\sin z}$ valid in a deleted neighborhood of $z = 0$.

Solution:
We notice that $\sin z = 0$ at $z = 0$, $z = \pm\pi$, ... An expansion of the function is thus possible in the deleted neighborhood of the origin. This domain

is described by $0 < |z| < \pi$. If you ask MATLAB for a Taylor series expansion of $1/\sin z$ about $z = 0$, you will get an error message—MATLAB recognizes that the expansion is not possible because a function must be analytic at its center of expansion z_0 for an expansion to be available. However, look at

$$g(z) = zf(z) = \frac{z}{\sin z} = \frac{z}{z - \dfrac{z^3}{3!} + \dfrac{z^5}{5!} - \cdots}$$

Here we have used the Maclaurin series of $\sin z$. If $z \neq 0$, we may divide the right side of the preceding to obtain

$$\frac{z}{\sin z} = \frac{1}{1 - \dfrac{z^2}{3!} + \dfrac{z^4}{5!} - \cdots}$$

from which it becomes clear that if we define $g(0) = 1$, we have removed the singularity of $g(z)$ at the origin since

$$\frac{1}{1 - \dfrac{z^2}{3!} + \dfrac{z^4}{5!} - \cdots}$$

is the reciprocal of a convergent power series (i.e., the reciprocal of an analytic function that is nonzero at the origin).

Let us ask MATLAB for the first six terms in the Maclaurin series of $g(z)$ with its singularity removed:

```
syms z b
b(z)=@(z)taylor(z/sin(z),z,0,'order',6)
```

The answer is

```
ans = (7*z^4)/360 + z^2/6 + 1
```

Notice that MATLAB realizes that $z/\sin(z)$ has a removable singularity at $z = 0$. Observe also that MATLAB writes the series in descending powers of z. In our example above where we expanded $1/\cosh z$ about $z = i$, the result was in ascending powers of $(z - i)$. A quirk of MATLAB (as of release 2015a) is that the function **taylor** will yield the Taylor series in ascending powers of $(z - z_0)$ if the expansion is not about $z_0 = 0$. For an expansion about zero, we obtain descending powers.

In the preceding expansion, we asked for six terms, but we seem to have gotten only three. Note that $z/\sin(z)$ is an even function of z. In our series, the first term is simply 1, the second is 0, the third contains the second power of z, the fourth vanishes, the fifth is the fourth power, while the sixth is absent. The terms with odd exponents of z vanish.

The infinite series of which we have obtained only the first three nonzero terms converges for $|z| < \pi$ to the function $\dfrac{z}{\sin z}$ when $z \neq 0$ and to 1 when $z = 0$. Dividing this series by z, we obtain the desired Laurent expansion:

$$\frac{1}{\sin z} = \frac{1}{z} + \frac{z}{6} + 7\frac{z^3}{360} + \dots \quad 0 < |z| < \pi$$

where we have placed the terms in their usual order of ascending powers.

The function $\dfrac{1}{\sin z}$ is analytic in the ring-shaped domain $\pi < |z| < 2\pi$, and a Laurent expansion of the function should be possible there. We can use the previously obtained Laurent series, valid for $0 < |z| < \pi$, to get us partway to the solution. The details can be found in W section 6.3 problem 41.

There are other ways to use the Taylor series function in MATLAB to obtain Laurent series. This is especially true when a function is analytic at infinity. Recall that a function $f(z)$ is analytic at infinity if the function $F(w) = f(1/w)$ is analytic in the w plane at $w = 0$. For example, $f(z) = \dfrac{1}{(z-1)(z-3)}$ is analytic at $z = \infty$ since $F(w) = \dfrac{1}{(w^{-1}-1)(w^{-1}-3)}$ is analytic at $w = 0$ (verify this), while $f(z) = z$ is not analytic at $z = \infty$ since $1/w$ is not analytic at $w = 0$.

Example 3.6

Expand $f(z) = \dfrac{z}{(z-1)(z+3)}$ in a Laurent series in powers of z, valid for $|z| > 3$.

Solution:
We verify that this function is analytic in the finite plane except at $z = 1$ and $z = -3$. Thus, if we are working with series having powers of z, a Taylor series is available for $0 \le |z| < 1$. A Laurent series can be found for $1 < |z| < 3$, and another for $|z| > 3$. The MATLAB function **taylor** will produce the Taylor series. Now

$$F(w) = f(1/z) = f(z) = \frac{w^{-1}}{(w^{-1}-1)(w^{-1}+3)} = \frac{w}{(1-w)(1+3w)}$$

is analytic at $w = 0$. Surprisingly, **taylor** in MATLAB will generate a Laurent series in powers of z. The series will be valid in the unbounded domain containing infinity, in this case, $|z| > 3$, as follows:

```
>> syms z c

c(z) = @(z) taylor(z./((z – 1). * (z + 3)),'ExpansionPoint',inf,'Order',10)

c(z) = 1/z – 2/z^2 + 7/z^3 – 20/z^4 + 61/z^5 – 182/z^6 + 547/z^7
        – 1640/z^8 + 4921/z^9
```

For our Laurent series, valid at $z = \infty$, we chose **inf** as the center of expansion. We have elected to obtain 10 terms, but only the odd powers 1 ... 9 are nonzero.

We see that the function **taylor** is in a sense a misnomer. If we apply **taylor** to a function of z that is analytic at infinity, *we obtain not a Taylor expansion but a Laurent expansion* containing powers of $\frac{1}{z^n}$, where $n \geq 0$ is an integer. Here is another example:

>> c(z) = @(z)taylor(exp(1./z),'ExpansionPoint',inf,'Order',5)

c(z) = 1/z + 1/(2*z^2) + 1/(6*z^3) + 1/(24*z^4) + 1

The above is the Laurent expansion (the first five terms) of $e^{1/z}$. Notice the rather strange ordering of the terms: $1/z$ is the first term, and $1/z^0$ is the last term. The actual Laurent expansion is $\sum_{n=0}^{\infty} \frac{1}{n!} \frac{1}{z^n}$. MATLAB does not yield the Laurent expansion of, for example, $ze^{1/z}$, because this function has a simple pole at infinity, as the reader should verify with the substitution $w = 1/z$.

In Example 3.6, we mentioned that the function $f(z) = \dfrac{z}{(z-1)(z+3)}$ has a Laurent expansion in the domain $1 < |z| < 3$. The methods we have been using with MATLAB will not produce this series.

However, MATLAB can still help us to obtain the series. We express $f(z)$ in the form

$$f(z) = \frac{\alpha}{z+3} + \frac{\beta}{z-1}$$

The first fraction is then expanded in a Taylor series in powers of z, valid for $|z| < 3$, while the second is expanded in a Laurent series in powers of z, valid for $|z| > 1$. The two series are added together to give us the desired Laurent series.

Although we should know how to obtain α and β in the above from the methods of elementary calculus, MATLAB can do it for us. At this point, you might wish to study the MATLAB function **residue** in the help folder. We proceed as follows:

>> [L M N] = residue([1 0], [1 2 −3])

$$L = \begin{bmatrix} 0.7500 \\ 0.2500 \end{bmatrix}$$

$$M = \begin{bmatrix} -3.0000 \\ 1.0000 \end{bmatrix}$$

N = []

Here is what was done. The denominator of our fractional expression is $(z - 1)(z + 3) = z^2 + 2z - 3$. This is a polynomial of second order whose coefficients, in descending powers, are given by in the row vector [1 2 –3]. The numerator of our fractional expression is the polynomial z whose coefficients we supply to MATLAB are in the form [1 0]. Again we go from highest to lowest power, with the zero arising because there is no constant (z^0) term in the numerator. The 1 is simply the coefficient of z in the numerator. Now L gives the coefficients in the partial fractions, while M supplies the location of the poles (i.e., information to make the denominators). Thus, the first fraction is $\dfrac{.75}{z+3}$, while the second is $\dfrac{.25}{z-1}$. The vector N here is empty. It arises in nonempty form if we are asked to reduce to partial fractions the ratio of two polynomials in which the degree of the numerator equals or exceeds that of the denominator, as is discussed later. The reader should check our answer—that is, verify that

$$\frac{.75}{z+3} + \frac{.25}{z-1} = \frac{z}{(z-1)(z+3)}$$

Now we expand the first fraction in a Taylor series and the second in a Laurent series, each in powers of z, and we add the two series, as follows:

```
syms z a b
a(z)=@(z)taylor(.75/((z+3)),'ExpansionPoint',0,'Order',4);
b(z)=@(z)taylor(.25/(z-1),'expansionpoint',inf,'order',4);
a(z)+b(z)
```

We chose to use four terms in each series.

The output is

ans = 1/(4*z) – z/12 + 1/(4*z^2) + z^2/36 + 1/(4*z^3) – z^3/108 + 1/4

Note the unorthodox grouping of the terms in $a + b$. Ordinarily, we would first state the terms with negative exponents of z (i.e., those in the series for b), while those in the series for a would be given second but in the order opposite to that given here.

Suppose the problem had been to expand

$$f(z) = \frac{z^3 + 2z^2 + i}{(z-1)(z+3)}$$

in a Laurent series valid for $|z| > 3$. We again seek a partial fraction decomposition, but note that we are dealing here with an improper rational expression—the numerator is of degree 3 and the denominator of degree 2. We proceed as follows:

```
>> [L M N]=residue([1 2 0 i],[1 2 -3])
```

$$L = \begin{bmatrix} 2.2500 - 0.2500i \\ 0.7500 + 0.2500i \end{bmatrix}$$

$$M = \begin{bmatrix} -3.0000 \\ 1.0000 \end{bmatrix}$$

$N = [1\ 0]$

Note that N is not empty.
The preceding means that the partial fraction decomposition of $f(z)$ is

$$f(z) = \frac{2.25 - .25i}{z+3} + \frac{.75 + .25i}{z-1} + z$$

The denominators of the two fractions are in the elements of M, while the corresponding numerators come from L. The vector N supplies the polynomial that must be added to the fractions—a consequence of $f(z)$ being an improper fraction. There are two elements in N that tell us that the polynomial has two terms, the first containing z and the second z^0. The coefficients are 1 and 0, respectively. We now expand each of the two fractions in Laurent series. We seek four terms in each series, for brevity. We do not seek to expand z, because it is already a one-term Taylor series.

```
syms z a b
a(z)=@(z)taylor((2.25-.25*i)/(z+3),'ExpansionPoint',inf,
  'Order',4)
b(z)=@(z)taylor((.75+.25*i)/(z-1),'expansionpoint',inf,'order',4)
a(z)+b(z)
```

The output of this program is

ans = 3/z + (−6 + i)/z^2 + (21 − 2*i)/z^3

Our final answer, the desired Laurent series for $|z| > 3$, requires our adding the term z to the above. Thus, we have

$$f(z) = \frac{z^3 + 2z^2 + i}{(z-1)(z+3)} = z + 3/z + (-6+i)/z^2 + (21-2*i)/z^3 + \cdots$$

for $|z| > 3$.

Exercises

1. Perform a calculation similar to Example 3.3 but compute sin(3 + 4i) using the Maclaurin series expansion of sin z. Compare the sum of series having 1, 2, ... 10 nonzero terms to the MATLAB determination of sin(3 + 4i). Recall that $\sin z = z - \dfrac{z^3}{3!} + \dfrac{z^5}{5!} - \ldots$

   How many terms would you need in your series to get agreement between the MATLAB value of sin(3 + 4i), and the series approximation, that is within 1%?

2. Consider the series $\dfrac{1}{1-z} = 1 + z + z^2 + z^3 + \ldots$. This series is uniformly convergent in the disc $|z| \le r$ if $r < 1$. If necessary, review the concept of *uniform convergence* in either W section 5.3 or S section 6.4. Because the series is uniformly convergent, we may integrate it term by term along a contour connecting $z = 0$ with $z = w$ in the complex plane. We require that $|w| < 1$ and that the contour lies in the above disc. Thus,

$$\int_0^w \frac{dz}{1-z} = \int_0^w dz + \int_0^w z\,dz + \int_0^w z^2\,dz + \ldots = w + \frac{w^2}{2} + \frac{w^3}{3} + \ldots \quad |w| < 1$$

   But notice also that $\displaystyle\int_0^w \frac{dz}{1-z} = -\text{Log}(1-z)\big]_0^w = \text{Log}\frac{1}{(1-w)}$. Thus,

$$\text{Log}\frac{1}{1-w} = w + \frac{w^2}{2} + \frac{w^3}{3} + \ldots \quad |w| < 1$$

   a. Take $w = \dfrac{i}{\sqrt{3}}$ in the preceding equation. By taking the real and imaginary parts of this equation, show that

$$\text{Log}\frac{\sqrt{3}}{2} = \frac{-1}{2\times 3} + \frac{1}{4\times 9} - \frac{1}{6\times 27} + \ldots$$

   and

$$\frac{\pi}{6} = \frac{1}{\sqrt{3}} - \frac{1}{3\times\sqrt{3}\times 3} + \frac{1}{5\times\sqrt{3}\times 9} - \ldots = \frac{1}{\sqrt{3}}\left[1 - \frac{1}{3\times 3} + \frac{1}{5\times 9} - \frac{1}{7\times 27} - \ldots\right]$$

   b. The last equation provides a method of computing π provided you know the value of $\sqrt{3}$. Multiply both sides of the preceding equation by 6, and compute π by using different finite series ranging from 1 to 10 terms, total. Use Example 3.3 as your model, and compare the result obtained from each series with the value that MATLAB chooses for π.

3. An apparent paradox. According to the fundamental theorem of algebra (see W section 4.6 and S section 5.2 and problem 5.10), the algebraic equation $a_0 + a_1z + a_2z^2 + \ldots + a_nz^n = 0$, where the a's are constants, n is a positive integer, and $a_n \neq 0$ has exactly n roots in the complex plane. Now suppose we approximate e^z by the first $n + 1$ terms in its Maclaurin series:

$$e^z \approx 1 + z + \frac{z^2}{2!} + \ldots + \frac{z^n}{n!}$$

It is apparent that the more terms we use in this Maclaurin polynomial approximation, the more places in the complex plane where the approximation has the value zero. However, $e^z \neq 0$ in the finite complex plane. Thus, it appears that the more terms we use in the Maclaurin expansion, the greater is the number of locations in the complex plane where the series cannot well approximate e^z. It can be proved that this paradox can be resolved by showing that the roots of $a_0 + a_1z + a_2z^2 + \ldots + a_nz^n = 0$, $a_k = \dfrac{1}{k!}$, move to infinity as $n \to \infty$. We confirm this numerically.

a. Write a MATLAB program that will find the roots of $1 + z + \dfrac{z^2}{2!} + \ldots + \dfrac{z^n}{n!}$ for $n = 1, 2, \ldots, 25$. These roots should be plotted as points in the complex plane and labeled with the number n from the corresponding equation. Observe that the roots do move farther from the origin with increasing n. Use the MATLAB function **roots**. You might want to open **help roots** in MATLAB to refresh your memory. Note that when entering the coefficients of the polynomial whose roots you are seeking, that numbers must be entered for the highest power first, and then descend in order.

b. Alter the preceding program so that it will find the root, for each value of n, that lies closest to the origin in the complex plane. Your new program should plot the magnitudes of these roots versus n.

4. a. Show that $\mathrm{Log}\, z = (z-1) - \dfrac{1}{2}(z-1)^2 + \dfrac{1}{3}(z-1)^3 - \dfrac{1}{4}(z-1)^4 + \dfrac{1}{5}(z-1)^5 - \ldots$ by direct use of Equations 3.9 and 3.10. State the center and radius of the domain in which the series converges to the function on the left.

b. Determine the first five coefficients by using the MATLAB function **diff** in the Symbolic Math Toolbox. Recall that the MATLAB function log is the principal value.

c. Determine the series in (a) by using the MATLAB function **taylor**.

d. Assume $z = x$ (is real). Obtain a plot comparable to Figure 3.3 in which you compute Log(x–1) and compare it with a five-term Taylor expansion (the first five nonzero terms) about $z = 1$. Consider the interval $0 < x < 3$. Comment on where the agreement is reasonably good and where it is not, and be sure to include discussion where the infinite Taylor series is valid.

e. Derive the formula $\tan^{-1}(a) + \tan^{-1}(b) = \tan^{-1}\left(\dfrac{a+b}{1-ab}\right)$, where a and b are real. This is easily done from the product $(1 + ia)(1 + ib) = 1 - ab + i(a + b)$ and the fact that the argument (or angle) of a product of complex factors is the sum of the arguments of each factor. Show from your result that $\tan^{-1}(1/3) + \tan^{-1}(1/2) = \pi/4$.

f. From the preceding, we have $4[\tan^{-1}(1/3) + \tan^{-1}(1/2)] = \pi$. Let $z = 1 + iy$ be the result of part (a). Derive the series $\tan^{-1}(y) = y - \dfrac{y^3}{3} + \dfrac{y^5}{5} - \dots$ State the restrictions on the real variable y for this series to be valid.

g. Prove that

$$\pi = 4\left[\left(\frac{1}{3} + \frac{1}{2}\right) - \frac{1}{3}\left(\frac{1}{3^3} + \frac{1}{2^3}\right) + \frac{1}{5}\left(\frac{1}{3^5} + \frac{1}{2^5}\right) + \dots + \frac{(-1)^{n-1}}{2n-1}\left(\frac{1}{3^{2n-1}} + \frac{1}{2^{2n-1}}\right)\dots\right]$$

Now, taking Example 3.3 as your model, use the preceding to compute π. Compare the result obtained from each finite series ($n = 1, 2, \dots$) with the value that MATLAB produces for π. Is this series more useful for computing π than the one in problem 2(b)?

5. a. Consider the principal branch of the function

$$f(z) = \frac{1}{(1-z)^{1/2}}$$

Where in the complex plane is this an analytic function? Where is this Maclaurin series expansion

$$\sum_{n=0}^{\infty} c_n z^n = \frac{1}{(1-z)^{1/2}}$$

valid? Give the center of the domain and its radius. Through the use of Equation 3.10, find a formula for c_n in terms of n.

b. Using the MATLAB function **diff**, show that the first six terms in the Maclaurin series as given by the formula you derived in

(a) agree with those obtained with having MATLAB generate the derivatives.

c. Confirm the results in (b) by using the MATLAB function **taylor**. Convert the expression found to an anonymous function of z using the @(z) notation.

d. Show that you can use your series to find $\sqrt{2}$ by suitable choice of z. Using this value of z in the above derived function handle, find what values you get for $\sqrt{2}$ by using the series having six terms found in (c) and compare this result with the MATLAB-generated sqrt(2). Use double precision. Now repeat this but use 50 terms. Are the results better with 50 terms?

3.3 Integration in the Complex Plane: Part 1, Finite Sums as Approximations

One reason that we have been paying so much attention to sums and limits is that these concepts provide the basis for understanding the most important and common operation in complex variable theory: contour integration. To perform a contour integration, we need a piecewise smooth curve (a contour) and a function, $f(z)$, to integrate along the contour. This function is often, but not always, analytic. The integral of $f(z)$ along the contour is—like all integrals—defined as the limit of a sum. In the case of a contour integral, we have this definition. The reader should refer to Figure 3.4.

We are here integrating the function $f(z)$ along the arc C in the complex plane. We integrate from point A to point B. The arc is divided into n sub-arcs. The endpoints of these arcs are $(x_1, y_1), (x_2, y_2), ..., (x_{n+1}, y_{n+1})$. Somewhere on each arc we choose just one point. The coordinates of these points are, respectively, $(x_1,y_1), (x_2,y_2), ..., (x_n,y_n)$, and for each of these points there are corresponding complex numbers $z_1 = x_1 + iy_1, z_2 = x_2 + iy_2, ..., z_n = x_n + iy_n$. Each arc is subtended by a vector chord creating complex numbers. Here, $\Delta z_1 = \Delta x_1 + i\Delta y_1$, $\Delta z_2 = \Delta x_2 + i\Delta y_2$, and so on. The vectors are shown in the figure. Thus, $\Delta x_k = X_{k+1}-X_k$ and $\Delta y_k = Y_{k+1}-Y_k$, where $k = 1, 2, ..., n$. We now define the integral

$$\int_A^B f(z)dz = \lim_{n\to\infty} \sum_{k=1}^n f(z_k)\Delta z_k \tag{3.15}$$

where all $\Delta z_k \to 0$ as $n \to \infty$. Note that the complex number $A = X_1 + iY_1$ and $B = X_{N+1} + iY_{N+1}$. We must think of the summation on the above right as arising from summing the following n products: the function $f(z)$ evaluated at a point on C multiplied by the complex number for the neighboring vector chord.

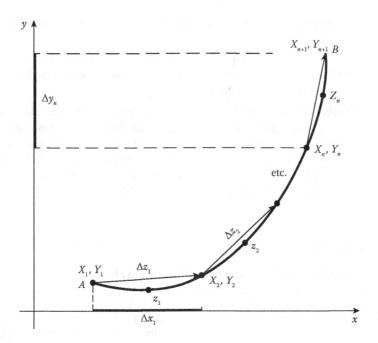

FIGURE 3.4
Defining a contour integral.

If we do not pass to the limit on the right, but instead use some finite value of n, we obtain only an approximation of the integral

$$\int_A^B f(z)\,dz \approx \sum_{k=1}^{n} f(z_k)\Delta z_k \qquad (3.16)$$

As n increases, the sum on the right ultimately becomes a better approximation to the integral on the left, although we cannot argue that the improvement is monotonic. For a particular $f(z)$ and C, we cannot say for certain that, for example, by going from $n = 10$ to $n = 11$ the sum on the right must be a better approximation to the integral. Note that the result on the right in Equation 3.16 depends to some extent on where we have placed each z_k on the corresponding subarc. However, in the limit as $n \to \infty$ and all $\Delta z_k \to 0$, this placement does not matter.

Let us investigate a particular integral whose value is known.

Example 3.7

Consider $\int_i^1 \cos z\,dz$, where the integral is along the arc C in the first quadrant $y = 1 - x^2$.

Approximate this integral by sums like that on the right side of Equation 3.16, where $n = 3$, and then compare the sum with the value of the integral obtained by using the antiderivative of cos z.

Solution:
We create the subarcs, arbitrarily, by making each x-axis projection of Δz_k, namely, Δx_k, to be of identical length. Figure 3.5 shows what happens with three such subdivisions.

Since x advances from 0 to 1 as we integrate, we have when $n = 3$ that $X_1 = 0$, $X_2 = 1/3$, $X_3 = 2/3$, $X_4 = 1$. The corresponding values of the Y coordinates are obtained from the equation of C—that is, $y = 1-x^2$. Thus, $Y_1 = 1$, $Y_2 = 8/9$, $Y_3 = 5/9$, $Y_4 = 0$.

Having found the endpoints of each subarc, we have that $\Delta z_1 = 1/3 - i/9$, $\Delta z_2 = 1/3 - i/3$, $\Delta z_3 = 1/3 - i5/9$ as some study of the figure should confirm. Notice that the sum of these three quantities is simply $1 - i$ which is a vector connecting the start of integration at $(1,0)$ with the finish $(0,1)$. This will *always* be the case with the Δz_k quantities—their sum will equal a vector going from the start to the finish of integration.

The quantities z_k must now be chosen. Recall that they must each reside somewhere on the corresponding subarc. We choose arbitrarily to place them in such a way that each x_k lies exactly between the two values of X_k defining the ends of the arc. The corresponding value of y_k is found from the equation of the contour C. Thus, $y_k = 1 - x_k^2$. We now have $x_1 = 1/6$, $y_1 = 35/36$, $x_2 = 1/2$, $y_2 = 3/4$, $x_3 = 5/6$, $y_3 = 11/36$. Our corresponding three-term sum on the right in Equation 3.16 is

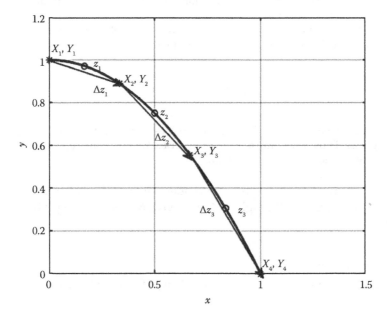

FIGURE 3.5
The arc C and vector chords.

$$\sum_{k=1}^{3} f(z_k)\Delta z_k = \cos(1/6 + i35/36)(1/3 - i/9) + \cos(1/2 + i3/4)(1/3 - i/3)$$
$$+ \cos(5/6 + i11/36)(1/3 - i5/9)$$

which is easily evaluated with MATLAB to be $0.8302 - 1.2061i$.

Using complex variable theory, one may evaluate the integral on the left in Equation 3.17

$$\int_i^1 \cos z \; dz$$

exactly (see W section 4.4 and S section 4.13).

Recall the following. Let $\int_{z_1}^{z_2} f(z)dz$ be an integral taken along a contour C connecting z_1 with z_2. Assume that C lies in a simply connected domain (one with no holes) and that in this domain we have found an analytic function $F(z)$ such that $dF/dz = f(z)$. Then the value of our integral is given simply by

$$\int_{z_1}^{z_2} f(z)dz = F(z_2) - F(z_1)$$

$F(z)$ is called the antiderivative of $f(z)$; if $f(z)$ is not analytic in a domain containing the contour of integration, then we cannot find an analytic $F(z)$ as it does not exist.

The function $\cos z$ is entire and is the derivative of another entire function $\sin z$. The path of integration is immaterial, and we have that

$$\int_i^1 \cos z \; dz = \sin 1 - \sin i$$

which we evaluate, to four decimal places, with MATLAB, as $0.8415 - 1.1752i$. Thus, with only a three-term series, we got a fair approximation to the "exact" value of the integral.

The following MATLAB program redoes the preceding example by means of an n term series on the right in Equation 3.16, where n ranges from 1 to 10 values.

Example 3.8

Repeat Example 3.7 but use series that range from 1 to 10 terms.

Solution:
```
nmax=input('max number of terms in series')
for(n=1:nmax)
```

```
m=n+1;% number of points needed on curve to define the n
% subarcs
X=linspace(0,1,m);% this gives
%the values of X needed
Y=1-X.^2;%gives values of Y on curve
delta_x=(X(2)-X(1))/2;
x=X(1:n)+delta_x;%the n values of x coordinate where we...
% evaluate f(z)on curve (near middle of chord)
y=1-x.^2; %the values of y corresponding to the above
z=x+i*y; % the complex values of z needed on the curve
fz=cos(z);%the n values of the function on the curve
yd=Y(2:m)-Y(1:m-1);% the vertical component of the vector
% chord
xd=2*delta_x;%the horizontal component of the vector chord
zdchord=xd+i*yd;% the vector chord complex representation
s=fz.*zdchord; %the product of the complex vector chord...
% and the function cos z evaluated nearby on the curve
series_sum(n)=sum(s);% the sum of those products
end
number_of_terms=[1:n]';
format long
approx_value=series_sum.';
disp('number of terms in series series sum')
disp([number_of_terms approx_value])
disp('sin(1)-sin(i)')
disp(sin(1)-sin(i))
```

The output of this program is

```
max number of terms in series10

nmax = 10

number of terms in series            series sum

1.000000000000000    0.741951831691765 - 1.530431115914932i
2.000000000000000    0.817101149113634 - 1.246464676977509i
3.000000000000000    0.830203869938418 - 1.206054884648359i
4.000000000000000    0.835034307863057 - 1.192415447540212i
5.000000000000000    0.837320720853477 - 1.186178843527661i
6.000000000000000    0.838576965669872 - 1.182810093770062i
7.000000000000000    0.839339429392264 - 1.180785102772347i
8.000000000000000    0.839836339025017 - 1.179473274644714i
9.000000000000000    0.840177956756501 - 1.178574997871836i
10.000000000000000   0.840422785323087 - 1.177933013899162i

sin(1)-sin(i)
   0.841470984807897 - 1.175201193643801i
```

Notice that the agreement between the above "exact" value and the value obtained by the series approximating the integral is surprisingly good. With only a five-term series, the disparity in both real and imaginary parts is less than 1%, and agreement continues to improve with the use of more terms.

As a check, we should observe that the results obtained from the three-term series agree with that obtained in the previous example.

The reader might wonder what the point of the preceding exercise is if we can evaluate the integral so easily by using the antiderivative, which quickly yields the exact result $\sin(1) - \sin(i)$. Of course, there is no point in determining the integral numerically in the way we just have. However, had the integrand been a function that is not analytic, for example, $f(z) = \cos|z|$ or $f(z) = \cos \bar{z}$, we cannot find an antiderivative, and a numerical determination is required.

The function $f(z) = \frac{\cos z}{z}$ is analytic except at $z = 0$ and is therefore analytic throughout a simply connected domain containing the path of integration like the one just employed. However, an antiderivative $F(z)$ cannot be found for this function in terms of the elementary functions of calculus, and a numerical integration is necessary unless one might have access to numerical tables (in this case, they do exist) yielding this integral.

MATLAB provides several functions that will perform numerical integrations, among them are **quad**, **quadl**, and **quadgk**. The last is among the most recent in MATLAB and of the most use to us here, because under certain circumstances, it permits integration between complex limits, as well as limits that can be infinite. It is so valuable that the next section is devoted to **quadgk**.

Exercises

1. a. Approximate the integral $\int_{0}^{1+i} \sin z\, dz$ along the contour $y = x^2$ by using a three-term series in the expression $\sum_{k=1}^{n} f(z_k)\Delta z_k$. Choose the vector chords Δz_k as they were done in Example 3.7 (i.e., they have identical projections on the real axis). Also, choose the x_k values to be identical to the value of x in the middle of each chord, as was done in Example 3.7.

   b. Explain why the exact value of the integral is $1 - \cos(1 + i)$, and compare this with the value found in (a).

   c. Write a MATLAB program that will approximate the integral by n terms where n varies between 1 and 10. Use the same criteria for Δz_k (uniform x-axis projections) as was done in Example 3.7 and the same criteria for choosing x_k as was done in part (a).

2. a. Consider the integral $\int_0^\pi \sinh z\,dz$ taken along the contour $y = \sin x$.

   Perform this by means of a numerical integration like that done in Examples 3.7 and 3.8. Use Δz_k that have identical projections along the x-axis. Use n vector chords to approximate the contour, where n varies from 1 to 10. Show your results for all 10 values. Now evaluate the integral exactly by using the antiderivative of sinh z. Are the results of your numerical integration real numbers? Explain why, ideally, they should be real.

   b. Use the above technique to approximate the integral by a sum with 100 terms. How different is this result from the one obtained with 10 terms? Consider especially the imaginary part.

3. Consider the integral $\int_{-i}^{i} \text{Log}\, z\,dz$ performed in the half plane Re$(z) \ge 0$

   along the half circle satisfying $x^2 + y^2 = 1$. Perform this by means of a numerical integration like that in Examples 3.7 and 3.8 but here use Δz_k that have identical projections along the y-axis. Again use n vector chords to approximate the contour where n has only the values 10, 100, and 1,000, in other words, n increases by factors of 10. The integrand is analytic in the given half plane and has an antiderivative that you can find in a table of integrals if you have forgotten it. Using this function, find the exact value of the integral and compare it with the "exact" value.

4. a. Consider the integral $\int_0^2 \dfrac{\sin z}{z}\,dz$ performed in the complex plane

   along $y = x(2 - x)$. Perform the integration numerically by following the technique in Examples 3.7 and 3.8, which uses values for Δz_k having identical projected lengths along the x-axis. If you evaluate sin z/z at those values of z whose x coordinates lie at the middle of each chord approximating the curve, then you will never have to ask MATLAB for the undefined sin $(0)/0$.

   Use series having 10, 100, and 1,000 terms and state your answers in double precision.

   b. The integrand is not analytic at $z = 0$ because it is undefined there. However, the integrand has a removable singularity at this point. Note that $\lim\limits_{z\to 0}\dfrac{\sin z}{z} = 1$ as can be verified with L'Hôpital's rule. We can remove the singularity in the integrand by defining the integrand as 1 when $z = 0$. This creates an integrand that is an entire function. The reader may wish to review "removable singularities" in W section 6.2 or S section 3.11. The integrand is thus an entire function, and we can deform the contour of integration into a straight line connecting $z = 0$ with $z = 2$. You may wish to review "deformation of contours" or "path independence" in W section 4.3 or S section 4.15. Thus, an integral having the same value as the one in (a) is $\int_0^2 \dfrac{\sin x}{x} dx$ on this line. This kind of integral turns up frequently in antenna engineering and optics where one encounters the function called the "sine integral" defined as $S_i(x') = \int_0^{x'} \dfrac{\sin x}{x} dx$. If you Google the words "sine integral, Si, and high accuracy," you will come to a website that gives values of this integral for different real values of x' to a high degree of accuracy. Thus, the results in part (a) should approximate $S_i(2)$. The site yields $S_i(2) = 1.60541297680269484858$. Compare this with the result obtained by using the series in part (a) with 1 million terms.

3.4 Integrations in the Complex Plane: Part 2, int and quadgk

MATLAB provides us with a wealth of methods for integration. We confine ourselves here to two of the most useful tools: **int** and **quadgk**. The first of these, **int,** will do symbolic integration. Ask it for the indefinite integral of cos x and it will return sin x. It will also perform definite integrals. If you ask it to integrate cos πx from 1/13 to 1/11, it will produce (sin(pi/11) − sin(pi/13))/pi. If possible, it does not produce decimal values for the results. However, if you ask **int** for the value of an integral which it cannot obtain by evaluating an antiderivative at the two limits, it can give you a numerical result based on numerical integration. You will also receive a warning that an "Explicit Integral" could not be found. To get the numerical result, you must use the command **vpa** as described below. The command **double** can also be used but is less reliable and will give the answer to exactly 16 decimal places. With **vpa** you have more flexibility and reliability.

The tool **quadgk** is only for definite integrals, which it evaluates numerically. The output of this function is a decimal number. The great utility of **quadgk** for us is that it will perform a contour integration in the complex plane provided that the contour is composed of straight lines. This last restriction might seem like a handicap, but we know that if our integrand is analytic, we can usually deform a curved contour into a straight line or into a contour composed of straight lines. In the case of both **int** (when used for definite integrals) and **quagk,** we can have one or both limits of integration set to infinity which we designate *inf*. The number *inf* is the real number, say x, as it increases without bound through positive values, $inf = \lim_{x \to \infty} x$, while $-inf$ is a real number x as it decreases without bound through negative values $-inf = \lim_{x \to -\infty} x$. One should note that like **quadgk**, **int** will do an integration in the complex plane, but the contour must be a single straight line and cannot be composed of a sequence of straight lines. For example, **quadgk** can integrate all the way around or part of the way around a triangle, while **int** would integrate only along one side.

At this point the reader should study both methods of integration, **int** and **quadgk**, by going to the **help** feature of MATLAB. In the remainder of this section, we see how **int** and **quadgk** can be used to verify results obtained with residue calculus or in some cases perform integrations that are possible with complex variable theory but require tedious mathematical manipulations.

Example 3.9

Find $I = \int\limits_{0}^{\infty} \dfrac{x^2 dx}{x^4 + 9}$ by using residue calculus and by using both **int** and **quadgk**.

Solution:

    a. With residues:

        Let us recall a theorem from complex variable theory (see W section 6.5 and S problems 7.15–7.17). Suppose we must evaluate the integral $\int\limits_{-\infty}^{\infty} \dfrac{P(x)}{Q(x)} dx$, where P and Q are polynomials in the variable x.

        Assume that the highest power of x in Q exceeds that in P by two or more, in other words, the degree of Q is at least two bigger than P, and we also agree that $Q(x) \neq 0$ for all real x. Then,

$$\int\limits_{-\infty}^{\infty} \frac{P(x)}{Q(x)} dx = 2\pi i \sum \text{Res}\left[\frac{P(z)}{Q(z)}\right] @ \text{ all the poles in the upper half plane} \quad (3.17)$$

        We will use *uhp* to abbreviate "upper half plane," and Res of course means "residue."

        In words: the integral above is $2\pi i$ times the sum of the residues of the function $P(z)/Q(z)$ at the poles of this function in the *uhp*. If we wish, we may instead sum the residues of the poles in the lower half plane. However, should we make that choice, we would need to put a *minus sign* in front of the two on the right side of the preceding equation.

        We were given an integration for I that has limits from 0 to infinity. Because of the even symmetry in the integrand, we have that $I = \int\limits_{-\infty}^{\infty} \dfrac{\frac{1}{2}x^2 dx}{x^4 + 9}$. Thus, in Equation 3.17, we take $P(z) = z^2/2$

and $Q(z) = z^4 + 9$. The equation $Q(z) = z^4 + 9 = 0$ has roots where $z = (-9)^{1/4} = \sqrt{3}e^{\pm i\pi/4}$ and $\sqrt{3}e^{\pm i3\pi/4}$. Two of these roots of $Q(z)$ lie in the upper half plane: $z = (-9)^{1/4} = \sqrt{3}e^{i\pi/4}$ and $\sqrt{3}e^{i3\pi/4}$. These roots result in $P(z)/Q(z)$ having simple poles since all four roots of $Q(z)$ are distinct (nonrepeated). Recall (see W section 6.3) that if $f(z) = g(z)/h(z)$ has a *simple pole* at a point z_0, where $g(z_0) \neq 0$, then the residue of $f(z)$ is

$$\text{Res}\, \frac{g(z)}{h(z)} @ z_0 = \frac{g(z_0)}{h'(z_0)} \tag{3.18}$$

This is one of the handiest formulas in complex variable theory. Thus, if we take $P(z) = g(z)$ and $Q(z) = h(z)$, the residues of $P(z)/Q(z)$ at $z = (-9)^{1/4} = \sqrt{3}e^{i\pi/4}$ and $\sqrt{3}e^{i3\pi/4}$ are, respectively,

$\dfrac{1}{2}z^2 / (4z^3) = 1/(8z)$ evaluated at these two points. The residues

are $\dfrac{1}{8\sqrt{3}}e^{-i\pi/4}$ and $\dfrac{1}{8\sqrt{3}}e^{-i3\pi/4}$. Summing these and applying

Equation 3.17, we have finally

$$I = \frac{2\pi i}{8\sqrt{3}}\left[\cos(\pi/4) - i\sin(\pi/4) + \cos(-3\pi/4) + i\sin(-3\pi/4)\right] = \frac{\pi}{2\sqrt{6}}$$

We note that the answer must be real, and it is.

b. To integrate with **int**, we use this MATLAB code:

```
syms x;
I=int((x.^2)./(x.^4+9),0,inf)
```

The output is

```
I = (pi*6^(1/2))/12
```

Observe that this is the same as $\pi\sqrt{6}/12 = \dfrac{\pi}{2\sqrt{6}}$, the result obtained by residues.

Notice the importance of the line of code:

```
syms x;
```

This tells MATLAB that x is a symbolic variable.

c. To integrate with **quadgk**, we use this code:

```
f=@(x) (x.^2)./(x.^4+9)
format long
I=quadgk(f,0,inf)
```

The output is

```
I = 0.641274915080932
```

In the preceding code, the line **format long** is optional. It gave us an output with 16 decimal places that we can use to compare with the output of part (b). If you use MATLAB to convert (pi*6^(1/2))/12 to a decimal expression and use long format, you will obtain the same result as in part (c). To make the conversion, use the line of code

double ((pi*6^(1/2))/12) or **vpa** ((pi*6^(1/2))/12, 16).

The number 16 here tells **vpa** to yield an expression with 16 decimal places. You can use other integers here.

Some comments:
Obviously, **quadgk** and **int** each have their uses. If you want to obtain an immediate numerical expression for use in other calculations, then **quadgk** would be your best bet. If you want an expression with numerical symbols to place inside an engineering or physics formula, you might prefer **int**. Of course, you can do the integration in the preceding example with residue calculus. However, if you were given the problem to find, for example, $\int_0^1 \frac{x^2 dx}{x^4+9}$, you could not quickly find the answer with residues. You might resort to an integral table or perhaps just use **quadgk** or **int** depending on your needs. The code would be like that used in Example 3.9 except that the upper limit of integration would no longer be **inf** but would be 1. There are symbolic problems in integration that cannot be done with **int** (i.e., it cannot do a closed-form expression for the result). See problem 2 that follows for example.

A problem such as $I = \int_0^{\infty} \frac{x^2 dx}{x^4+x^2+3+i}$ can in principle be done with residue calculus, but you would be faced with the messy job of locating the roots of $z^4 + z^2 + (3 + i)$ in the complex plane. Here you are much better off with either **int** or **quadgk**, depending on your goal. A similar integral is considered in exercise problem 5.

A problem like $\int_0^3 e^{i(\sinh z)^{1/2}} dz$ will pose a difficulty for **int** as the following code shows:

```
syms z
I=int(exp(i*sqrt(sinh(z))),0,3)
```

Which yields the following warning:
Warning: Explicit integral could not be found.
However, an additional line of code:

```
vpa(I,10)
```
yields this answer, in numerical form, correct to 10 decimal places.
ans = 2.15702077*i + 0.1021467801

3.4.1 Integration in the Complex Plane with MATLAB

The great utility of **quadgk**, as we have mentioned, lies in its ability to do integrations in the complex plane along contours composed of straight lines. If we ask MATLAB for **quadgk**(*fun, a, b*) where *fun* is expressed as a function handle using the α notation, and *a* and *b* are numbers that can be complex, then the integral is evaluated numerically along a straight line connecting *a* and *b* in the complex plane. Here is an example. Notice that when using the anonymous function, we do not have to tell MATLAB that the argument of α is a symbol. This is assumed. But if you forget and use **syms**, the program will still run.

Example 3.10

Find $I_1 = \int\limits_{(1-i)}^{(1+i)} e^{iz}\, dz$ by using **quadgk**, and verify the answer with complex variable theory.

Solution:
The required code is simply as follows:

```
syms z
f=@(z) exp(i*z)
format long
I=quadgk(f,1-i,1+i)
check_answer=(exp(i*(1+i))-exp(i*(1-i)))/i
```

Whose output is

I = −1.977795411525730 + 1.269927829569473i

check_answer = −1.977795411525731 + 1.269927829569473i

   Notice the "check answer" portion of the output. We recall that since e^{iz} is an analytic function, we may simply evaluate the integral using the antiderivative of this function. Thus, $I = \dfrac{e^{iz}}{i}\Big|_{1-i}^{1+i} = \dfrac{e^{i(1+i)} - e^{i(1-i)}}{i}$. Observe that the agreement between the two methods is nearly perfect. Had the given integrand been $e^{\bar{z}}$, which is not analytic, we could not have checked the results of our numerical integration so easily, as this function does not have an antiderivative.

   A note is appropriate here about the use of infinity (**inf**) in the limits of **quadgk** if we are integrating in the complex plane. One can use this as one of the limits *provided that the other limit is real*. One can also use - **inf** if the other limit is real. You can use **−inf** and **inf** together. We cannot really use infinity as a limit unless we stick to the real axis, although complex integrands are permissible when integrated on the real axis.

3.4.2 Waypoints

Using **quadgk**, we can integrate along a contour composed of straight lines. The *waypoints* vector, placed in **quadgk**, tells **quadgk** where the endpoints of these straight lines lie. For example, suppose we wish to integrate $\cos z/z$ from $1 + i$ to $-3i$ using this path of successive straight lines: from $1 + i$ to $2i$ and then along the line from $2i$ to $-1 + 2i$, and from there to $-1 - i$, and finally from $-1 - i$ to $-3i$. (See the solid lines with arrows in Figure 3.6.) We write code in which *waypoints* is a row vector whose elements are the points where the contour abandons its old straight line path and starts on a new one. We *do not include* in waypoints the starting points and endpoints in the numerical integration. These points still appear as the *limits of integration*. The numerical entries in waypoints are the points on the path where the tangent of the overall contour of integration is discontinuous; we enter these points as complex numbers in waypoints from left to right in the order in which they are encountered on the path. If you elect to use *waypoints*, you may not use infinity (i.e., **inf** or **−inf**) in the waypoints vector or in either the starting point or endpoint of the integration if any of the elements in waypoints is not real, or if any of the limits of integration are not real.

Here is the required code to solve the problem described above:

```
format long
syms z
f=@(z)cos(z)./z;
IA=quadgk(f,1+i,-3i,'waypoints',[2i,-1+2i,-1-i])
```

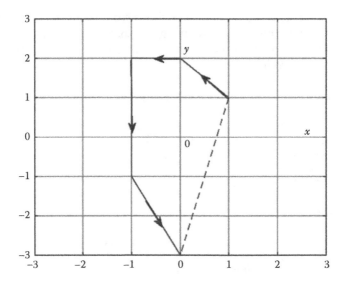

FIGURE 3.6
An illustration of "Waypoints."

The output is

```
IA = 4.078219914209672 + 4.4251398468664735i
```

Notice (refer to the figure and the solid lines) that the integration begins at 1 + *i* and ends at –3*i* and that the waypoints are 2*i*, –1 + 2*i*, and –1 – *i*.

We can check this result. If we integrated along a straight-line path connecting 1 + *i* with –3*i* by using I=quadgk(f,1+i,-3i,), we would not obtain the above value for IA. This is because the operation we are doing depends on the path of integration. The function cos *z*/*z* has a singularity (a simple pole) at *z* = 0 but is otherwise analytic. We cannot deform the integration along the original path (the solid line in Figure 3.6) into the dotted-line path without forcing the contour to pass through the singularity.

Suppose, however, we were to take the result IA from the above and add it to the integral of cos *z*/*z* taken along the straight line going from –3*i* to 1 + *i*. The sum of these two integrals would be the value of the integral $\oint \frac{\cos z}{z}dz$ taken in the positive (counterclockwise) direction around the closed contour that is the polygon shown in Figure 3.6.

We can do this integration with residues, because cos *z*/*z* has a simple pole at *z* = 0, which is enclosed. The result should be $2\pi i\,\text{Res}[\frac{\cos z}{z}@z=0]$, where Res is an abbreviation for residue. Using Equation 3.18, we have that the residue is cos 0 = 1. So the result of our integration around the closed contour in Figure 3.6 is 2*πi*.

Let us add our integral IA obtained above from **quadgk** to the integral of cos *z*/*z* taken along the dotted line in Figure 3.6. We call the latter integral IB. The sum of these should equal 2*πi* and would provide a check on the value of IA. Here is our code:

```
format long
syms z
f=@(z)cos(z)./z;
IA=quadgk(f,1+i,-3i,'waypoints',[2i,-1+2i,-1-i])
IB=quadgk(f,-3i,1+i)
total_is= IA +IB
two_pi_i=2*pi*i
```

Whose output is

```
IA = 4.078219914209672 + 4.4251398468664735i

IB = -4.078219914209674 + 1.858045460314853i

total_is = -0.000000000000002 + 6.283185307179587i

two_pi_i = 0 + 6.283185307179586i
```

Observe that the agreement between the sum of these two integrals and the desired $2\pi i$ is excellent, and any disparity is due to round-off errors.

Rather than integrate around the quadrilateral by combining two integrals, we can perform the integration with a single **quadgk** statement. We take the beginning *and* final limits of integration as $1 + i$ and use the remaining corners as waypoints. The code is

```
format long
syms z
f=@(z)cos(z)./z;
IC=quadgk(f,1+i,1+i,'waypoints',[2i,-1-i,-3i])
```

which yields

```
IC = -0.000000000000020 + 6.283185307179271i
```

This agrees with the total sum of the two prior integrals, IA plus IB, up to 13 decimal places. We would have been happier if the real part above had turned out to be zero, but rounding errors prevented this.

Exercises

1. Evaluate $\int_0^1 e^{i\sin x} \cos x \; dx$ by the following three methods and compare the results
   a. By using the antiderivative of the integrand
   b. By using **int**
   c. By using **quadgk**

2. Evaluate $\int_0^1 e^{i\sin x} \; dx$. Note that, unlike the previous problem, an antiderivative of the integrand does not exist in terms of standard functions. Determine this integral by means of
   a. **int**

      and
   b. **quadgk** and compare the results.

3. Evaluate the integral $I = \int_{-\infty}^{\infty} \dfrac{dx}{x^4 - i}$ by these methods:
   a. Using residue calculus
   b. Through the use of **int** (obtain the result both in symbols and in a decimal expression)

c. Through the use of **quadgk**.

d. Verify that the three answers above are numerically the same (to within round-off errors).

4. Using **int**, evaluate the integral $I(a) = \int_{-\infty}^{\infty} \frac{dx}{x^2 + a^2}$, where $a \neq 0$ is an arbitrary complex number. If necessary, read the documentation for **int** to see how to evaluate an integral containing a symbolic parameter, which in this case is a. Briefly, you must mention that parameter in the **syms** statement. You might wish to make your answer look more attractive by using the command **pretty**. Now take $a = \rho e^{i\phi}$, where ϕ is the principal angle, $-\pi < \phi \leq \pi$ and ρ is positive. Compare your result with that obtained by residues. What values of a must be avoided if we are to avoid singularities on the path of the integration? How is this reflected in your MATLAB result?

5. Evaluate $I = \int_{0}^{\infty} \frac{dx}{x^4 + x^2 + i}$ by using the following:

a. The function **int**. Convert your symbolic result to a decimal one using both **vpa** and **double**. How different are they? The preceding integration can be done using residues, but it is tedious (as can be seen by your answer, in symbols, to this part), and you would have to find the zeros of $z^4 + z^2 + i$ in the complex plane and then find the residues at each of these poles. We do this in part (c).

b. The function **quadgk**. Compare your results to those in part (a).

c. If we are determined to do this problem with residues, MATLAB can assist us.

    First, find the location of all the zeros of $z^4 + z^2 + i$ by using MATLAB's **roots**. Explain why all the poles of $1/(z^4 + z^2 + i)$ are simple. Observe that the integral is $I = \pi i \sum \text{Res} \frac{1}{z^4 + z^2 + i}$ @ poles in uhp. Prove that the residue at a pole is $1/(4z^3 + 2z)$ evaluated at the pole. Using MATLAB, find the residues at all the poles and sum those that arise in the upper half plane. Compare your final answer with those obtained in parts (a) and (b).

    Hint: If vrt is the column vector containing the roots of the above polynomial, then uu=imag(vrt)>0 is a column vector having ones where the corresponding roots are in the uhp and zeros where they are in the lower half plane (lhp).

6. MATLAB can compute residues. Suppose you have a rational expression consisting of two polynomials in z:

$$f(z) = \frac{b_n z^n + b_{n-1} z^{n-1} + \ldots + b_0}{a_m z^m + a_{m-1} z^{m-1} + \ldots + a_0}$$

Then MATLAB can find the location of the poles of this expression as well as the residue at each pole. At this point, you should read the documentation for **residue** in the help folder. Briefly, you employ the statement

$$[r,p,k]=\text{residue}(b,a)$$

where b is the row vector consisting of the coefficients of the polynomial in the numerator. These are entered in order of descending powers, just as is done in using **roots**. Be sure to enter a zero if a term is missing. For example, if the numerator is $4z^3+2z+i$, then $b = [4\ 0\ 2\ i]$. In a similar way, a is a row vector describing the denominator.

The outputs of **residue** are the vector r, which is the residue of the expressions at the poles, and the vector p, which is the location of the poles (stated in the same order as the residues are given). If $m > n$, then the vector k has no entries. If $m \le n$, then k gives the coefficients of a polynomial of degree $n-m$. The polynomial is the remainder after $f(z)$ is resolved into partial fractions. The coefficients are given in the usual order, just described. Notice that **residue** works only for rational expressions (ratios of polynomials).

a. For the function $f(z) = \dfrac{3z^2 + 2iz + (1+i)}{5z^4 + iz^3 + i}$, use MATLAB to find all the poles and state the residue at each pole.

b. Using these residues, find $\displaystyle\int_{-\infty}^{\infty} \dfrac{3x^2 + 2ix + (1+i)}{5x^4 + ix^3 + i}\,dx$. Remember to use only the residues in the uhp or, if you prefer, only the lhp.

c. Check your answer to (b) by using **int**, being sure to convert your answer to a decimal form to facilitate comparison with the answer from (b).

7. Recall from complex variable theory that if we are given the integral

$$\int_{-\infty}^{\infty} \frac{P(x)}{Q(x)} e^{ivx}\,dx$$

where $v > 0$ and the degree of $Q(x)$ exceeds that of $P(x)$ by 1 or more,

then $\displaystyle\int_{-\infty}^{\infty} \frac{P(x)}{Q(x)} e^{ivx}\,dx = 2\pi i \sum \text{Res}\,\frac{P(z)}{Q(z)} e^{ivz}$ @ all poles in the uhp.

If $v < 0$, we would use the poles in the lower half plane and replace 2 with −2.

a. Using residues, find

$$\int_{-\infty}^{\infty} x \frac{e^{i2x}}{(x-1)^2+1}dx$$

b. Evaluate this integral using **int** in MATLAB.

c. Evaluate the same integral using **quadgk**. Notice that the answer you have given is erroneous as can be seen by comparing your answer with the result of part (a). The function **quadgk** some-times fails to evaluate an integral where the integrand is oscil-latory, dies out slowly, and one or both limits are infinite. Try integrating between –1,000 and 1,000 and see if this helps. Notice that a difficulty did not arise in part (b) because **int** uses the anti-derivative to evaluate the integral.

d. Using the result of (a) or (b), find

$$\int_{-\infty}^{\infty} x \frac{\cos 2x}{(x-1)^2+1}dx$$

and

$$\int_{-\infty}^{\infty} x \frac{\sin 2x}{(x-1)^2+1}dx$$

You might wish to review W section 6.6 or S section 7.4.

8. a. Using **quadgk**, perform the integration $\int_{1-i}^{1+i} \cosh z \, dz$ along the straight-line segment connecting the two limits.

   b. Integrate these same two points by using the broken-line path going from $1-i$ to $-1-i$ then to $-1+i$ and from there to $1+i$. Use waypoints. Compare your answer to (a).

   c. Check your answer by using the antiderivative of the integrand and compare all three results.

9. a. Repeat part (a) above but use as the integrand $\cosh z/z$.

   b. Repeat part (b) above but use the new integrand.

   c. Explain why in parts (a) and (b) you obtained differing results.

   d. Reconcile the difference between the results by using a residue.

10. a. Consider

$$\oint \frac{\cos(\pi z)}{\sin(\pi z)} dz$$

where the integral is done on the contour $|z|=3/2$. Explain why we can get the same result if we integrate instead in the positive direction around the square having corners at $z = \pm\frac{3}{2}$ and $z = \pm\frac{3}{2}i$.

b. Perform the given integration by using residues.

c. Check your answer by using **quadgk** and the square contour. Employ waypoints. Are the results the same?

d. By using **quadgk**, do the integration

$$\oint \frac{\cos(\pi|z|)}{\sin(\pi z)} dz$$

around the square. Explain why we cannot do this problem with residue calculus.

11. Compute $\int_{-i}^{i} \text{Log}\, z \, dz$ using **quadgk** and **waypoints**. Use the following two different paths:

a. Straight lines connecting $-i$, 1, and i.

b. Straight lines connecting $-i$, -1, and i.

c. Discuss the fact that the answers to (a) and (b) are not the same. Note that you are using the principal value of the Log.

d. Derive both of the answers analytically by using the antiderivative of the log.

    Hint: For part (b), notice that the imaginary part of log changes discontinuously as you cross the branch cut at $z = -1$. Thus, you should break your integration into two integrals: one from $-i$ to $-1-i*$eps (at the bottom of the cut) and the other from -1 to i. Notice that using principal values in MATLAB $\log(-1-i*$eps$) \approx -i\pi$ and it takes $\log(-1)=i\pi$.

12. a. Using **quadgk** and waypoints, find

$$I = \oint \frac{\sin z}{\sinh z \,(z^2 + 1)(z + 10i)} dz$$

taken around the circle $|z-1|= 4$.

    Hint: Deform your contour into a box that encloses the same poles as the circle.

b. Check your answer to (a) by using residue calculus.

4

Harmonic Functions, Conformal Mapping, and Some Applications

Harmonic functions and conformal mapping are two of the richest and most interesting topics in complex variable theory. It is our good fortune that harmonic functions, which are always the real and imaginary parts of analytic functions, can be used to describe two-dimensional configurations of fluid flow, heat transfer, and electrical fields. Historically, conformal mapping achieved importance as a means of solving two-dimensional problems in these branches of engineering and physics. Although numerical methods on modern computers have rendered this application of less importance, conformal mapping still provides us with solutions to certain canonical problems that can be used to verify the correctness of computer-generated solutions. This is analogous to the practice of learning to perform integrations in elementary calculus courses even though, for example, the MATLAB® Symbolic Mathematics Toolbox will perform them for you. One must know what kinds of answers to expect from a computer and to have a means of checking them.

In Chapter 2, we learned some graphical techniques for studying functions of a complex variable. These included three-dimensional plots as well as the use of contour lines. Conformal mapping provides another means of visualizing the properties of a function of a complex variable. In this case, we work in just two dimensions and use two separate planes.

4.1 Introduction

Let $z = x + iy$ and $w = u + iv$ be complex variables. If $w = f(z)$ describes a function of the complex variable z and is defined at the point z_0, we can say that this function *maps* the point z_0 into the point w_0 where $w_0 = f(z_0)$.

We can display this fact using two planes—the z and w planes. The points $z_0 = x_0 + iy_0$ and $w_0 = u_0 + iv_0$ might be plotted as * or some other symbol. We say that w_0 is the *image* of z_0 under the given transformation $w = f(z)$.

We might also say that z_0 is the image of w_0 under an inverse transformation. Suppose, for example, that $z = g(w)$ yields z when w is known. Here $g(w)$ is said to be the *inverse transformation* of $f(z)$. We have $z_0 = g(w_0)$. We assume in the preceding discussion that we are dealing with well-behaved, single-valued functions.

For example, suppose $w = f(z) = \log z$ and $z_0 = 1 + i$ so that its image is

$$w_0 = \text{Log}(1+i) = \text{Log}\sqrt{2} + i\frac{\pi}{4} \approx .3466 + i\frac{\pi}{4} = u_0 + iv_0. \text{ Incidentally, } z = g(w) = e^w$$

and $z_0 = 1 + i = e^{w_0} = e^{\text{Log}\sqrt{2}+i\pi/4}$.

We display the mapping with the code:

```
clf
z=(1+i);
w=log(z);
figure(1);
plot(z,'*');grid on;xlabel('x');ylabel('y');
%the above plots the point 1+i in z plane,
% and labels the axes.
text(1.02,1.02,'z_0')
% above labels the above point, we use 1.02, 1.02
%and not 1,1, so as to keep the letter zo off the star *
text(1,1.6,'\fontsize {12} The z Plane')
figure(2);
plot(w,'*');grid on;xlabel('u');ylabel('v')
%above plots log(1+i) in w plane.
text(real(w)+.05,imag(w)+.05,'w_0');
%note the use of .05 in the above to keep
%the wo off the *
%labels the above point
text(0,1.7,'\fontsize{12} The w Plane')
```

The output is as shown in Figure 4.1.

Suppose we wish to map a large number of points from one plane to another. We might designate them as A, B, C … (we obviously have 26 of them if we need that), and their images as *A, B, C* …. We resort here to a bold italic font. Alternatively, we might simply use a, b, c or some other such variation in lieu of A, B, C.

To understand how to effectively use letters, realize that we may convert any letter of the alphabet, regarded as a character, to a unique numerical value using a convention called the ASCII standard representation. The capital letters A, B, C, …, Z have ASCII numbers 65, 66, 67, …, 90. Similarly, the lowercase letters a, b, c, …, z have ASCII numbers 97, 98, 99, …, 122.

We may manipulate characters as if they were numbers by operating on the corresponding number in the ASCII representation. For example, we may begin with the letter A in a string and produce the letter D by adding

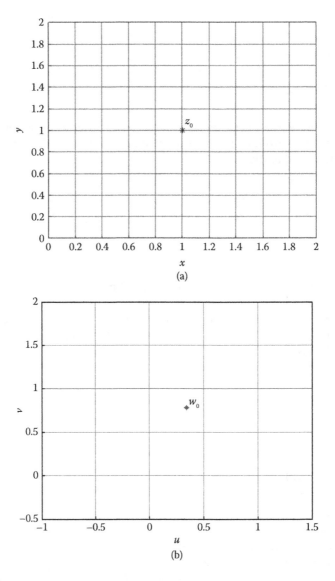

FIGURE 4.1
(a) The point for $1 + i$ and (b) the principal value of $\log(1 + i)$.

3 to the ASCII number for A (which is 65) and then converting the result (which is 68) back to a character which will be D. We convert from the ASCII number to the actual letter of the alphabet by use of the operation **char**. Here is how we might convert a letter A to letter D. Recall the use of single quotes to make a character, '...'.

```
>> uu='A'
uu = A
>> vv=uu+3
vv = 68
>> char(vv)
ans = D
```

Here is how we might go from the lowercase letter z to the lower case u. Recall that there are four letters *between* u and z and that the ASCII number for u is smaller than that for z.

```
>> tt='z'
tt = z
>> LL=tt-5
LL = 117
>> char(LL)
ans = u
```

Now consider the 11 points on the line $x = 1$ given by $z = 1, 1 \pm 1i, 1 \pm 2i, \ldots 1 \pm 5i$. We will identify them with the letters A ... K (starting with $1 - 5i$ and proceed upward along $x = 1$ to $1 + 5i$). Using the mapping $w = \log z$, we will plot not only the given values of z but their images in the w plane and label them as *A ... K*.

This is accomplished with the following code:

```
clf
y=linspace(-5,5,11);
x=1;
z=1+i*y;
w=log(z);
figure(1)
   plot(real(z),imag(z),'.');grid on;
%plots a dot for each value used in
%z plane
   xlabel('x');ylabel('y');title('z plane');grid on
figure (2)
plot(real(w),imag(w), '.');axis equal;
%plots a dot at each value of logz
xlabel('u');ylabel('v');title('w plane')
for n=1:length(y)
     letter=char(n+64);
%the above creates the letters A, B,...
     figure(1);
     text(1.02,imag(z(n)),letter,'FontSize',10);
     % the above plots the letters A,B,
     %note that using 1.02 instead of 1 in the above
     %keeps the text off the dot in the plot.
     %a similar thing is done for figure(2)

      figure(2)
```

```
text(real(w(n))+.02,imag(w(n)),letter,'FontSize',10, ...
   'FontAngle',...
   'italic','FontWeight','bold');
      % the above plots in the w plane the
  % italicized bold A, B, C...
      % which are the images of the points in the z plane
   hold on
end;grid on
```

This yields the output as shown in Figures 4.2 and 4.3.

Every point that we consider in the z plane is marked with a dot and a letter A through K. Its image in the w plane is also marked with a dot and a corresponding letter *A* through *K* in boldface italic.

If we use plot(real(z),imag(z)) in place of plot(real(z),imag(z),'.'), we obtain a continuous line connecting the points we have chosen to map. A similar change can be made to plot(real(w),imag(w),'.') to obtain a continuous line.

Sometimes, instead of wanting to map a selection of discrete points from one plane to another, we wish to map a curve, generally one that is smooth or piecewise smooth. Since a curve contains an infinite number of points,

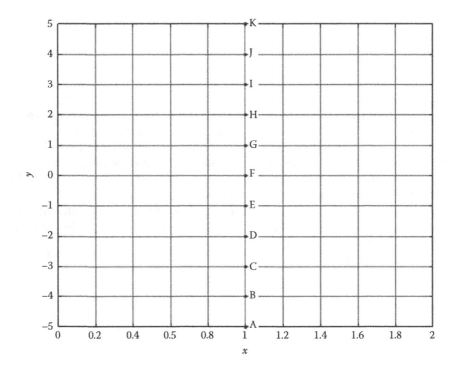

FIGURE 4.2
Some points for values of $1 + iy$.

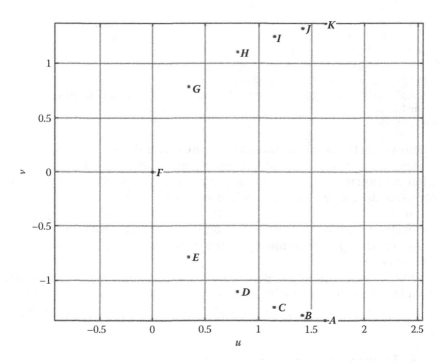

FIGURE 4.3
Principal value of the logarithm for numbers in preceding figure.

we cannot map them all. The usual procedure is to map a sufficient number of points from the curve in the z plane into the w plane—the number of points is large enough so that if MATLAB connects them together the resulting curves in either plane resemble the usually smooth ones that we are considering. If in doubt, overestimate the number of points you think you might need. The only drawback will be increased required time for the computation, and this is often not a consideration.

Let us consider two intersecting curves and again map them with the transformation $w = \log z$. These will be the circle $|z| = 2$ and the line segment $x = \text{Re}(z) = 1, -1 \le y \le 1$.

As a first attempt, we proceed as follows with this code:

```
clf
y=linspace(-3,3,101);
x=1;
z1=1+i*y;
w1=log(z1);
th=linspace(-pi,pi,100);
z2=2*exp(i*th); % yields 100 points on circle of radius 2
w2=log(z2);
```

```
figure(1)
 plot(real(z1),imag(z1),'linewidth',2);hold on
 plot(real(z2),imag(z2),'linewidth',1);grid on;axis equal
xlabel('x');ylabel('y');title('z plane');grid on
figure (2)
plot(real(w1),imag(w1),'linewidth',2);hold on
plot(real(w2),imag(w2),'linewidth',1);
grid on;axis equal
xlabel('u');ylabel('v');title('w plane')
y=linspace(-3,3,6);th=linspace(-.9*pi,pi,6);
```

The output is as shown in Figures 4.4 and 4.5.

The line $x = 1$ in the z plane is the heavier line in Figure 4.4, and the circle $|z| = 2$ is the thinner line in the same figure. The image of $x = 1$ is the heavier line in Figure 4.5, while the image of the circle is the heavier line in the same figure. To see the effect of using too few data points for your mapping, try changing the line of code above from y=linspace(-3,3,101); to y=linspace(-3,3,11);. You will get a jagged result.

What is lacking in the above two figures is any indication of how points from one curve are mapped into another. We can remedy this by labeling some points on each of the two curves in Figure 4.4, using the letters A, B, C, ...

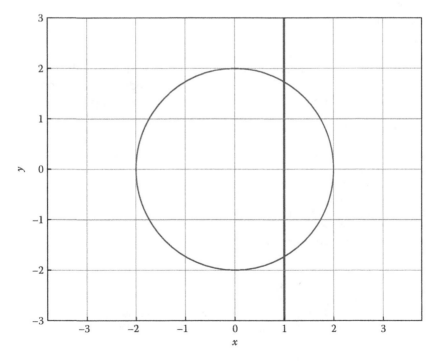

FIGURE 4.4
Circle $|z| = 2$ and Line $Re(z) = 1$.

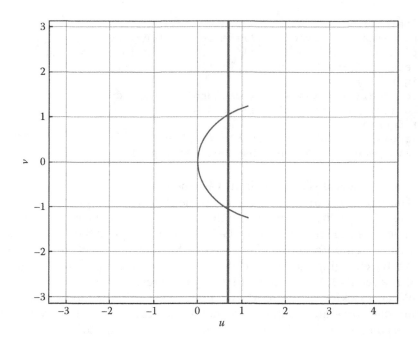

FIGURE 4.5
Images of the preceding curves when mapped by Log z.

and a, b, c, ... while indicating their image points with bold italic versions of
these letters. We add on this code to the above code. The additional lines
provide an idea of how each curve is deformed by the mapping process:

```
y=linspace(-3,3,6);th=linspace(-.98*pi,.98*pi,6);
z1=1+i*y; z2=2*exp(i*th);
%we will use fewer data points for the letters
%on the curves than were needed to make the curves
 for n=1:length(y)
     letter1=char(n+64);
% above creates the letters A, B,...
     letter2=char(n+96);
%above creates letters a,b, ...
      figure(1);
      text(1,imag(z1(n)),letter1,'FontSize',10);
%plots the letters A,B,
     % in the z plane on x=1;
     text(real(z2(n)),imag(z2(n)), letter2,'FontSize',10)
     % above plots the letters a,b, on the circle in
      % the z plane
     w1(n)=log(z1(n));
     w2(n)=log(z2(n));
      figure(2);
```

```
text(real(w1(n)),imag(w1(n)),letter1,'FontSize',10, 'FontAngle',
    ...'italic','FontWeight','bold');
% the above puts in the w plane the italicized bold A,B,..
 % which are the images of the points A,B,.. in the z plane
text(real(w2(n)),imag(w2(n)),letter2,'FontSize',10,'FontAngle',
    ...'italic','FontWeight','bold');
%the above puts in the w plane the italicized bold
%a,b..which are the
   %images of a,b,c in the z plane
       hold on
end;
```

With these new lines appended to the preceding program, the output is as shown in Figures 4.6 and 4.7.

Notice the line of code: th=linspace(−.98*pi,.98*pi,6);. We do not set limits on *th* (the angle used to locate points of the circle in Figure 4.6) from −pi to pi. Otherwise in Figure 4.6, the points marked *a* and *f* would lie on top of one another because of the line of code z2=2*exp(i*th), which would produce two identical values for z2;.

We see how the vertical line $x = 1$ in Figure 4.6 is bent into the bow shape in Figure 4.7 that is symmetrical about the real axis. We also see how the circle of radius 2 is deformed into a line segment. It is not hard to show (try this)

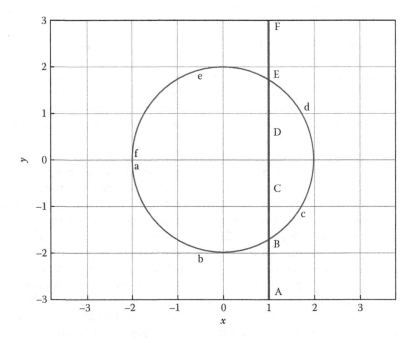

FIGURE 4.6
Circle $|z| = 2$ and Line $\mathrm{Re}(z) = 1$.

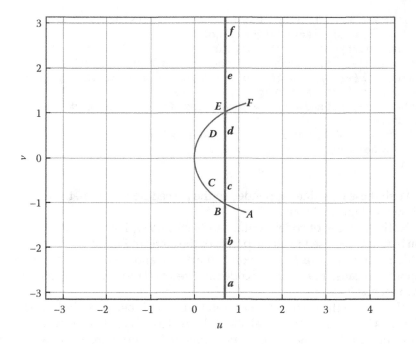

FIGURE 4.7
Images of the preceding curves when mapped by Log z.

that the segment lies along the line $u = \text{Log } 2$, $-\pi < v \leq \pi$, because MATLAB uses the principal value of the log.

Consider the intersection between the circle and the straight line in Figure 4.6 near the point marked E. The reader should show that the intersection is at $x = 1, y = \sqrt{3}$. With a little high school geometry, he or she should see that the acute angle between the tangent to the circle and the vertical line is 60°. See Figure 4.8 (detail).

If you study Figure 4.9, you will find that the intersection between the images of the curves shown in the detail appears to also be about 60°.

It is in fact exactly 60°; notice that in each case the curve with the lowercase letters is displaced 60° counterclockwise from the one with uppercase letters. What we have here is an illustration of the *conformal property* in complex variable theory (see W section 8.2, S section 8.4): if $w = f(z)$ is analytic in a domain D, and if C_1 and C_2 are two curves intersecting at a point z_0 in D, then the images of these two curves in the w plane will intersect in such a way that their angle of intersection (at $w_0 = f(z_0)$) will be the same as that between C_1 and C_2, provided that $f'(z_0) \neq 0$. Not only is the magnitude of the angle preserved, but so is its *sense*. For example, if we must rotate C_1 clockwise to stretch it onto C_2 (using the smallest possible angle), then this must be true of their respective images. We now see where the word *conformal* comes from

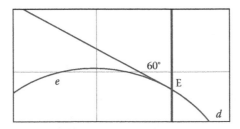

FIGURE 4.8
Detail of an intersection.

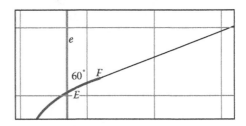

FIGURE 4.9
Detail of the images of the intersection for curves in preceding figure.

in "conformal mapping": the angular intersection of intersecting curves in one plane conforms in magnitude and direction to the angular intersection of their images in the other. To repeat, conformal mapping assumes that an analytic function is used for the mapping and that its first derivative is nonvanishing at a point of intersection where angle preservation must be preserved.

A nonconstant analytic function will always map a domain into a domain (see W section 8.3). Suppose all the points in a domain D in the z plane are mapped into the w plane by means of an analytic function $w = f(z)$. Recall that because the function is analytic, it is single valued. The result of the mapping is a domain D' in the w plane. If z_1 and z_2 are points in D, and the equation $f(z_1) = f(z_2)$ is only satisfied in D if $z_1 = z_2$, then we say that the mapping from D to D' is *one-to-one*. Examples of mappings that are and are not one-to-one are given in Exercises 2 and 3.

In Figure 4.6, there is a domain whose boundaries are defined by the line segment BCDE and the arc containing the letters c and d. With the aid of Photoshop™, we show this domain heavily shaded in Figure 4.10. We can see how this domain is mapped into the w plane by studying how its boundary points are mapped into the w plane. Studying Figures 4.6 and 4.7 leads us to conclude that the dark-shaped area is mapped into the dark domain shown in Figure 4.11.

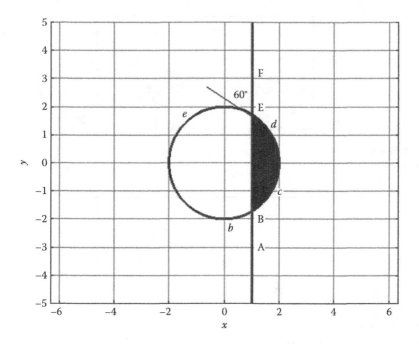

FIGURE 4.10
A domain in the z plane.

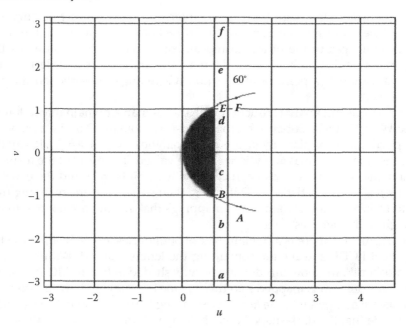

FIGURE 4.11
The image of the domain in the preceding figure under the transformation $w = \text{Log } z$.

For confirmation, note that the point $z = 3/2$, which lies in the domain shown in Figure 4.10, has an image lying at $w = \text{Log}(z) \approx .4055$, which is seen to lie in the domain depicted in Figure 4.11.

Exercises

1. a. Consider the unit circle $|z| = 1$. Construct this circle in the z plane by choosing 100 points uniformly spaced around the circle, placing them in the z plane, and having MATLAB connect them with a smooth curve. Be sure to use axes of equal scales to see the circle. Suppose you had used only the nine points at $\arg(z) = 0, 45, 90, 135, \ldots 315, 360$ degrees, what kind of curve would MATLAB give you, and how well would it approximate a circle?

   b. Using the 100 points employed above, map the circle into the w plane using the transformation $w = e^z$.

   c. Place the nine points described in part (a) on the circle generated with 100 points in part (a). Label these points with the letters A, B, C, ...

   d. Using MATLAB, plot the images of these points with the letters (italic) *A, B, C,* ... on the figure in part (b).

2. Consider the rectangular-shaped domain in the z plane whose boundary points are on the four straight-line segments: $y = 0$, $-1 \leq x \leq 1$; $y = \pi$, $-1 \leq x \leq 1$; $x = \pm 1$, $0 \leq y \leq \pi$.

   a. Using $w = e^z$, map these four line segments into the w plane by means of MATLAB. Do this by mapping 100 points from each segment. Use a heavy line width (e.g., 3) for the boundary in each plane.

   Using MATLAB, place a different capital letter of the alphabet at each of the four corners of the domain in the z plane, and place one at the midpoint of each line segment. Put corresponding bold italic letters at the images of these points in the w plane.

   b. The interior points of the above rectangle form a domain. How is it mapped into the w plane by the given transformation?

   c. Explain why the mapping from the z plane to the w plane is one-to-one—that is, if we choose any point in the domain found in part (b), for example, $f(z_1)$, it has exactly one image in the domain given in the z plane (i.e., there is only one possible value for z_1).

   d. Suppose we change the rectangle in the z plane to $y = 0, -1 \leq x \leq 1$; $y = 2\pi, -1 \leq x \leq 1$; $x = \pm 1$, $0 \leq y \leq 2\pi$, and use the same mapping.

Is the mapping again one-to-one? Into what domain is the interior of the rectangle mapped?

e. Suppose we replace 2π by 3π in the above. Is the mapping again one-to-one?

3. a. Consider the domain in the z plane whose boundary is the circle $|z - 1 - i| = 1$. The domain is the interior of this circle. Using MATLAB, produce a plot of the image of this boundary under the transformation $w = z^2$. Use at least 100 points to make this circle. Place at least four points uniformly spaced on the circle, designating them with capital letters of the alphabet, and put corresponding lowercase letters at the images of these points in the w plane.

b. Is there a one-to-one mapping between the domain described above and the domain created by the given transformation? Explain.

c. If we change the circle to $|z - 1 - i| = 2$, is the mapping one-to-one? Explain.

4. Consider the bounded domain in the z plane whose boundaries are the lines $y = 0, .1 \le x \le 1; x = 0, .1 \le y \le 1; |z| = .1, 0 \le \arg z \le \pi/2; |z| = 1, 0 \le \arg z \le \pi/2$.

This domain is mapped with the transformation $w = z^i$, where the principal value is used.

a. Using MATLAB, plot the boundaries of the domain in the z plane and the boundaries of the image in the w plane. In the z plane, place letters A, B, C... at these points: $z = .1, .5, 1, e^{i\pi/4}, i, .5i, .1i, .1 e^{i\pi/4}$, and put corresponding italicized letters at the images of these points in the w plane.

b. Prove that the mapping between the two domains is one-to-one.

5. A domain in the z plane consists of all the points lying between two squares with a common center. The centers of both squares lie at $z = 1 + i$. The outer square has sides parallel to the x–y axes and has side length 2. Two sides are on the axes. The inner square has sides parallel to the same axes and has a side length of one half.

a. Using the transformation $w = z^3$, map these two squares into the w plane with MATLAB. Place letters A, B ... H on each of the eight corners and place bold or italic versions of these same letters at their images in the w plane. We suggest using a small font size (e.g., 6) in the w plane.

b. Study the mapping in part (a). Where is it not conformal, and why?

c. Is the mapping of the domain from the z plane to the w plane one-to-one? Explain.

4.2 The Bilinear Transformation

One of the most commonly used conformal mappings in engineering is the bilinear transformation given by

$$w(z) = \frac{az+b}{cz+d} \tag{4.1}$$

(See W section 8.4 and S section 8.10.)

Here a, b, c, d are constants. Notice that if $\dfrac{a}{c} = \dfrac{b}{d}$, this would reduce to w being a constant and so the mapping is not conformal. The same would be true if we solve the above for $z(w)$, the inverse transformation. Then z becomes a constant. To avoid these possibilities, we require that $ad \neq bc$. You should easily prove that this inequality guarantees that $w'(z) \neq 0$ for all finite z (i.e., the mapping is everywhere conformal).

Among the simpler special cases of bilinear transformations are such examples as $w = 2iz$, $\; w = 1/z$, $\; w = \dfrac{z-i}{z+i}$. The inverse of a bilinear transformation is a bilinear transformation (i.e., if you were to solve Equation 4.1 for z, you would obtain z in terms of a bilinear expression in w). Although this is a high school level exercise, we perform it here so as to demonstrate the **solve** function in MATLAB.

The program is

```
z= solve('w=(a*z+b)/(c*z+d)','z')
'a pretty answer for z='
pretty(z)
```

The output is

```
z = -(b - d*w)/(a - c*w)
a pretty answer for z=
```

$$-\frac{b-dw}{a-cw} \tag{4.2}$$

which is obviously also a bilinear transformation. Note that if you run the above program, you will receive a warning message (not shown here) that the result is valid only if $ad \neq bc$.

You may avoid this warning by using, if you wish,

```
solve('w=(a*z+b)/(c*z+d)','z','IgnoreAnalyticConstraints',true)
```

to the right of the equals sign in the first line of code in lieu of

```
solve('w=(a*z+b)/(c*z+d)','z').
```

The most useful property of the bilinear transformation is this: If an infinite straight line in the z plane is mapped by means of the transformation (Equation 4.1), its image in the w plane is found to be either an infinite straight line or a circle. If a circle in the z plane is similarly mapped, the result in the w plane will be either a circle or an infinite straight line. This is proved in all standard texts (see, e.g., W section 8.4). By an infinite straight line, we mean one along which we can travel in two opposite directions and arrive at infinity. Thus, $x + y = 1$ is infinite, but $x + y = 1$, $y > 0$ is not infinite but could be called semi-infinite. We will use the term "straight line" to mean an *infinite straight line*.

Since the inverse of a bilinear transformation is still a bilinear transformation, it means that a circle or straight line in the w plane, obtained by mapping some curve in the z plane by means of a bilinear transformation, must have arisen from a straight line or circle in the z plane.

If the coefficients in a bilinear transformation a, b, c, d are real in Equation 4.1, then it has the property that $w(\bar{z}) = \bar{w}(z)$. This means that points that are conjugates of each other in the z plane will have image points in the w plane that are conjugates of one another under this mapping. Thus, a circle whose center is on the real axis in the complex z plane will have as its image in the w plane either another circle with center on the real axis or an infinite line parallel to the imaginary axis. An infinite line parallel to the imaginary axis in the z plane will be mapped into either a circle in the w plane, with center on the real axis, or an infinite line parallel to the imaginary axis.

Example 4.1

Let us verify some mappings using the transformation $w = \dfrac{z-1}{z+i}$ and MATLAB. We will map, by means of this transformation, these two circles: $|z| = 1$ and $|z - 1 - i| = 2$.

Solution:
Before beginning, notice that as we approach the point $z = -i$, the function $w(z) \to \infty$. Now $z = -i$ is on the circle $|z| = 1$. Thus, at least one point of the circle has an image at infinity, which means that the image of the entire circle must be an infinite straight line. Using MATLAB to plot points from the z to the w plane, we must be careful to avoid $z = -i$. Otherwise, we will be dividing by zero and will receive an error message. Observe, too, that the given circle passes through $z = 1$, where $w = 0$. Thus, the given line must pass through the origin in the w plane.

The circle $|z - 1 - i| = 2$ does not pass through $z = -i$. Thus, w is never infinite on this locus, and the image of the circle must be another circle.

The code is

```
th=2*pi*linspace (0,1,1000);
z=exp(i*th);
figure (1);
```

```
plot(real(z),imag(z));axis equal;grid on
xlabel('x');ylabel('y');title 'Unit Circle z plane'
w=(z-1)./(z+i);
figure (2)
plot(real(w),imag(w));grid on;axis equal;
xlabel('u');ylabel('v');
z=1+i+2*z;
figure(3)
plot(real(z),imag(z));axis equal;grid on
xlabel('x');ylabel('y');title 'Circle |z-1-i|=2'
w=(z-1)./(z+i);
  figure(4);
  plot(real(w),imag(w));grid on;axis equal;
  xlabel('u');ylabel('v');
```

The resulting output is shown in Figures 4.12 through 4.15.

There are some things to note here. We have used 1,000 data points on the two circles in the z plane in order to map these circles into the w plane. This may seem excessive, but in the case of the transformation applied to the circle $|z - 1 - i| = 2$, it is not. This is because the mapping function $w = \dfrac{z-1}{z+i}$ changes rapidly as we approach $z = -i$. If we expect to obtain a smooth curve from the mapping, we must obtain enough image points in the w plane to yield a circle.

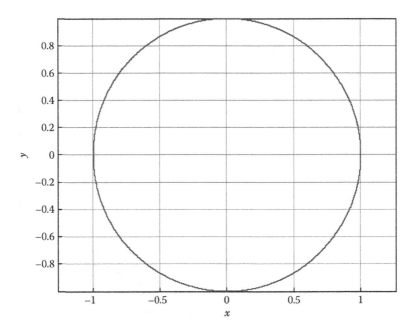

FIGURE 4.12
Unit circle, z plane.

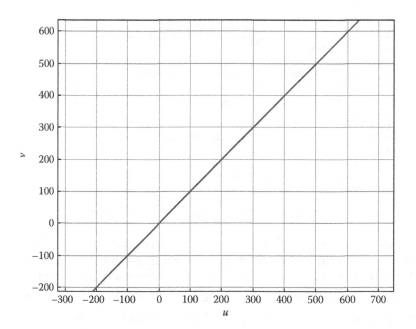

FIGURE 4.13
Image in the w plane of the unit circle for $w = \dfrac{(z-1)}{(z+i)}$.

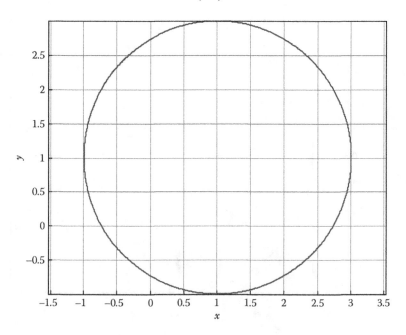

FIGURE 4.14
Circle $|z - 1 - i| = 2$.

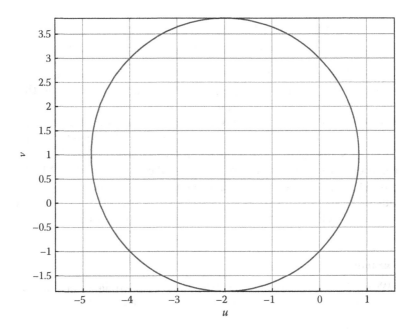

FIGURE 4.15

Image in the w plane of the circle $|z - 1 - i| = 2$ for $w = \dfrac{(z-1)}{(z+i)}$.

We noted that $w = \dfrac{z-1}{z+i}$ goes to infinity at $-i$ and that the unit circle passes through this point. On the unit circle, we have $z = \exp(i\theta)$ $0 \le \theta \le 2\pi$. By not using $\theta = 3\pi/2$, we can avoid evaluating w at this point. This is satisfied by the line of code th=2*pi*linspace(0,1,1000); . In fact, we can replace 1,000 with any even integer and avoid the troublesome point, as the reader should verify.

4.2.1 The Cross-Ratio

Let the points z_1, z_2, z_3, z_4 lie in the complex plane, and let their images, created by means of a bilinear transformation, be w_1, w_2, w_3, w_4. Then the "cross-ratio" of these points is preserved (see W section 8.4 and S section 8.10), which means

$$\frac{(z_1 - z_2)(z_3 - z_4)}{(z_1 - z_4)(z_3 - z_2)} = \frac{(w_1 - w_2)(w_3 - w_4)}{(w_1 - w_4)(w_3 - w_2)}$$

Let us take one of these points, say z_4, to be the arbitrary variable z having image w. Substituting these values in the preceding, we have

$$\frac{(z_1 - z_2)(z_3 - z)}{(z_1 - z)(z_3 - z_2)} = \frac{(w_1 - w_2)(w_3 - w)}{(w_1 - w)(w_3 - w_2)} \qquad (4.3)$$

The preceding can be solved for either $z(w)$ or $w(z)$ as we wish.

Notice that the values z_1, z_2, z_3 must differ from each other, and the same goes for their corresponding images w_1, w_2, w_3. (Recall that the bilinear transformation is one-to-one.) If one of the quantities on the left, say z_3, is infinite, we replace the quotient of the two terms (in this case $(z_3 - z)/(z_3 - z_2)$) involving this quantity by the number 1. The same technique is used on the right if there is an infinite quantity there. We can write a MATLAB program in which one supplies z_1, z_2, z_3 as well as the desired images w_1, w_2, w_3. The output from the program can be either $w(z)$ or $z(w)$, or both. The matter of what to do when a quantity is infinite is taken up in problem 2 in the exercises.

Example 4.2

Write a MATLAB program in which you enter the finite values z_1, z_2, z_3 and w_1, w_2, w_3 and receive back the bilinear transformation $w(z)$.

Solution:
The following code will work:

```
w=solve('(w1-w2)*(w3-w)/((w1-w)*(w3-w2))=(z1-z2)*(z3-z)/
    ((z1-z)*(z3-z2))','w');
zvalues= input(' z1 z2 z3 as row vector elements')
z1=zvalues(1);
z2=zvalues(2);
z3=zvalues(3);
wvalues= input(' w1 w2 w3 as row vector elements')
w1=wvalues(1);
w2=wvalues(2);
w3=wvalues(3);
% the following line substitutes the numerical values into
    the formula %for w, in place of the symbols
w=subs(w,{'w1','w2','w3','z1','z2','z3'},{w1,w2,w3,z1,
    z2,z3});
simplify(w)
factor(w)
```

Some things to note include that the line of code using **solve** will yield the symbolic solution for w. This employs the symbols for the six values z_1, z_2...w_1, w_2.... The line containing the function **subs** places the numerical values in these symbols. The function **factor** will help us recognize that we have a bilinear transformation. Notice, however, that **simplify** does not always live up to its name. If you attempt **simplify** $((z - i)/(z^2 + 1))$, MATLAB will not factor the denominator, divide the common factor, and give you back $1/(z + i)$. Thus, the fact that such a function is a bilinear transformation is not evident. You must be prepared for such outcomes. Sometimes **factor** can be helpful in these situations. Notice also that the function **solve** results in the output of the program warning you that we cannot find $w(z)$ if the points to be mapped in either plane are not numerically distinct. The warning can be eliminated by adding 'IgnoreAnalyticConstraints', true in the solve expression as discussed above.

Let us try out the preceding program. Taking $z_1 = 0$, $z_2 = 2$, $z_3 = 1 + i$ with corresponding images $w_1 = -1, w_2 = 1/3, w_3 = \dfrac{1}{5} + \dfrac{2i}{5}$, we have finally for w the output

    ans = (z − 1)/(z + 1)
    ans = [z − 1, 1/(z + 1)]

These two answers are consistent; the second simply factors the first.

You can check this formula by installing the previous values of z_n and verifying that the desired values of w_n are obtained.

Because this bilinear transformation contains only real coefficients, we know that it will map a circle, in the z plane, with center on the x axis into either a circle with center on the u axis or into an infinite line perpendicular to that axis. The circle with center at $z = 1$ having radius 1 passes through the given points $0, 2, 1 + i$. The image of this circle in the w plane passes through the three points $w_1 = -1, w_2 = 1/3, w_3 = \dfrac{1}{5} + \dfrac{2i}{5}$ and has center at the average of -1 and $1/3$ (i.e., at $w = -1/3$). Suppose we want to see how the transformation thus obtained maps the circle $|z - 2i| = 1$. We add on these lines of code to the preceding:

```
th=linspace (0,2*pi,100);
z=2*i+exp(i*th);% creates the points on the circle, center at 2i
w=subs(w,{'z'},{z});% puts the values we need into w
figure (1)
plot(real(w),imag(w));axis equal
```

For the values of $z_1, z_2 \ldots w_1, w_2 \ldots$ specified above, the output of this would be a circle.

Because a nonconstant analytic function maps a domain into a domain, it should be possible to find a bilinear transformation that maps the real axis in the z plane into the unit circle in the w plane in such a way that the image of the upper half of the z plane is the interior of that circle. Notice that we can rewrite Equation 4.1 as

$$w(z) = \frac{a}{c}\left(\frac{z+b/a}{z+d/c}\right) \tag{4.4}$$

where we must assume that a and $c \neq 0$. If c were zero in Equation 4.1, we could not argue that $z = \infty$ has an image at a finite point in the w plane. The limit of the right side in the above expression as $z \to \infty$ is a/c. As we want this to lie on the unit circle in the w plane, we can take $\dfrac{a}{c} = e^{it}$, where t is any real number. Suppose $z = x$ is real. We want the image for such a point to have magnitude 1.

Then we obtain from Equation 4.4

$$|w(z)| = |e^{it}| \left(\frac{|x+b/a|}{|x+d/c|} \right) = \left(\frac{|x+b/a|}{|x+d/c|} \right) = 1$$

Now we have $|x + b/a| = |x + d/c|$ that can only be satisfied if $b/a = d/c$ or $b/a = \overline{(d/c)}$. We must reject the former as unworkable (see Equation 4.4), as the transformation reduces to a constant. Thus, with the latter, and $p = -b/a$, we have for the conformal mapping that transforms the real axis into the circle $|w| = 1$,

$$w(z) = e^{it} \left(\frac{z-p}{z-\bar{p}} \right) \quad t \text{ is real}$$

or

$$w(z) = m \left(\frac{z-p}{z-\bar{p}} \right) \qquad |m| = 1 \qquad (4.5)$$

How do we know that the space above the axis is mapped into the interior of the circle? If we take p as any point in the z plane satisfying $\text{Im}(p) > 0$, we see from the above that it is mapped into the origin of the w plane—the center of the circle. Because domains are mapped into domains by the above, we conclude that the transformation in Equation 4.5, with $\text{Im}(p) > 0$, maps the upper half plane into the domain bounded by the unit circle and the boundary $y = 0$ into the unit circle.

Example 4.3

Write a MATLAB program that will accomplish the following. It is to produce a bilinear transformation that will transform the real axis in the z plane into the unit circle of the w plane. The program will request a particular point in the upper half plane that will be mapped into the origin in the w plane. Finally, the program will request a particular point on the real axis z that is to have a specified image on the unit circle in the w plane.

Solution:

The following program will work:

```
% the line below solves for w(z)
clear
m=solve('w=m*(x-p)/(x-conj(p))','m');
```

```
p=input('point in upper z plane mapped to w =0');
x=input('point on real axis mapped to a place on unit circle');
w=input('image of above pt. on unit circle');
m=subs(m, {'w','x','p'},{w,x,p})
'above requires |m|=1'
  syms w z;
  w=m*(z-p)/(z-conj(p))
```

Here we try the program out. We will ask that $z = i$ be mapped into $w = 0$ and that $z = 1$ be mapped into $w = e^{i\frac{\pi}{4}}$, which, as required, is on the unit circle.

Here are the inputs, outputs, and the result:

point in upper z plane mapped to w = 0   i
point on real axis mapped to a place on unit circle   1
image of above pt. on unit circle exp(i*pi/4)

m = 2^(1/2)*(−1/2 + i/2)

above requires |m|=1

The output is

$w = (2\hat{~}(1/2)*(z - i)*(-1/2 + i/2))/(z + i)$

Exercises

1. a. In Example 4.1, using MATLAB, we mapped the unit circle $|z| = 1$ into the w plane by means of the transformation $w = \dfrac{z-1}{z+i}$. We obtained what appears to be the infinite line $v = u$. Prove mathematically that this line is indeed obtained.

    Hint: In the given expression for $w(z)$, multiply numerator and denominator by the conjugate of the denominator. We can generate the given circle in the x–y plane with $z = e^{i\theta}$, where $0 \le \theta \le 2\pi$. Substitute this expression for z in the equation just obtained, and show that if $w = u + iv$, then u and v are such that $v = u$ which is the equation of a straight line through the origin in the w plane. It should be obvious that the line passes through infinity when z goes through $-i$.

   b. Using MATLAB, plot in the w plane the image of the circle $|z - 1 - i| = 1$ under the transformation $w = 1/z$. The original circle has center at $z = 1 + i$. Does its image have center at $1/(1 + i)$? Where is its center? Figure this out analytically and confirm it from your plot.

This is something to think about: If a bilinear transformation
maps a circle into a circle, does it always map the center of the
first circle into the center of the second one?

2. a.   Consider the cross-ratio in which the variable points z and w are
images of each other but the point z_3 lies at infinity, the other
points being finite. In this limit, the cross-ratio Equation 4.3
becomes

$$\frac{(z_1 - z_2)}{(z_1 - z)} = \frac{(w_1 - w_2)(w_3 - w)}{(w_1 - w)(w_3 - w_2)}$$

Write a MATLAB program in which you supply at a prompt
the five numerical values w_1, w_2, w_3, z_1, z_2. The program should
then solve for w as a function of z, using the **solve** function.
Use the function **simplify** to simplify your answer.

   b.   Use the program you just obtained to find one transformation that maps the infinite line $x + y = 1$ onto the unit circle.
Explain why there is more than one transformation that will
do this.

3. a.   A fixed point of a mapping is one in which some value of z,
say z_j, has the same numerical value as its image w_j. Thus, a
fixed point for a bilinear transformation must satisfy $z = \frac{az+b}{cz+d}$.
Write a MATLAB program in which you supply at a prompt
the numerical values of a, b c, d, and you will get back the fixed
points. You should use the function **solve**. Note that in general
there are at most two fixed points. When will there be no fixed
points? When will there be one fixed point? When will there
be an infinity of fixed points? Use your program to determine
the two fixed points of $w(z) = \frac{(z-i)}{z+1}$. Note that if your answer
is two values of z, include the line of code eval(z) to see their
numerical values.

   b.   Check your numerical answer to part (a) by showing that for the
two values found, we have $w(z) = z$.

4. A bilinear mapping of a bilinear mapping is still bilinear. This is
easily proved. Let

$$f(z) = \frac{az+b}{cz+d} \quad \text{and} \quad f_1(z) = \frac{ez+f}{gz+h}$$

which are each bilinear. Then

$$f(f_1(z)) = \frac{a\dfrac{az+b}{cz+d}+b}{c\dfrac{az+b}{cz+d}+d} = \frac{(a^2+bc)z+bd+ab}{(ac+dc)z+bc+d^2}$$

which is also a bilinear transformation.

Obviously, $f(f(z))$ and $f(f(f(z)))$, and so on, are bilinear transformations.

a. Write a MATLAB program in which you enter at a prompt the four numerical values a, b, c, d. Now let $f_0 = f(z)$ and $f_1 = f(f(z))$, $f_2 = f(f(f(z)))$, and so on. The program should produce as output $f_k(z)$ for $k = 1, 2, \ldots, 10$. Test your program with these functions:

$$f(z) = \frac{z+1}{z-1} \quad \text{and also} \quad f(z) = \frac{z+i}{z-2i}$$

b. Explain how your results show for all $k \geq 1$ that with $f(z) = \dfrac{z+i}{z-i}$, we have $f_k(z)$ can be only three possible functions, even though we have investigated only the first 10 functions. What are these functions?

c. Use your program to find how many possible functions are generated if you begin with $f(z) = \dfrac{z-1}{z+1}$? What are these functions?

5. a. Consider Equation 4.5. How should this equation be modified, if at all, to map the real z axis into the unit circle of the w plane with the *lower* half of the z plane being mapped onto the interior?

b. Consider Equation 4.5. How should this equation be modified to do the following: the imaginary axis in the z plane gets mapped into the unit circle in the w plane while the right half of the z plane is mapped onto the interior of that circle? Write a program comparable to that in Example 4.3 that will do the following: the program requests a point in the right half of the z plane that is to be mapped to the origin of the w plane. The program requests a point on the imaginary axis in the z plane that is to be mapped into a point, also specified by the user, on the unit circle in the w plane.

c. Use the above program to find the bilinear transformation that maps the imaginary axis into the unit circle, and with the point

$z = 1 + i$ being mapped to the origin. The point $z = 2i$ is to be mapped into $w = -1$.

d. Using the above transformation, map the line segments $y = x$, $0 \le x \le 10$ and $y = -x$, $0 \le x \le 10$ into the w plane. Use a heavier line for the first mapping so that you can distinguish the two image curves. At what angle should these images intersect, and why?

4.3 Laplace's Equation, Harmonic Functions: Voltage, Temperature, and Fluid Flow

Let $f(z) = u(x, y) + iv(x, y)$ be a function that is analytic in some domain D. Then it is easily proved (see W section 2.5 or S section 3.4) that both real and imaginary parts satisfy Laplace's equation:

$$\frac{\partial^2 \Omega}{\partial x^2} + \frac{\partial^2 \Omega}{\partial y^2} = 0 \tag{4.6}$$

in this domain—that is, we may put either $u(x, y)$ or $v(x, y)$ in place of Ω in the above and find that the equation is true. For example, $f(z) = z^2 = x^2 - y^2 + i2xy$ is an entire function, and it is readily seen that both $u(x, y) = x^2 - y^2$ and $v = 2xy$ will satisfy Equation 4.6 in any domain. Functions satisfying Laplace's equation in a domain are said to be *harmonic functions*. A function not satisfying Equation 4.6 throughout some domain *cannot* be either the real or imaginary part of an analytic function in a domain.

If a function $\Omega(x, y)$ does satisfy Equation 4.6, then there exists an analytic function of the form $f_1(z) = \Omega(x, y) + iv(x, y)$. Given $\Omega(x, y)$, there are procedures for finding $v(x, y)$ (see W section 2.5 and S problem 3.7). We show how MATLAB can help do this in the next section where we learn how knowledge of $v(x, y)$ as well as $f_1(z)$ might be useful. We say that $v(x, y)$ is the *harmonic conjugate* of $\Omega(x, y)$. Note that there also exists another analytic function of the form $f_2(z) = u(x, y) + i\Omega(x, y)$.

That both the real and imaginary parts of an analytic function are harmonic has enormous utility for the solution of *two-dimensional* physical problems in electrostatics, heat transfer, and fluid mechanics. In these problems, the sources of electric fields, heat, and fluid flow are infinitely long structures in a direction perpendicular to the $x-y$ plane. The properties of these sources do not vary along the dimension perpendicular to the plane, and no physical property is assumed to change in time.

The space surrounding these sources is assumed to be filled with a material that will do at least one of the following: support an electric field, permit the conduction of heat, and sustain a flowing fluid. The material is of infinite extent in any direction perpendicular to the complex plane. The complex plane is assumed to show a cross section of the material and is perpendicular to any sources that are present. The behavior of all physical quantities is assumed to be identical in all planes parallel to the complex plane.

For an example of two-dimensional heat conduction, imagine that all of space is filled with a material that conducts heat. An infinitely long, hot tube is maintained at a fixed elevated temperature in this material. The tube passes perpendicularly through the complex plane. Then the temperature in this material depends only on the variables x and y, and we would expect the temperature in the material to fall as we move farther from the tube.

We can complicate the heat flow configuration described above, adding other tubes, each with its own temperature, or ones in which the cross sections of the tubes are not necessarily circular.

Further complications, in various physical situations, might involve the use of a material surrounding the sources that is bounded except in directions normal to the x–y plane. We might require, depending on the physics, that temperature, fluid flow, or electric fields satisfy certain prescribed conditions on the boundaries of the material.

All of these three kinds of physical situations are analyzed by means of harmonic functions to which we give the symbol $\phi(x, y)$. In the case of electrostatics, this harmonic function is called the *electrostatic potential,* or sometimes the voltage, in fluid mechanics it is known as the *velocity potential,* while for heat conduction it is called simply the *temperature.* The use of the potential in fluid problems comes with three caveats: like water, the fluid cannot be compressed nor are there present in the fluid *vortices* (or whirlpools)—the fluid is *irrotational.* This term is explained in all standard books on fluid mechanics. Also, there is assumed to be no internal friction among the moving fluid particles.

From $\phi(x, y)$, we can, in electrostatics, obtain the components of the electric field $E_x(x, y)$ and $E_y(x, y)$. These components give the x and y directed components of force on a unit (1 Coulomb) test charge placed at (x, y). In fluid mechanics, we can obtain from $\phi(x, y)$ the components of the fluid velocity vector $V_x(x, y)$ and $V_y(x, y)$ describing the motion of a particle of fluid at (x, y). In heat conduction, we can use $\phi(x, y)$ to get the components of the heat flux density vector $Q_x(x, y)$ and $Q_y(x, y)$ at a point. The vector gives the direction of heat flow and the rate of flow at (x, y).

The equations involved are as follows:

For electrostatics,

$$E_x\left(x,y\right)=\frac{-\partial\phi}{\partial x} \quad E_y\left(x,y\right)=\frac{-\partial\phi}{\partial y} \tag{4.7}$$

For fluid flow,

$$V_x(x,y) = \frac{\partial \phi}{\partial x} \quad V_y(x,y) = \frac{\partial \phi}{\partial y} \tag{4.8}$$

For heat conduction,

$$Q_x(x,y) = \frac{-k\partial \phi}{\partial x} \quad Q_y(x,y) = \frac{-k\partial \phi}{\partial y} \tag{4.9}$$

Here k is a constant peculiar to the heat-conducting medium and is called its *thermal conductivity*. If you are familiar with the concept of the *gradient* or the del operator, ∇, you see that

$$\mathbf{E} = -\nabla \phi = -\text{grad } \phi \tag{4.10}$$

$$\mathbf{V} = \nabla \phi = \text{grad } \phi \tag{4.11}$$

$$\mathbf{Q} = -k\nabla \phi = -k \text{ grad } \phi \tag{4.12}$$

The vector $\mathbf{E}$, for example, which has components $E_x(x, y)$ and $E_y(x, y)$, is "minus the gradient of ϕ"; it is the vector whose components are $\frac{-\partial \phi}{\partial x}$ and $\frac{-\partial \phi}{\partial y}$. Note that some authors would put minus signs after the equals signs in Equations 4.8 and 4.11. There is no unanimity here.

MATLAB is useful to us here in several ways. First, given a function $\phi(x, y)$, how can we tell if it is harmonic? For a simple function, say $\phi(x, y) = x^2 - y^2$, it should be evident that Equation 4.6 is satisfied with $\phi = \Omega$, and by contrast, $\phi(x, y) = x^2 + y^2$ does not satisfy the equation and is not harmonic. Verify this. Thus, the first of these can serve as a voltage, velocity potential, or temperature function while the second cannot. Suppose we are given a more complicated function $\phi(x,y) = e^{e^x \cos y} \cos(e^x \sin y)$. It is not obvious if this is harmonic, although the question is readily resolved with the following code:

```
syms x y
g=exp(exp(x)*cos(y))*cos(exp(x)*sin(y));
A=diff(g,x,2)+diff(g,y,2)
```

The output is:

```
A = 0
```

Notice that the third line of code:

```
A=diff(g,x,2)+diff(g,y,2)
```

sums together the second x and second y derivatives of the function $\phi(x,y) = e^{e^x \cos y} \cos(e^x \sin y)$. The result is found to be zero, which proves that this function is harmonic in any domain of the complex plane. This function was constructed by a choice of $\phi(x, y)$ as the real part of the function $e^{(e^z)}$, which must of course be harmonic. The reader should compute the real and imaginary parts of this function.

Suppose we proceed as above and try to determine whether the function

$$\phi(x,y) = \frac{-2xy}{\left(x^2 + y^2\right)^2}$$

is harmonic. The code

```
syms x y
gg=(-2*x*y)/(x^2+y^2)^2;
B=diff(gg,x,2)+diff(gg,y,2)
```

has as its output

B = (48*x*y)/(x^2 + y^2)^3 − (48*x^3*y)/(x^2 + y^2)^4 − (48*x*y^3)/(x^2 + y^2)^4

Since, unlike in the previous example, we do not simply get zero, we might conclude that this is not a harmonic function. Wrong. The additional line of code, which simplifies the above, B=simplify(B) has as its output B = 0 which confirms that we have a harmonic function, one that the reader should verify is the imaginary part of $f(z) = \dfrac{1}{\left(x+iy\right)^2}$. We had to simplify our result to see this. We should, to be safe, try to simplify our result in such calculations, to determine if we have a harmonic function. However, before concluding that the function $\phi(x,y) = \dfrac{-2xy}{\left(x^2 + y^2\right)^2}$ is harmonic everywhere, take note of the fact that it is undefined at $z = 0$ and therefore is not differentiable at the origin. MATLAB ignores this. Thus, the correct conclusion is that this function is harmonic in any domain *not containing the origin*.

In the preceding, we began with a real function $\phi(x, y)$ and used MATLAB to determine whether the function is harmonic. Alternatively, we might begin with an analytic function $f(z)$ and extract its real or imaginary part. Each must be harmonic in any domain where $f(z)$ is analytic. Often the problem of extracting $u(x, y) + iv(x, y)$ from $f(z)$ is simple. For example,

$$f(z) = 1/z = 1/(x+iy) = \frac{x}{x^2+y^2} - i\frac{y}{x^2+y^2}$$

whose real and imaginary parts

$$\frac{x}{x^2+y^2} \quad \text{and} \quad \frac{-y}{x^2+y^2}$$

are harmonic in any domain not containing the origin.

But suppose we have a more complicated function like $f(z)=e^{\cos(z^2)}$. This is an entire function whose real and imaginary parts are harmonic everywhere. The following code shows how to extract the real and imaginary parts $u(x, y)$ and $v(x, y)$ and confirms in addition that these parts are harmonic:

```
%getting real part and imaginary part
clear
syms x y real
   z=x+i*y;
w=exp(cos(z^2))
u=real(w)
v=imag(w)
%check on harmonic behaviour
u_check=diff(u,x,2)+diff(u,y,2);
v_check=diff(v,x,2)+diff(v,y,2);
u_check=simplify(u_check)
v_check=simplify(v_check)
```

The output is

w = exp(cos((x + y∗i)^2))
u = exp(cosh(2∗x∗y)∗cos(x^2 − y^2))∗cos(sinh(2∗x∗y)∗sin(x^2 − y^2))
v = −exp(cosh(2∗x∗y)∗cos(x^2 − y^2))∗sin(sinh(2∗x∗y)∗sin(x^2 − y^2))
u_check = 0
v_check = 0

Notice an important line of code:

```
syms x y real
```

This tells MATLAB that *x* and *y* are *real*. Without this line, we could not use the real(*w*) and imag(*w*) operations to extract the real and imaginary parts of *w* because MATLAB would assume that *x* and *y* are complex.

We have chosen to check that the real part of *w*, which we call *wr(x, y)*, is harmonic through the inclusion of the line of code u_check=diff(u,x,2)+diff(u,y,2).

The output of zero, which is the sum of the second *x* and *y* derivatives of *wr(x, y)*, confirms that we have a function that is harmonic everywhere. A similar check is made on the imaginary part of *w*.

Incidentally, the functions $u(x, y)$ and $v(x, y)$ obtained from the preceding code might be more comfortably written as

$$u(x,y)=e^{\left(\cosh(2xy)\cos\left(x^2-y^2\right)\right)}\cos\left(\sinh(2xy)\sin\left(x^2-y^2\right)\right)$$

and

$$v(x,y) = -e^{\left(\cosh(2xy)\cos\left(x^2-y^2\right)\right)} \sin\left(\sinh\left(2xy\right)\sin\left(x^2-y^2\right)\right)$$

Sometimes we might find that having supplied MATLAB with a function *f(z)*, we discover that the function is sufficiently complicated that MATLAB cannot determine explicit expressions for the real and imaginary parts. The following example will suffice:

```
clear
syms x y real
syms z
z=x+i*y
A0=sin(z)
A0r=real(A0)
A1=sin(1/z)
A1r=real(A1)
A2=sin(z*sin(1/z))
A2r=real(A2)
A3=sin(z^2)
A3r=real(A3)
```

Whose output is

z = x + y*i
A0 = sin(x + y*i)
A0r = cosh(y)*sin(x)
A1 = sin(1/(x + y*i))
A1r = cosh(y/(x^2 + y^2))*sin(x/(x^2 + y^2))
A2 = sin(sin(1/(x + y*i))*(x + y*i))
A2r = real(sin(sin(1/(x + y*i))*(x + y*i)))
A3 = sin((x + y*i)^2)
A3r = cosh(2*x*y)*sin(x^2 - y^2)

Refer to A2r above. Note that MATLAB was unable to present the real part of sin(z sin(1/z)) as an explicit real function of *x* and *y*. In the case of the other functions sin(z), sin(1/z), sin(z²), we were successful.

To deal with the case of sin(z sin(1/z)), we can break the operation into steps for MATLAB, first obtaining the real and imaginary parts of sin(1/z) and entering that result into the argument of sin(z sin(1/z)). It will be found that the following will produce an explicit function for the real part of sin(z sin(1/z)):

```
clear
syms x y real
```

```
syms z
z=x+i*y
A1=sin(1/z)
A1r=real(A1);
A1i=imag(A1);
A2=sin(z*(A1r+i*A1i))
A2r=real(A2)
```

The output is

A2r = cosh(x*cos(x/(x^2 + y^2))*sinh(y/(x^2 + y^2)) − y*cosh(y/(x^2 + y^2))*sin(x/(x^2 + y^2)))*sin(x*cosh(y/(x^2 + y^2))*sin(x/(x^2 + y^2)) + y*cos(x/(x^2 + y^2))*sinh(y/(x^2 + y^2)))

Using MATLAB, you can readily show that the preceding is a harmonic function in any domain not containing the origin.

In Section 2.2 we saw how the properties of a real function of the two real variables, like $\phi(x, y)$, can be studied through a set of contour plots. In electricity, if the harmonic function $\phi(x, y)$ describes the potential in space, then such contour plots are the loci on which the voltage is constant. These are the traces in the x–y plane of the surfaces of constant voltage, often referred to as equipotential surfaces. Similarly, if a temperature is given by $\phi(x, y)$, then the contour plots of $\phi(x, y)$ are isotherms, or surfaces of constant temperature. Finally, in fluid mechanics, we can also plot and speak of equipotential surfaces.

To review this subject, consider for example an infinitely long, uniformly charged, filament passing perpendicularly through the complex plane at the point (x_0, y_0). The electric potential (voltage) created in the surrounding material is given by

$$\phi(x,y) = C \operatorname{Log} \frac{1}{\sqrt{(x-x_0)^2 + (y-y_0)^2}} \tag{4.13}$$

Here C is a real constant that depends on the density of electric charge along the filament as well as the electrical properties of the surrounding, nonconducting, medium.[*] If $z = x + iy$ and $z_0 = x_0 + iy_0$, we have also that

$$\phi(x, y) = C \operatorname{Log}(1/|z - z_0|) = -C \operatorname{Log}(|z - z_0|) = -C \operatorname{Re}[\operatorname{Log}(z - z_0)] \tag{4.14}$$

One should use MATLAB (as was done earlier) to quickly verify that $\phi(x, y)$ is a harmonic function in any domain not containing (x_0, y_0).

---

[*] In this case, $C = \rho_L/(2\pi\varepsilon_0)$, where ρ_L is the charge per unit length on the filament and ε_0 is a constant found in handbooks called "the permittivity of free space."

Electrical engineers like to model electric power lines as two charged filaments separated by some distance. Each filament carries an equal and opposite charge. There is an electric field surrounding the filaments.

Example 4.4

An electrical transmission line is being approximated as two oppositely charged filaments. The positive filament passes through $z = 1$ while the negative one passes through $z = -1$. Each filament is perpendicular to the complex plane. The potentials created by the filaments can be obtained from Equations 4.13 and 4.14. We take $z_0 = 1$ or $z_0 = -1$ as needed. Thus,

$$\phi_1(x,y) = -\text{Log}\sqrt{(x-1)^2 + y^2} = -\text{Log}|z-1| = -\text{Re}\big(\text{Log}(z-1)\big)$$

and

$$\phi_2(x,y) = \text{Log}\sqrt{(x+1)^2 + (y)^2} = \text{Log}|z+1| = \text{Re}\big(\text{Log}(z+1)\big)$$

These arise from the charges passing through $z = 1$ and $z = -1$, respectively. The constant C has been set equal to 1. The combined potential is $\phi(x, y) = \text{Re}(-\text{Log}(z - 1) + \text{Log}(z + 1))$. We can plot the equipotentials for this function using the familiar MATLAB function **contour** as follows:

```
x=linspace(-4,4,100);
y=linspace(-4,4,100);
n=linspace(-2,2,5);
[x,y]=meshgrid(x,y);
z=x+i*y;
 w=-log(z-1)+log(z+1);
wr=real(w);
 [c,h]= contour(x,y,wr,n);colorbar;
    xlabel('x');ylabel('y');
    title('Contour Lines for Log|z+1|-Log|z-1|')
    grid
    clabel(c,h);axis([-2.5 2.5 -2.5 2.5])
```

This generates Figure 4.16.

Figure 4.16 is in black and white (gray scale). If you run the above code, you will obtain a more attractive color figure and a color bar.

As mentioned in our presentation of Equations 4.7 through 4.9, the gradient of the potential $\phi(x, y)$ is useful. In the present case, it can give us the components of the electric field, but it can also yield temperature and

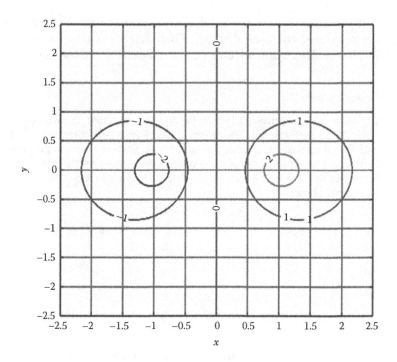

FIGURE 4.16
Contour lines for $\log|z+1| - \log|z-1|$.

the heat flux density in other situations. The electric field is the vector having components given in Equations 4.7 and 4.10.

MATLAB provides a convenient graphical means for visualizing the gradient vector. This involves using the MATLAB functions **gradient** and **quiver**, and the reader should read the documentation for these in the help folder. Let the function $\phi(x, y)$ be approximated by numerical values at the intersections of a rectangular grid, perhaps one generated by **meshgrid**. The MATLAB function **gradient** uses the differences between values of $\phi(x, y)$ at adjacent points to compute numerical values of the components of the gradient at the same grid points. For example, at some point with coordinates $x_j, y_k,$

$$\frac{\partial \phi}{\partial x} \approx \frac{\phi\left(x_{j+1}, y_k\right) - \phi\left(x_j, y_k\right)}{\Delta x}$$

which approximates the x directed component of the gradient of ϕ. Unless told otherwise, MATLAB takes the spatial distance between the values used to compute such an expression as unity so that $\Delta x = 1$, which means $\dfrac{\partial \phi}{\partial x}$ is approximated as $\phi(x_{j+1}, y_k) - \phi(x_j, y_k)$. The reader should keep in mind that we are obtaining a numerical approximation, perhaps a crude one, by using this function.

More often, we simply want a graphical picture of the gradient of $\phi(x, y)$, not numerical values. This is accomplished by our using, in addition to the above, the MATLAB function **quiver**. The entries giving the components of the gradient at each intersection of the grid in the x–y plane are used by **quiver** to generate pictorial arrows signifying the gradient. These arrows are placed on a plot at the intersections of the grid. The length of the arrow is proportional to the strength of the gradient vector, while the arrow's direction is the direction of the gradient vector. Following is a simple example.

Example 4.5

Let $\phi(x, y) = x^2 - y^2 = \text{Re}(z^2)$, a harmonic function. The following code illustrates the use of **gradient** and **quiver** to show us a picture of the gradient of the function. You might wish to study **quiver** in the help folder.

```
clf
v=linspace(-3, 3, 100);
[x,y]=meshgrid(v);
f=(x.^2-y.^2);
  figure (1)
  n=linspace(-2,2,5);
[c,h]=contour(x,y,f,n);axis equal
clabel(c,h);colorbar
hold on
%following uses coarse spacing for gradient
%which is desirable
v=linspace(-3, 3, 10);
[x,y]=meshgrid(v);
f=x.^2-y.^2;
  [gx,gy]=gradient(f);
  quiver(x,y,gx,gy);
  title('contour lines and gradient for x^2-y^2')
```

The output is as shown in Figure 4.17.

There are several matters to notice here. The gradient vector at each location is normal to the contour of constant ϕ passing nearest to that location. The vector points toward the nearest other contour on which ϕ has a more positive numerical value than on the first. The length of the vector is inversely proportional to the spacing of adjacent contours, which is because the narrower the spacing, the more rapidly is ϕ changing as we move normal to the contours.

Another important matter is the code. We use

```
v=linspace(-4, 4, 100)
[x,y]=meshgrid(v);
```

to plot the contours of constant ϕ. However, to plot the gradient, we employ

```
v=linspace(-4, 4, 10)
[x,y]=meshgrid(v);
```

The 100 was replaced by 10.

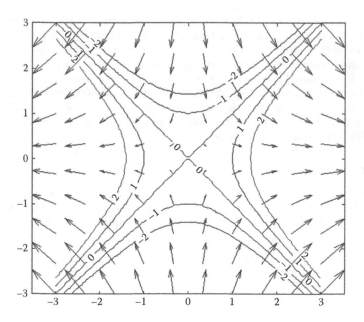

FIGURE 4.17
Contour lines and gradient for $x^2 - y^2$.

Notice that we are using a *coarser* grid to plot the gradient (compare 100 versus 10). Had we used as fine a grid to plot the gradient as was done with the surfaces of constant ϕ, we would have obtained a plot from **quiver** so densely filled with arrows as to be unintelligible. The reader should verify this for him- or herself. The plot shown above is in black and white, but the code will generate a more attractive color picture.

To find the numerical values of the gradient of ϕ, we are best served by using the formula for grad ϕ rather than the line of code [gx,gy]=gradient(f), which would yield a result based on an approximate evaluation of the derivatives with respect to x and y. In the present problem, the gradient is a vector having components $\dfrac{\partial(x^2 - y^2)}{\partial x} = 2x$ and $\dfrac{\partial(x^2 - y^2)}{\partial y} = -2y$. These are readily evaluated on any grid of points by means of MATLAB. Observe that both components of the vector vanish at the origin, which explains the absence of a vector in the above figure at this point.

Example 4.6

Using **quiver**, plot the negative of the gradient, for the equipotentials shown in Figure 4.16. We are using the negative of the gradient because the electric field is generally the vector with components −∇ϕ, where ϕ(x, y) is the electrostatic voltage.

The following code will work. We simply add these lines to those used in Example 4.4:

```
hold
v=linspace(-2.5, 2.5, 10);
[x,y]=meshgrid(v);
z=x+i*y;
w=-log(z-1)+log(z+1);
wr=real(w);
  [gx,gy]=gradient(-wr);
quiver(x,y,gx,gy,.75);
```

This generates the output shown in Figure 4.18.
Notice the line of code:

```
quiver(x,y,gx,gy,.75);
```

The .75 scales down the length of the arrows to three quarters of their normal length. Without this number, the arrows are longer and crowd each other in the figure. Observe that the vectors representing the electric field emanate out from the positive charge on the right and head toward the negative one on the left, as is to be expected. The arrows strike the equipotentials at right angles.

Your plot will be in color and will have a color bar. We give here only a gray-scale version.

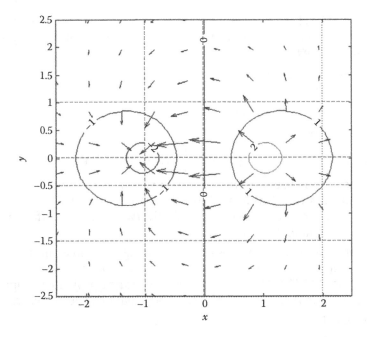

FIGURE 4.18
Electric field for $-\log|z-1| + \log|z+1|$.

Notice also that we use `[gx,gy]=gradient(-wr);`. We do not apply `-gradient(wr)`, which would result in an error message—an idiosyncrasy of MATLAB, which I am told will be remedied.

Figure 4.18 has interpretations beyond electrostatics. Imagine the filament on the right (passing through $x = 1$, $y = 0$) as a source of heat and the one on the left (passing through $x = -1$, $y = 0$) as a sink for heat—a place where heat is absorbed. The arrows indicate the direction of heat flow through a heat-conducting medium surrounding the filaments, while the length of the arrows is a measure of the intensity of that flow.

A similar interpretation is available in fluid mechanics: because of the absence of the minus sign in Equation 4.11 (compare with its neighboring equations), the filament on the left emits fluid uniformly in all directions and fills some unbounded container of liquid, while the filament on the right is a sink for this flow—it receives all the fluid sent by the filament on the left. The arrows above are opposite to the direction of fluid flow, while their length is proportional to the density of flow.

Exercises

1. Using MATLAB, determine whether the following functions are harmonic. If a function is harmonic in general except at certain points, state where these points are located.

   a. $\cos(x^2 + y^2) \cos(2xy)$

   b. $\cos(x^2 - y^2) \cos(2xy)$

   c. $\text{Log}(\sinh^2 x + \sin^2 y)$

   d. $\text{Log}(\cosh^2 x + \sin^2 y)$

   e. $e^{-\tan^{-1}(y/x)} \sin\left(\frac{1}{2}\text{Log}(x^2 + y^2)\right)$, where the principal value of the arctan is used. In MATLAB you will have to use atan(y/x) and not atan2(y,x), as MATLAB will not do symbolic differentiations involving the latter function. As the two functions differ only by a constant, their derivatives are identical.

2. With $z = x + iy$, express the following functions in the form $f(z) = u(x, y) + iv(x, y)$, finding $u(x, y)$ and $v(x, y)$ by means of MATLAB, and use MATLAB to show that both $u(x, y)$ and $v(x, y)$ satisfy Laplace's equation in a domain. State where the equation is not satisfied.

   a. $\cos(1/z)$

   b. $\cos(\cos z)$

   c. $1/(z - i)^5$

d. $(z + z^{-1})^5$

e. $\sinh(\sin z^2)$

3. An electrostatic potential in space is being created by two identical positively charged filaments. One filament passes through $z = 1$, while the other passes through $z = -1$. Each filament is perpendicular to the complex plane. The potentials created by the filaments are from Equations 4.13 and 4.14.

$$\phi_1(x,y) = -\text{Log}\sqrt{(x-1)^2 + y^2} = -\text{Log}|z-1|$$

and

$$\phi_2(x,y) = -\text{Log}\sqrt{(x+1)^2 + (y)^2} = -\text{Log}|z+1|$$

where the constant C has been set equal to 1. The combined potential is given by

$$\phi(x,y) = -\text{Log}\sqrt{(x-1)^2 + y^2} - \text{Log}\sqrt{(x+1)^2 + y^2}$$

a. Plot the equipotentials created by this pair.

Suggestion: Use the equipotentials –2, –1, 0, 1, 2. Use a grid with 100^2 points satisfying $|x| \leq 3$, $|y| \leq 3$.

b. On the above figure, plot the negative of the gradient of the potential by using **grad** and **quiver**.

Suggestion: Use a much coarser grid than that used in part (a), otherwise you will have too many arrows to be visible. You may also wish to shorten the length of the arrows. Make sure the arrows (which are proportional to the electric field) are normal to the nearest equipotentials and point from higher to lower potentials.

4. A strip of electric charge is of width 2. The density of charge on the strip is uniform. The strip, of infinite length, is perpendicular to the x–y plane and occupies the space $y = 0$, $-1 < x < 1$. It can be shown that the electrostatic potential created by the strip is of the form $C \, \text{Re}[(z - 1) \log(z - 1) - (z + 1) \log(z + 1)]$, where C is a real constant, and $z = x + iy$.

The value of the real constant C is directly proportional to the density of electric charge on the strip.

a. Taking $C = 1$, plot the equipotentials created by this charged strip. Suggested values for the equipotentials are $\phi = -4, -3, -2, -1, 0$. Use a grid like that described in problem 3(a).

b. In the above figure, plot the negative of the gradient of the potential. Use the suggestions in problem 3(b).

4.4 Complex Potentials, Cauchy–Riemann Equations, the Stream Function, and Streamlines

In the preceding section, we discussed a harmonic function $\phi(x, y)$ which can be interpreted as giving the voltage, temperature, or velocity potential in various two-dimensional configurations. Suppose this to be the real part of the analytic function $\Phi(x, y)$ or $\Phi(z)$. Thus, we put

$$\Phi(x, y) = \phi(x, y) + i\psi(x, y) \tag{4.15}$$

Here $\psi(x, y)$ is a real function that is known as the *stream function* because of its connections to fluid mechanics, although the term is used in electricity and heat transfer as well. Thus, ψ and ϕ are *conjugate functions*. The former can be as useful as the latter, as we see later. The function $\Phi(x, y)$ is called the *complex electric potential* (in electrostatics) or *complex temperature* (in the case of heat flow) and the *complex velocity potential* (for fluid flow). Suppose we know either ϕ or ψ. Can we use one to find the other?

4.4.1 Cauchy–Riemann Equations

Recall that all standard textbooks on complex variable theory treat his subject. (See W section 2.5 and S section 3.5.) In familiar notation, we are given the analytic function $f(z) = u(x, y) + iv(x, y)$, where

$$\partial u/\partial x = \partial v/\partial y \tag{4.16}$$

$$\partial u/\partial y = -\partial v/\partial x \tag{4.17}$$

As a simple example, to refresh the reader's memory, we consider the function $u(x.y) = y + x^2 - y^2 + x$. Putting $\Omega(x, y) = u(x, y)$ in Equation 4.6, we readily see that this function is harmonic. We now seek $v(x, y)$ such that $f(z) = u(x, y) + iv(x, y)$ is analytic. We have from Equations 4.16 and 4.17 that

$$2x + 1 = \partial v/\partial y \tag{4.18}$$

$$-2y + 1 = -\partial v/\partial x \tag{4.19}$$

We integrate Equation 4.18 with respect to the variable y and Equation 4.19 with respect to the variable x. We also swap the locations of the minus sign in Equation 19 and get Equations 4.20 and 21:

$$v = 2xy + y + c_1(x) + C_2 \tag{4.20}$$

$$v = 2xy - x + c_2(y) + C_4 \tag{4.21}$$

Here $c_1(x)$ is a function that can depend on x but not on y, while $c_2(y)$ is a function that can depend on y but not on x. C_2 and C_4 are true constants. One should verify the above results by substituting back into Equations 4.18 and 4.19.

Now Equations 4.20 and 4.21 should both be valid expressions for v. The difference between them must be identically zero. Subtracting the second from the first gives us the expression we call *difference*:

$$\textit{difference} = y + x + c_1(x) - c_2(y) + C_2 - C_4$$

Setting *difference* equal to zero, we see that we must take $c_1(x) = -x$, $c_2(y) = y$ as well as $C_2 = C_4$.

The number $C_2 = C_4$ is simply a true constant, independent of x and y, and its value can only be found if the numerical value of $v(x, y)$ is known at some value of (x, y). Thus, using our derived values in either Equation 4.20 or Equation 4.21 gives

$$v = 2xy + y - x + C_2$$

Here we accomplish the same thing with MATLAB:

```
clear all
syms y x v ux uy v1(y) v2(x) c1(x) c2(y) difference;
u=y+x^2-y^2+x;
ux=diff(u,x);
uy=diff(u,y);
%below solves du/dx=dv/dy
v1=dsolve(ux==diff(v1),y)+c1(x);
%below solves du/dy=-dv/dx
v2=dsolve(uy==-diff(v2),x)+c2(y);
v1=simplify(v1)
v2=simplify(v2)
difference=v1-v2;
difference=simplify(difference)
```

The reader might wish to study the MATLAB **help** for **diff** and **dsolve**. Notice that the lines of code ux=diff(u,x) and uy=diff(u,y) compute $\dfrac{\partial u}{\partial x}$ and $\dfrac{\partial u}{\partial y}$. The next two lines of code, which employ **dsolve**, solve $\dfrac{\partial u}{\partial x} = \dfrac{\partial v}{\partial y}$ and $\dfrac{\partial u}{\partial y} = \dfrac{-\partial v}{\partial x}$. We add on "constants" c_1 and c_2, which can depend on x and y, respectively. We derive two expressions for $v(x, y)$, which we call v_1 and v_2, respectively. We form the difference of these expressions. The output of the program is

```
v1 = C2 + y + c1(x) + 2*x*y
v2 = C4 - x + c2(y) + 2*x*y
difference = C2 - C4 + x + y + c1(x) - c2(y)
```

Studying the above line, we see that if the difference in our two expressions for v is to be zero throughout a region of the x–y plane, then `c1(x)=-x` and `c2(y)=y`. We may install these in the expression for either `v1` or `v2`. The constant `C2=C4` remains undetermined. We have just found the value of $v = 2xy + y - x + C2$ to within a constant. In this particular problem, C2 and C4 are the numerical values of $v(0, 0)$. In the exercises, you will do some MATLAB problems similar to the above where the computer program can save us some labor.

A program similar to the above can be written if we are supplied with the imaginary part of an analytic function and are seeking the real part. This is left to the exercises.

Recall that in textbooks in complex variable theory, the Cauchy–Riemann equations are derived by equating two equally valid expressions for the derivative of a function of a complex variable. Thus, with $f(z) = u(x, y) + iv(x, y)$, we have both $f'(z) = \dfrac{\partial u}{\partial x} + i\dfrac{\partial v}{\partial x}$ and $f'(z) = \dfrac{\partial v}{\partial y} - i\dfrac{\partial u}{\partial y}$. We can eliminate $\dfrac{\partial v}{\partial x}$ in the first of these by means of a Cauchy–Riemann equation (Equation 4.17) and obtain $f'(z) = \dfrac{\partial u}{\partial x} - i\dfrac{\partial u}{\partial y}$. If we take $\Phi(z) = \phi(x, y) + i\psi(x, y)$ as our analytic function, we have $\Phi'(z) = \dfrac{\partial \phi}{\partial x} - i\dfrac{\partial \phi}{\partial y}$, and finally

$$\overline{\Phi'(z)} = \frac{\partial \phi}{\partial x} + i\frac{\partial \phi}{\partial y} \tag{4.22}$$

We have noted that if $\phi(x, y)$ is an electrostatic potential, then the x and y components of the electric field in space are given, respectively, by (see Equation 4.7) $E_x(x,y) = -\dfrac{\partial \phi}{\partial x}$ and $E_y(x,y) = -\dfrac{\partial \phi}{\partial y}$. Thus, the negative of the expression in Equation 4.22 is a complex quantity whose real and imaginary parts are, respectively, the x and y components of the electric field.

We have

$$E_x(x,y) + iE_y(x,y) = -\overline{\Phi'(z)} \tag{4.23}$$

which should show the utility of the complex potential in electrostatics.

In a similar way, we may use the complex velocity potential of fluid mechanics to obtain a complex number whose real and imaginary parts are the vector components of the fluid velocity:

$$V_x(x,y) + iV_y(x,y) = \overline{\Phi'(z)} \tag{4.24}$$

For this complex potential $\Phi(z)$, the real part $\phi(x, y)$ is the velocity potential describing fluid flow.

Finally, the complex temperature of heat flow has this property:

$$Q_x(x,y) + iQ_y(x,y) = -k\overline{\Phi'(z)} \tag{4.25}$$

The real and imaginary parts on the left are the components of the heat flux density vector, and k is the thermal conductivity of the material in use.

Of what use is the stream function $\psi(x, y)$—the imaginary part of the potential—in each of these three cases? Recall that for the analytic function $\Phi(z) = \phi(x, y) + i\psi(x, y)$, if we generate a set of curves $\phi(x, y) = \alpha_1, \alpha_2, \ldots$, and another set of curves $\psi(x, y) = \beta_1, \beta_2, \ldots$ (where the α and β values are real constants), the two sets of curves will meet at right angles—they are orthogonal. The orthogonality property can break down at a point where $\Phi'(z) = 0$. Recall also that in electrostatics the electric field at each point in space is at right angles to the curve of constant potential (the curve of constant ϕ) passing through that point. Thus, at each point in space, the electric field vector must be tangent to the curve of constant ψ passing through that point. The statement breaks down only at those places where the electric field vanishes—here we cannot assign it a direction.

In a similar fashion, in fluid mechanics, the curves along which the stream function maintains constant values, the "streamlines," are the curves that are tangent to the direction of fluid flow. The curves do not by themselves show the actual sense of flow (i.e., the direction of motion along the streamline)—this must be computed from either Equation 4.24 or Equation 4.8. However, we can frequently tell the direction along the curve by a simple physical argument. Finally, in heat conduction problems, the streamlines—the curves of constant ψ—are the curves that are tangent to the direction of heat flow.

In fluid mechanics, we require that any impermeable boundary be coincident with a streamline. Let us look at the following case.

The analytic function $\Phi(z) = z$, $z = x + iy$ is not a very interesting complex velocity potential. We have from Equation 4.24 that $\overline{\Phi'(z)} = 1 = V_x + iV_y$ so that $V_x = 1$ and $V_y = 0$. Thus, fluid is moving uniformly in the positive x direction with uniform velocity. The stream function $\psi(x, y) = y$ and so the streamlines are simply the lines on which y assumes fixed real values, for example, $y = 0, \pm1, \pm2, \ldots$. Suppose we add to this analytic function the analytic function $1/z$. Thus, $\Phi(z) = z + 1/z$, where we exclude the origin. Notice that as $z \to \infty$, $\Phi(z) \to z$, which means that the flow is uniform at infinity. Let us ask MATLAB to generate the streamlines for this potential. This code will do that:

```
clf;clear
x=linspace(-6,6,200);
y=linspace (-6,6,200);
n=linspace (-2,2,9);
[x,y]=meshgrid(x,y);
```

```
z=x+i*y;
  w=z+1./z;
  wi=imag(w);
  [c,h] = contour(x,y,wi,n);colorbar;
  axis equal
  xlabel('x');ylabel('y'); grid
    clabel(c,h);
```

This output is as shown in Figure 4.19.

The arrows that appear in Figure 4.19 were not generated by the code but were added later as will be explained. Do not be alarmed that we cannot read the numbers inside the unit circle—they are not needed.

It appears that the contour on which $\psi = 0$ is satisfied both on the unit circle and on the line $y = 0$, $-\infty < x < \infty$. This can be confirmed as follows:

$$\Phi(z) = z + 1/z = x + iy + 1/(x + iy) = x + iy + \frac{(x - iy)}{x^2 + y^2}$$

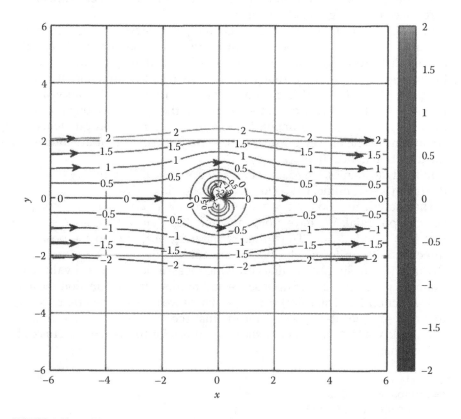

FIGURE 4.19
Streamlines for $z + \frac{1}{z}$.

Thus, $\psi(x,y) = \operatorname{Im}\Phi(x,y) = y - \dfrac{y}{\left(x^2+y^2\right)}$. If we equate the preceding to zero, we find that the equation is satisfied by either $y = 0$ or $x^2 + y^2 = 1$ (which is $|z| = 1$) as the figure suggests. Thus, the unit circle is a streamline of this given complex potential.

The complex velocity for this flow is found from Equation 4.24. With the given potential, we have $V_x + iV_y = 1 - 1/\bar{z}^2$. We see that as $z \to \infty$, $V_x \to 1$ and $V_y \to 0$. There is a flow of fluid that moves from left to right. It begins as a uniform flow and ends the same way. Knowing the direction of the flow, we added arrows to some of the streamlines.

The contour $|z| = 1$ is a streamline, which means that flow is everywhere tangent to this circle. We may place a rigid barrier in coincidence with this circle and not affect the flow of fluid in and outside this barrier. Of course, no fluid will get inside the barrier. Thus, we may interpret the streamlines outside the unit circle in Figure 4.19 as showing the flow around what might be an infinitely long wooden log of unit radius that has been placed inside a uniform stream of infinite extent. Under these circumstances, the streamlines shown inside the unit circle in this same figure have no physical meaning, and the fact that we cannot read the numbers on these lines is of no consequence. If you want to know the actual fluid velocity at any point outside the log, you can use

$$V_x + iV_y = 1 - 1/\bar{z}^2 = 1 - \frac{\left(x+iy\right)^2}{\left(x^2+y^2\right)^2} \tag{4.26}$$

to establish the components V_x and V_y.

If both components of the velocity were to vanish at a point, then we could not assign a direction to the fluid velocity vector. We cannot say that it is tangent to the streamline. In such cases, the tangent to the streamline is not defined. Studying Equation 4.26, we see that if $\bar{z}^2 = 1$, which means $z = \pm 1$, the fluid velocity is zero. Studying Figure 4.19, we see that indeed we cannot establish the tangent to the streamline as this point—the streamline on $y = 0$ branches off into two different directions. In fluid mechanics, such points where the velocity vanishes are called "stagnation points" as the fluid is nonmoving (stagnant).

Let us turn our attention to electrostatics and create a new potential by taking the one from the previous example and multiplying it by i. The code is essentially the same as above except that we replace `w=z+1./z;` with `w=i*(z+1./z);`.

We have also generated a plot of a unit circle, by means of the lines of code added at the end of the program:

```
hold on
axis([-2 2 -2 2])
th=linspace(0,2*pi,1000);
    z=exp(i*th);
    plot(z,'linewidth',2)
```

We have added arrows to the streamlines after the plot has been formed. Figure 4.20 presents the result.

Our complex potential is now

$$\Phi(z) = \phi(x,y) + i\psi(x,y) = iz + i/z = -y + ix + i/(x+iy) = -y + ix + \frac{(y+ix)}{x^2+y^2} \quad (4.27)$$

If $|z| \gg 1$, we have approximately that $\Phi(z) \approx iz = -y + ix$. Thus, $\phi(x,y) =$ Re($\Phi(z)$) $\approx -y$. Recalling that $E_x = \dfrac{-\partial\phi}{\partial x}$ and $E_y = \dfrac{-\partial\phi}{\partial y}$, we realize that, as we move far from the origin, the electric field asymptotically tends to $E_x = 0$ and $E_y = 1$. The field is uniform and pointing in the positive y direction. In this way, we were able to assign arrows to the streamlines in Figure 4.20.

From Equation 4.27, we have that the electrostatic potential $\phi(x,y) = \text{Re}\,\Phi(z) = -y + \dfrac{y}{x^2+y^2}$. Observe that on the unit circle, $x^2 + y^2 = 1$, we have that $\phi(x,y) = 0$. Thus, the unit circle is a locus of constant potential, and we would expect that the streamlines strike this circle at right angles. Figure 4.20 would appear to confirm this except at $z = \pm 1$. The failure of orthogonality can be explained if we notice that $\Phi'(z) = i(1 - 1/z^2)$

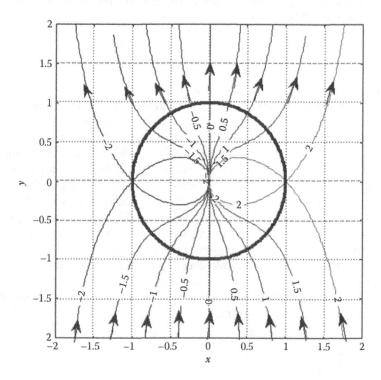

FIGURE 4.20
A plot of a unit circle and the streamlines of arrows.

is zero if $z = \pm 1$. Recall that $E_x(x,y) + iE_y(x,y) = -\overline{\Phi'(z)}$. Thus, the electric field vanishes at $z = \pm 1$, and we cannot assign it a direction or find a tangent to the streamlines at these points.

If an electric field strikes a perfect electrical conductor, the intersection must be orthogonal. We can thus insert a perfectly conducting metal tube of unit radius into the configuration of Figure 4.20. The axis is perpendicular to the z plane and passes through the origin. We may regard this as a metal tube placed into an electric field that is uniform and parallel to the y axis at infinity. The field will not penetrate the tube, the field inside is zero, and we should ignore the streamlines inside the unit circle that appear in the plot of Figure 4.20. The streamlines outside the tube show the direction of the electric field. The numerical values of the field can be found from $E_x(x,y) + iE_y(x,y) = -\overline{\Phi'(z)}$.

Of what use is the numerical value of the stream function? Consider a piecewise smooth, closed curve C such as that in Figure 4.21. Suppose $\Phi(z) = \phi(x, y) + i\psi(x, y)$ is a complex analytic potential describing a configuration in heat flow, fluid mechanics, or electrostatics. Let C' be a segment of C with α and β the values of z at the extremes of the segment. We encounter α before β in moving counterclockwise. Let

$$\Delta\psi = \psi(\alpha) - \psi(\beta) \tag{4.28}$$

If α and β were to lie on the same streamline, this expression would be zero, because ψ is constant along a streamline. Now it can be shown

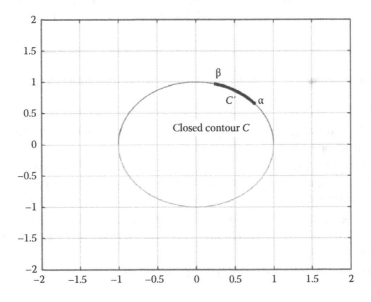

FIGURE 4.21
A contour for studying ψ.

(see W chapter 8 and its appendix) that in heat conduction, the expression $H = k\Delta\psi$ gives the flux of heat flow (the rate at which energy is transported by heat conduction) across a surface of unit length (perpendicular to the x–y plane) whose edge is this segment C'. Positive flow is from the interior to the exterior of the domain whose boundary is C. Recall that k is the heat conductivity of the medium in use. In fluid mechanics, the expression $G = -\Delta\psi = \psi(\beta) - \psi(\alpha)$ is the rate at which fluid moves across the same surface (assumed permeable) from interior to exterior. Finally, in electrostatics, the formula $F = \varepsilon\Delta\psi$ gives the flow of electric flux lines across the surface. Here, ε is a constant, called the permittivity, that describes the electrical properties of the medium in which the electric field exists. Note that all of the preceding statements involving Equation 4.28 presume that ψ varies continuously as we move along the surface in question.

Let us look at the configuration of Figure 4.19. If the cylinder were not present, then the flow of fluid would be uniform and given by the potential $\Phi(z) = x + iy$. How much fluid flows through a rectangular surface like the one described above whose edge is the line segment $x = 1$, $-1 \le y \le 1$? Now following the notation of Equation 4.27, we have $\alpha = (1, -1)$ and $\beta = (1, 1)$. Since $\psi = y$, we have $\Delta\psi = -1 - 1 = -2$. The rate of flow of fluid is $G = -\Delta\psi = 2$. This might be measured in such units as liters per second or gallons per minute, depending on the system in use. If the cylinder of unit radius shown in Figure 4.19 is now placed in the fluid, so as to impede the flow, we have shown that $\psi(x,y) = \text{Im}\,\Phi(x,y) = y - \dfrac{y}{\left(x^2 + y^2\right)}$. Evaluating ψ at points at α and β, we have $-\frac{1}{2}$ and $\frac{1}{2}$, respectively. Thus, $G = -\Delta\psi = 1$. The flow through the surface has been reduced from 2 to 1 by this impediment.

Exercises

Solve the following problems by means of suitable MATLAB programs. All of the functions given are harmonic.

1. If the real part of an analytic function is $\sin(x^2/2 - y^2/2 + x)\cosh(xy + y)$, find the imaginary part. Evaluate any arbitrary constant in your answer by assuming that the sought-after function is zero at the origin.

2. If the real part of an analytic function is $y + \arctan(y/(x - 1))$, find the imaginary part. Evaluate any constant by assuming that the function is zero at $z = 0$.

3. If the real part of an analytic function is $x^2 - y^2 + 2 + \dfrac{x^2 - y^2}{\left(x^2 + y^2\right)^2}$, find the imaginary part. Evaluate any constant by assuming that the function is zero at $z = 1$.

4. If the imaginary part of an analytic function is

$$\sin\left(\frac{1}{2}\log\left(x^2+y^2\right)+y+1\right)\cosh\left(\arctan\left(\frac{y}{x}\right)-x\right),$$

find the real part to within an additive constant.

5. If the imaginary part of an analytic function is

$$e^{\arctan(y/x)}\cos\left(\text{Log}\sqrt{x^2+y^2}\right)+1+2y-x$$

find the real part. Assume that the real part equals 1 at $z = 1$.

6. In the previous section, Example 4.5, we used **gradient** and **quiver** in MATLAB to create arrows pointing in the direction of the gradient of the function $u(x, y) = x^2 - y^2$. The gradient at each point in space must be normal to the surface, on which u is constant, passing through that point. The gradient must therefore be tangent to the streamline passing through that point. Determine the stream function for this function $u(x, y)$ and plot the streamlines on the same figure as the vectors for the gradient. Verify that the arrows from **quiver** are along the streamlines.

7. Suppose fluid motion is caused by a flow that is uniform at infinity and in the x direction. Suppose that there are present, normal to the complex plane, a line source of fluid and a corresponding sink of identical strength. The source and sink penetrate the x axis. The streamlines for this overall arrangement form interesting shapes. One in particular forms an oval shape known as a Rankine Oval. Consider the complex velocity potential given by $\Phi(z) = \phi(x, y) + i\psi(x, y) = V_\infty z - C \text{ Log}(z - 1) + C \text{ Log}(z + 1)$. Here V_∞ is the velocity of the fluid at infinity, while C (a positive real) is the strength of the fluid source at $z = -1$. There is a corresponding sink at $z = 1$.

   a. Take $V_\infty = 1$ and $C = 1$. Using MATLAB, plot the streamlines of this flow. A suggestion for a nice plot: let the values of ψ on the plot be the nine integers satisfying $-4 \le \psi \le 4$.

   Consider $|x| \le 5$ and $|y| \le 5$.

   b. Study your plot. What is special about the curve on which $\psi = 0$? This is called a Rankine Oval. Other such ovals are obtained by adjusting the separation of source and sink, the strengths of these, C, and the value of V_∞.

   c. Consider a rectangular surface that freely allows the passage of fluid through itself. The surface is normal to the x axis and is described by $x = 2, 0 \le y \le h$, where h is real. The surface is one unit long in the direction normal to the z plane. Write a MATLAB program, using the stream function, that plots the rate of fluid flow (the flux of fluid) through this surface as h varies from 0 to 4.

4.5 Mapping, Dirichlet, and Neumann Problems and Line Sources

4.5.1 Dirichlet Problems

(See W chapter 8.5 and S chapter 9.)

Suppose a domain D in the x–y plane represents a region in which there is an electric field or the conduction of heat. The region is the cross section of a material extending to infinity. For the moment we assume that the electrical voltage or the temperature is known on the boundaries of the domain. The electrostatic potential or temperature, $\phi(x, y)$, inside D must satisfy Laplace's equation and agree with the given boundary conditions on the boundary points of D. Such problems are often solved through the method of conformal mapping. They are known as Dirichlet-type boundary value problems.

The domain D in the x–y plane is mapped, using a conformal transformation $w = f(z) = u(x, y) + iv(x, y)$, into the u, v plane. The new domain is D_1. The boundaries are mapped into this plane as well. Thus, the point (x_0, y_0) on the boundary of D is mapped into $u(x_0, y_0) + iv(x_0, y_0) = u_0 + iv_0$ on the boundary of D_1. Refer to Figure 4.22. The values that $\phi(x, y)$ assumes at each boundary point of D are assigned to the image of that point in the w plane. Thus, the values of $\phi(u, v)$ are known at the boundary points of D_1.

Suppose the shape of the domain D_1 is simpler or more familiar than D. We find or recognize a solution to the problem, call it $\phi_1(u, v)$, that is harmonic in the w plane and satisfies the values of $\phi_1(u, v)$ that were specified at the boundary points. Let us take $\phi_1(u, v)$ and replace u and v by their explicit expressions in terms of x and y. Thus, $\phi_1(u, v) = \phi_1(u(x, y), v(x, y)) = \phi(x, y)$. The function $\phi(x, y)$ is a harmonic function. A solution of Laplace's equation *remains a solution* of this equation when transferred from one plane to another by an analytic function (i.e., by a conformal transformation).

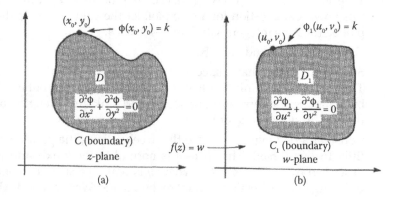

FIGURE 4.22

A domain D (a) and its image D_1 (b).

This follows because an analytic function of an analytic function is analytic. This function provides the solution to our problem—it is harmonic in the domain D and satisfies the given boundary conditions.

In Neumann problems, which often occur in fluid mechanics where there is a rigid boundary, or in heat transfer configurations involving insulation, we again seek a function $\phi(x, y)$ that satisfies Laplace's equation in a domain. However, in this case the component of the gradient of $\phi(x, y)$ that is normal to the boundary of the domain (written as $\frac{\partial \phi}{\partial n}$, where n stands for distance along the normal) must take on certain prescribed values—often zero. The method described above still applies—if we can solve this problem in the w plane with the required boundary conditions on $\frac{\partial \phi}{\partial n}$ satisfied at the image of the original boundary, we can transform this solution back into the z plane (with our analytic transformation).

Before demonstrating the power of the method, let us solve some simple problems in the w plane without resorting to the preceding technique. These results will be of use when we perform that method. Consider the complex potential:

$$\Phi(w) = -\frac{i}{\pi} \operatorname{Log} w \tag{4.29}$$

which is analytic everywhere except on the negative real axis and the origin. Separating into real and imaginary parts, we have

$$\Phi(w) = -\frac{i}{\pi} \operatorname{Log} w = \frac{\arg(w)}{\pi} - \frac{i}{\pi} \operatorname{Log}|w| = \phi(u, v) + i\psi(u, v) \tag{4.30}$$

We will confine our attention to the half space $v \geq 0$ with the origin excluded. Now

$$\phi(u, v) = \frac{1}{\pi} \arg(w) = \frac{1}{\pi} \tan^{-1} \frac{v}{u} \tag{4.31}$$

where we use the principal value of the argument of w. Also,

$$\psi(u, v) = -\frac{1}{\pi} \operatorname{Log}|w| = \frac{1}{2\pi} \operatorname{Log}\left|\frac{1}{u^2 + v^2}\right| \tag{4.32}$$

Notice that the harmonic function $\phi(u, v) = \arg(w) = \frac{1}{\pi} \tan^{-1} \frac{v}{u}$ is continuous in the upper half of the w plane and equal to zero on the positive u axis and equals 1 on the negative real axis. It is undefined at the origin.

The equipotential surfaces are simply rays extending out from the origin, the edges of the surfaces on which arg(w) is constant. A plot of this is presented in Figure 4.23.

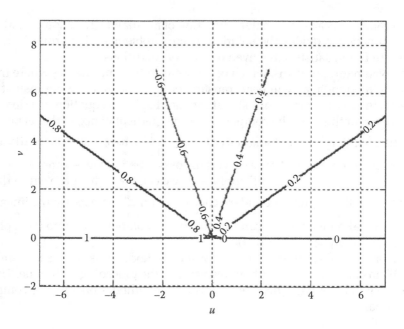

FIGURE 4.23

The contour lines for real $\mathrm{Re}\left(\dfrac{-i}{\pi}\mathrm{Log}\,w\right)$ in the upper half plane.

Figure 4.23 is obtained (almost) from this code:

```
clf
x=linspace (-7,7,100);
y=linspace (0, 7,100);
[x,y]=meshgrid(x,y);
z=x+i*y;
phi=-i*log(z)/(pi);
phir=real(phi);
phim=imag(phi)
values=linspace(0 ,1,6);
figure (1)
[g,h]=contour (x,y,phir,values,'linewidth',2);
clabel(g,h);
grid
xlabel('u');ylabel('v');axis equal
x=linspace (-7,7,100);
y=linspace (0, 7,100);
[x,y]=meshgrid(x,y);
z=x+i*y;
phi=-i*log(z)/(pi);
phir=real(phi);
phim=imag(phi)
```

```
values=linspace(0 ,-1,6);
figure (2)
[g,h]=contour (x,y,phir,values,'linewidth',2);
clabel(g,h);
grid
xlabel('u');ylabel('v');axis equal
```

If you run the above code, you see (with Release 2015a) that MATLAB fails to correctly label the ray $y = 0, x > 0$ as being the contour on which our function is zero. We have added in this zero by editing Figure 4.23. I am told that this weakness in MATLAB will likely be fixed in a future version.

The imaginary part of the complex potential is

$$-\frac{1}{\pi}\text{Log}|w| = \frac{1}{2\pi}\text{Log}\left|\frac{1}{u^2 + v^2}\right| \tag{4.33}$$

The streamlines are the lines or edges of the surfaces on which $u^2 + v^2$ is constant. These are circles. Since we are focusing on the upper half of the w plane, we are considering semicircular arcs. We can generate the streamlines using the above code and get the curves as shown in Figure 4.24. Note that we must use the *imaginary* part of the complex potential and also use different numbers for the values assumed by the stream function on the contour lines.

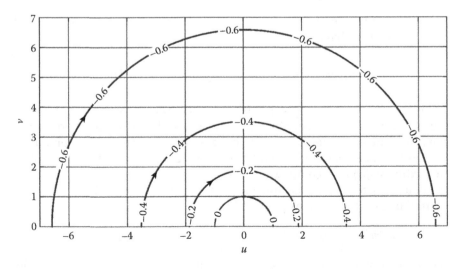

FIGURE 4.24

The contour lines for $\text{Im}\left(\frac{-i}{\pi}\text{Log } w\right)$ in the upper half plane.

Comments:

1. Refer to Equations 4.10 through 4.12. In electrostatics we have that the electric field is given by $E = -\nabla\phi(x, y)$. Because of the minus sign, the arrows on the streamlines, if they are to show the direction of the electric field, should point from higher to lower values of electric potential. Thus comparing the previous two figures, we see that arrows placed on the streamlines, showing the direction of the electric field, should be in the clockwise direction. This is also the case for the heat flux density vector, while arrows showing the direction of fluid flow would be oppositely directed because there is no minus sign. We have placed a few arrows on the figure, using the plot editor, assuming this is a physical configuration in electrostatics or heat flow.

2. In a region of space that has no electric sources (i.e., charge) or sources of heat, the highest and lowest values of $\phi(x, y)$ (the electric potential or the temperature) will occur on the boundaries. (See W section 4.6, problems 13 and 14.) Thus, we might think of Figure 4.23 as a region in the upper half space $v \geq 0$. It is bounded below by a plane whose right half is at $\phi = 0$ and whose left half is at $\phi = 1$. (These are temperatures or voltages.) Note that in the upper half space we can say that $0 \leq \phi \leq 1$, and the results confirm that. This kind of reasoning is useful in choosing the values of $\phi(x, y)$ to be plotted in the **contours** command in MATLAB. We can avoid seeking contours on which $\phi(x, y)$ assumes values outside of the interval described by the potentials or temperature on the boundary. This argument does not apply when there are sources (e.g., charges) present.

If we were to use $w - 1$ in lieu of w in the previous complex potential, Equation 4.29, we have

$$\Phi_1(w) = \frac{-i}{\pi}\mathrm{Log}(w-1) = \frac{\arg(w-1)}{\pi} - \frac{i}{\pi}\mathrm{Log}|w-1| = \phi_1(u, v) + i\psi_1(u, v)$$

We would obtain equipotentials and streamlines identical to those in Figures 4.23 and 4.24, except that the patterns would be displaced one unit to the right. The equipotentials would be rays emanating from the point $u = 1$, $v = 0$, while the streamlines would be semicircles centered at this same point. Regarding $\Phi_1(w)$ as a complex electrostatic potential, the voltage $\phi_1(u, v)$ on the boundary $v = 0$, $u > 1$ is now 0 while $\phi_1(u, v)$ for $v = 0$, $u < 1$ is 1. Similarly, for the complex potential

$$\Phi_2(w) = \frac{-i}{\pi}\mathrm{Log}(w+1) = -\frac{\arg(w+1)}{\pi} - \frac{i}{\pi}\mathrm{Log}|w+1|,$$

the equipotentials for $\phi(u, v)$ are rays starting at $u = -1$, $v = 0$, while the streamlines are semicircles with centers at this point. The potential on the boundary $v = 0$, $u > -1$ is now 0, while the potential for $v = 0$, $u < -1$ is 1.

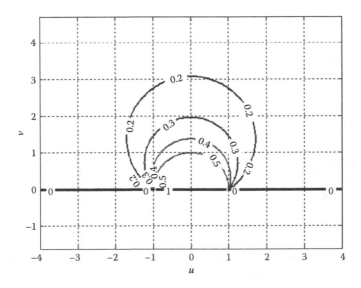

FIGURE 4.25

The contour lines for $\mathrm{Re}\left(\dfrac{-i}{\pi}\mathrm{Log}\,(w-1)+\dfrac{i}{\pi}\mathrm{Log}\,(w+1)\right)$ in the upper half plane.

Consider the subtraction $\Phi_0(w) = \Phi_1(w) - \Phi_2(w) = \phi_0(u, v) + i\psi_0(u, v)$. Thus,

$$\Phi_0(w) = -i\frac{\mathrm{Log}\,(w-1)}{\pi} + i\frac{\mathrm{Log}\,(w+1)}{\pi} \tag{4.34}$$

$$\phi_0 = \frac{1}{\pi}\Big[\arg(w-1) - \arg(w+1)\Big] \tag{4.35}$$

$$\psi_0(w) = -\frac{\mathrm{Log}\,|(w-1)|}{\pi} + \frac{\mathrm{Log}\,|(w+1)|}{\pi} \tag{4.36}$$

To find the conditions satisfied by $\phi_0(u, v)$ on the line $v = 0$, we find the difference between the conditions satisfied by the real parts of the complex potentials $\Phi_1(w)$ and $\Phi_2(w)$. We find that $\phi_0(u, v) = 0$ for $v = 0$, $u < -1$, and $\phi_0(u, v) = 0$ for $v = 0$, $u > 1$. Notice that $\phi_0(u, v) = 1$ for $v = 0$, $-1 < u < 1$. Let us use MATLAB to plot the resulting equipotentials and streamlines in the space above the real axis.

Figure 4.25 shows the equipotential surfaces, in the space $v \geq 0$. If we imagine that the earth is maintained at ground potential (zero volts) except for a conducting strip of width 2, centered at the origin, and if this strip is kept at 1 volt, the above figure indicates the electric potential above the earth. A similar interpretation can be made involving temperature above the earth if this is a problem in heat transfer. The code we used is as follows:

```
clf
clear
u=linspace (-4,4,100);
v=linspace (0, 4,100);
```

```
[u,v]=meshgrid(u,v);
w=u+i*v;
phi=-i*log(w-1)/(pi)+i*log(w+1)/(pi);
phir=real(phi);
phim=imag(phi);
[g,h]=contour (u,v,phir,[ .2 .3 .4 .5],'linewidth',2);
clabel(g,h);
xlabel('u');ylabel('v');axis equal
grid
```

Notice the line of code:

```
[g,h]=contour (u,v,phir,[.2 .3 .4 .5],'linewidth',2);
```

The elements in the vector [.2 .3 .4 .5] are the numerical values chosen that apply on each equipotential surface. They were selected after some experimentation to produce a useful plot. As mentioned above, we seek values for the elements lying on or between 0 and 1.

The contours of potential 0 and 1 were plotted separately and given extra width. The reader should be able to supply the additional code.

The streamlines created by the complex potential $\Phi_0(w)$ are plotted with a similar code and are shown in Figure 4.26.

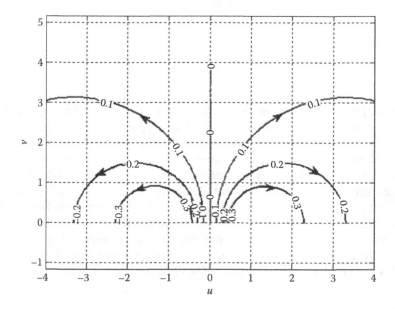

FIGURE 4.26

The contour lines for $\mathrm{Im}\left(\dfrac{-i}{\pi}\mathrm{Log}\,(w-1)+\dfrac{i}{\pi}\mathrm{Log}\,(w+1)\right)$ in the upper half plane.

Assuming this is a problem in electrostatics, we have added arrows to the streamlines in accordance with Figure 4.26. The electric field points in a direction from higher to lower potential.

4.5.2 A Dirichlet Problem

Example 4.7

Using conformal mapping, find an expression for the electrostatic potential in the space $y \geq 0$, $x \geq 0$ subject to the boundary conditions shown in Figure 4.27. Use MATLAB to plot the equipotentials and the streamlines.

We can think of the above as showing the cross section of a piece of sheet metal that has been bent into a 90° angle. The sheet metal is kept at an electrostatic potential of zero volts. However, a section of the metal along both the x and y axes is maintained at 1 volt. The potential is not defined at $x = 1$ on $y = 0$, or at $y = 1$ at $x = 0$.

Solution:

Consider the conformal mapping $w = z^2$ applied to the configuration shown in Figure 4.27. This would bend the boundary into the straight line $v = 0$, $-\infty < u < \infty$. The image of a portion of the boundary, the L-shaped conductor maintained at 1 volt in the z plane, is now the line segment in the w plane, $v = 0$, $-1 < u < 1$. We assign to this the same potential,

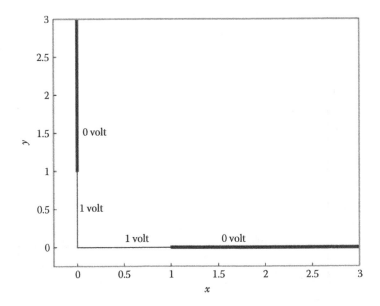

FIGURE 4.27
An electrostatic Dirichlet problem.

the numerical value of 1, that its image had in the z plane. The remainder of the real axis in the w plane is assigned the value its image had in the z plane: zero. The situation is shown in the w plane, as shown in Figure 4.28.

The portion of the z plane in the first quadrant is now the region $v \geq 0$.

The problem of finding the potential above this plane has been solved. It is the one we just did and whose results are shown in Figures 4.25 and 4.26. We can transfer this result into the z plane by substituting z^2 for w in the set of Equations 4.34 through 4.36.

Thus,

$$\Phi_0(w) = -i\frac{\text{Log}(z^2-1)}{\pi} + i\frac{\text{Log}(z^2+1)}{\pi} \tag{4.37}$$

$$\phi_0 = \frac{1}{\pi}\left[\arg(z^2-1) - \arg(z^2+1)\right] \tag{4.38}$$

$$\psi_0(w) = -\frac{\text{Log}\left|(z^2-1)\right|}{\pi} + \frac{\text{Log}\left|(z^2+1)\right|}{\pi} \tag{4.39}$$

We can write MATLAB programs to plot the equipotentials and streamlines. For example, the equipotentials are as shown in Figure 4.29.

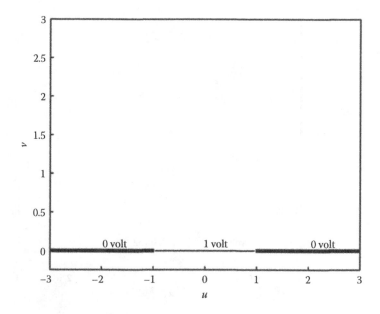

FIGURE 4.28
A transformation of the previous figure.

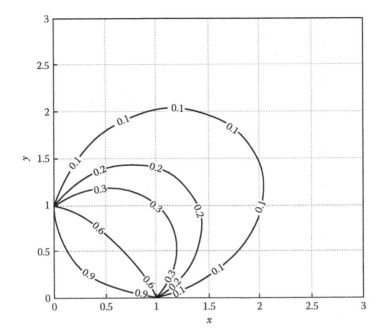

FIGURE 4.29
The equipotentials for Figure 4.27.

Figure 4.29 is obtained from the following code:

```
clf
x=linspace (0,3,100);
y=linspace (0,3,100);
[x,y]=meshgrid(x,y);
z=x+i*y;
phi=-i*log(z.^2-1)/(pi)+i*log(z.^2+1)/(pi);
phir=real(phi);
phim=imag(phi);
values=[0 .1 .2 .3 .6 .9 1];
[g,h]=contour (x,y,phir,values,'linewidth',2);
clabel(g,h);
xlabel('x');ylabel('y');axis equal
grid
```

We notice that the equipotential at .9 volts nearly has the same shape as the nearby L-shaped conductor at 1 volt. Some trial and error was used before assigning the numbers in the line of code for values, in order to get an interesting plot. The same goes for the code used for streamlines.

We can plot the streamlines for this configuration using nearly the same code as above. We use phim in the above code (the imaginary part of the complex potential) to get the contour lines, which are the streamlines. The result is as shown in Figure 4.30.

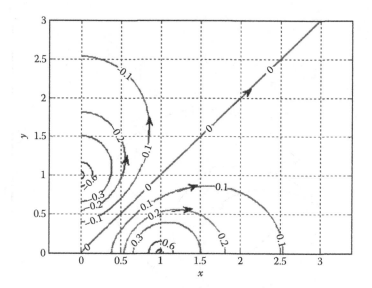

FIGURE 4.30
The streamlines for Figures 4.27 and 4.29.

We have chosen values of ψ ranging between –1 and 1 for our contour values. We used the line of code: values=[-1-.6-.3-.2-.1 0.1.2.3.6 1];.

We have added arrows to the plot, placing them on the streamlines so that they point in the direction of the electric field (i.e., going from higher to lower potential). There is a streamline (along $y = x$) that goes from the corner (at zero potential) to infinity.

A problem like the above can sometimes be done more easily if we use the polar coordinates r, θ, where now $z = re^{i\theta}$. This could be the case, for example, if the 90° angle in Figure 4.27 was changed to some other angle, say 45° or 60°. To work in polar coordinates, we use the following code which is written here for the case of a 90° bend. Note the values used for theta:

```
clf
r=linspace (0,3,100);
theta=linspace (0,pi/2,100);
[r,theta]=meshgrid(r,theta);
z=r.*exp(i*theta);
phi=-1i*log(z.^2-1)/(pi)+1i*log(z.^2+1)/(pi);
phir=real(phi);
phim=imag(phi);
values=[ 0 .1 .2 .3 .6 .9 1];
x=real(z);y=imag(z);
[g,h]=contour (x,y,phir,values,'linewidth',2);
clabel(g,h);
xlabel('x');ylabel('y');axis equal
grid
```

This should produce the same output as we see in Figure 4.29.

4.5.3 A Neumann Problem

Example 4.8

Fluid flows down into a closed channel as shown in Figure 4.31. The fluid flow is V_o in the positive direction along the bottom wall at $x = 0$, $y = 0 +$.

Find the velocity potential for this configuration, and plot the stream-lines for the flow.

Solution:

The fluid cannot penetrate the walls of the channel. Hence, the velocity vector V can have no component along the normal to the walls at any point. The velocity vector is given by $V = \text{grad } \phi$, where ϕ is the velocity potential for the fluid flow. Thus, the gradient of ϕ will have no component normal to the walls of the closed channel. This is ensured if the walls of the channel coincide with streamlines because the velocity vector is tangent to the streamlines. This is part of our boundary conditions, and because it involves the normal component of $V = \text{grad } \phi$, it is a Neumann problem. The remainder consists of our finding a velocity potential $\phi(x, y)$ that will satisfy the prescribed condition on the velocity at the center of the bottom of the channel.

If we can find a conformal mapping that will straighten the walls of the channel into a line in the w plane (e.g., into the real axis), we have simplified the problem. A table of conformal mappings (see, e.g., S section 8.4) is useful in discovering this transformation. We end up employing the following:

$$w = u + iv = \sin\frac{\pi}{2}z = \sin\frac{\pi}{2}x\cosh\frac{\pi}{2}y + i\cos\frac{\pi}{2}x\sinh\frac{\pi}{2}y \qquad (4.40)$$

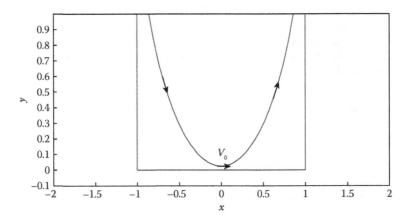

FIGURE 4.31
Flow into and out of a channel.

Notice that for $x = \pm 1$, $y \geq 0$, which holds for points on the left- and right-hand boundaries, $\cos \frac{\pi}{2} x = 0$. Points here are mapped onto the part of the u axis where $|u| \geq 1$. The bottom of the channel, where $y = 0$, $-1 \leq x \leq 1$, is mapped onto the line $v = 0$, $-1 \leq u \leq 1$. Notice that $dw/dz = \frac{\pi}{2}\cos(\pi z/2)$ vanishes at $z = \pm 1$, which means that the transformation is *not* conformal at these two points. This enables the transformation to map the U-shaped boundary into a straight line $v = 0$, $-\infty \leq u \leq \infty$.

Because the boundary is a streamline in the z plane, it follows that the u axis in the w plane must also be a streamline. The simplest nonconstant potential, satisfying such a condition, which is analytic in the upper half of the w plane is

$$\Phi(w) = \phi(u, v) + i\psi(u, v) = Aw = A_p \, (u, iv) \tag{4.41}$$

where A is any real number. The stream function Av vanishes on the u axis. For reasons that will soon be apparent, we choose $A = \frac{2}{\pi} V_0$. Using this value, and combining Equations 4.40 and 4.41, we have for the complex potential in the z plane,

$$\Phi(z) = V_0 \frac{2}{\pi} \sin \frac{\pi}{2} z = V_0 \frac{2}{\pi}\left(\sin \frac{\pi}{2} x \cosh \frac{\pi}{2} y + i \cos \frac{\pi}{2} x \sinh \frac{\pi}{2} y \right) \tag{4.42}$$

From Equation 4.24 and the above, we find that the fluid velocity in the channel is

$$\overline{\Phi'(z)} = V_0\left(\overline{\cos \frac{\pi}{2} z} \right) = V_0\left(\cos \frac{\pi}{2} x \cosh \frac{\pi}{2} y + i \sin \frac{\pi}{2} x \sinh \frac{\pi}{2} y \right)$$

$$= V_x(x,y) + iV_y(x,y)$$

As required, at $x = 0$, $y = 0+$, we have that $V_x = V_0$ and $V_y = 0$. (This explains our choice of A.) A plot of the streamlines can be obtained from MATLAB code. We have chosen to take $V_x = V_0 = 1$ at the bottom of the channel. The program is

```
clf
x=linspace (-1.1,1.1,100);
y=linspace (-.1, 3,100);
[x,y]=meshgrid(x,y);
z=x+i*y;
phi=2/pi*sin(pi/2*z);% this is the complex potential
phim=imag(phi);
values=[0 .5 1 3 5 10 15 20];
[g,h]=contour (x,y,phim,values,'linewidth',2);
clabel(g,h);
xlabel('x');ylabel('y');xlim([-2 2]);ylim=([0 3]);
 axis equal
grid
```

The choice of values of ψ on each streamline, given by the line of code values=[0 .5 1 3 5 10 15 20], were determined by some experimentation to produce the attractive and useful plot given in Figure 4.32. Some arrows have been added to indicate the direction of fluid motion.

Comment:
The reader should note this subtlety. At the point $x = 0$, $y = 0+$, the complex velocity vector is simply V_0. The image of this point in the w plane is $w = 0$. The complex fluid velocity vector in the w plane is $\bar{\Phi}'(w) = \frac{2}{\pi} V_0$. These are numerically different and illustrate a general principle: Although the complex potential assumes identical numerical values at a point in the z plane and at the image of that point in the w plane, the same cannot be said in general for the magnitude (or direction) of the complex velocity vector at points that are images of each other. This is because in the w plane, we have a fluid velocity of $\overline{\frac{d\Phi}{dw}}$, while in the z plane it is

$$\overline{\frac{d\Phi(w(z))}{dz}} = \overline{\frac{d\Phi(w)}{dw}\frac{dw}{dz}}$$

and in general these are not the same.

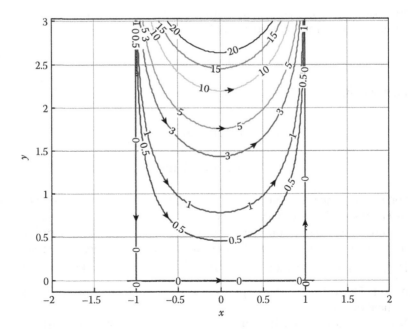

FIGURE 4.32
Streamlines for flow into a closed channel.

Further Comment:
We were fortunate in doing this problem, which began in the z plane, to find an explicit expression for *w*, as a function of *z*, that would map our geometry in the z plane into a more tractable one in the *w* plane. Sometimes we can find a function of *z*, call it *z*(*w*), that will change a simple geometry in the *w* plane into a more complicated but useful geometry in the z plane. However, the inverse of this function *w*(*z*) may not be readily available from known algebraic and transcendental functions. Such an instance occurs in problems 5 and 6 of the exercises. However, we can often still trace the equipotentials and streamlines in the given problem (in the z plane) using a technique shown in problem 5 and, as shown in problem 6, obtain the fluid velocity by using the **solve** command in MATLAB to obtain numerical values of *w* for specified values of z.

4.5.4 Line Sources and Conformal Mapping

In the previous section, we discussed line sources of electric flux, heat energy, or fluid flow.

If such a source passes though the point $z = z_0$ in the complex plane, its most general form for the resulting complex potential is

$$\Phi(z) = A \operatorname{Log} \frac{a}{(z - z_0)} \tag{4.43}$$

Here the constants A and a are nonzero reals, and a is positive. We will call A the strength of the line source. With

$$\Phi(z) = \phi(z) + i\psi(z)$$

We have that

$$\phi(z) = \phi(x, y) = A \log \frac{a}{|z - z_0|} \tag{4.44}$$

$$\psi(z) = \psi(x,y) = -A \arg(z - z_0) \tag{4.45}$$

For convenience, we have used principal values throughout the above. This is not strictly necessary but is usually done. As noted in W (section 8.7), the strength of the line source in electrostatics, heat conduction, or fluid flow can be related to A by means of Equation 4.28. For example, a line charge carrying ρ_L coulombs per meter gives $A = \frac{\rho_L}{2\pi\varepsilon_0}$, where ε_0 is a known numerical constant. There are similar expressions giving A if the line source is shedding heat or fluid.

Note from Equation 4.44 that the constant a establishes the distance from the line charge, $|z - z_0|$, at which the potential ϕ falls to zero. This number is sometimes chosen arbitrarily and is related to the units in use.

Suppose the line source, in the z plane, resides near a boundary on which the value of ϕ is specified (a Dirichlet problem) or the boundary is required to be a streamline (a Neumann problem). Because of the presence of the boundary, we must assume that the complex potential created by the line source is no longer identical to what the source would create were it isolated. We can sometimes solve such a problem through a conformal mapping into the w plane where $w = f(z)$ creates a simpler or more familiar boundary. The line source is transformed into a line source passing through $w_0 = f(z_0)$. The *strength of the line source* measured by A *will remain the same* as in the original problem. This is because (see W section 8.7) the change in ψ measured along a curve in the z plane and measured along the image of that curve in the w plane will be identical. Example 4.9 presents the technique.

Example 4.9

Consider the line charge (or "line source") that produces the complex electric potential $A \operatorname{Log} \dfrac{1}{z - ih}$, $h > 0$, when the line is in an infinite uniform material; we take A as positive real which means that the charge is positive. The source passes through $x = 0$, $y = h$. The source is placed h units above an infinite "grounded" conducting plane as shown in Figure 4.33. The x axis is the cross section of the plane that is at the potential (or voltage) $\phi = 0$. We use $\Phi(x, y)$ to describe the actual complex potential that now exists above the plane. The real potential $\phi(x, y)$ must

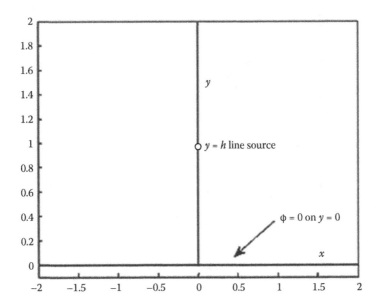

FIGURE 4.33

A line source at $y = h$ above a grounded plane.

vanish on the plane. Find the complex potential above the plane and use MATLAB to sketch the equipotentials and stream lines.

Solution:
Suppose we do a conformal mapping that will have the line source at $z = ih$ pass through the origin in the w plane. We map the infinite line $y = 0$ into the unit circle in the w plane. The symmetry thus obtained is one in which we might find a solution to the problem of finding $\Phi(w)$ in the w plane.

Referring to the previous section on bilinear transformations, we find that one such transformation is

$$w = \frac{z - ih}{z + ih} \tag{4.46}$$

Note that there is no unique transformation that will accomplish this; we can multiply the right side of the above by a complex number of the form $e^{i\omega}$ (where ω is any real) and still have a valid transformation. You should convince yourself that this mapping maps the upper half of the z plane into the interior of the unit circle. Note that as required, the image of $z = ih$ is $w = 0$. With this transformation, we have Figure 4.34.

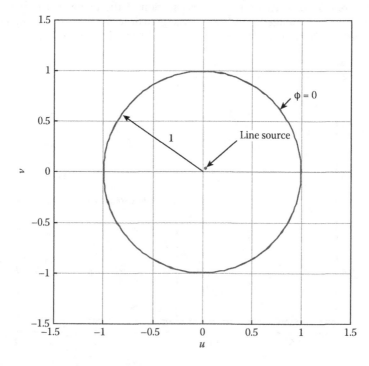

FIGURE 4.34
A conformal transformation of the previous figure.

A line source now appears at the center of a cylinder. The cylinder is maintained at the same potential as its z-plane image: the x axis. Thus, on the cylinder, $\phi(u,v) = \phi(|w| = 1) = 0$. Because the strength of the line source, A, is preserved, we can say that the complex potential created by the line is of the form $\Phi(w) = A \operatorname{Log} \dfrac{a'}{w} = A \operatorname{Log} \dfrac{a'}{|w|} - iA \arg(w) = \phi(u,v) + i\psi(u,v).$

To make the potential $\phi(u,v)$ vanish at $|w| = 1$, we take $a' = 1$. Thus,

$$\Phi(w) = A \operatorname{Log} \frac{1}{w} = A \operatorname{Log} \frac{1}{|w|} - iA \arg(w) = \phi(u,v) + i\psi(u,v) \quad (4.47)$$

Combining the preceding with Equation 4.46, we have for the complex potential in the z plane:

$$\Phi(z) = A \operatorname{Log} \frac{z+ih}{z-ih} = A \operatorname{Log} \left| \frac{z+ih}{z-ih} \right| + iA \arg \frac{z+ih}{z-ih} \quad (4.48)$$

Let us use the above to plot the equipotentials and streamlines in the upper half space of the z plane. Below the line $y = 0$, there is no electric field. The infinite grounded conducting plane prevents any penetration into $y < 0$. We take $A = 1, h = 1$ for our plots.

Thus, the following code generates the lines of constant ϕ. We did some experimentation to discover that values of $\phi = $ 0, .1, .5, 1, 2, 5 produce good plots. We have placed an o at the location of the line source: (1,0). We present the code and results below:

```
clf
x=linspace (-3,3,100);
y=linspace (-.1,5,100);
[x,y]=meshgrid(x,y);
z=x+i*y;
phi=log((z+i)./(z-i));
phir=real(phi);
values=[0 .1 .5 1 2 5];
[g,h]=contour (x,y,phir,values,'linewidth',2);hold on
clabel(g,h);
xlabel('x');ylabel('y');axis equal
plot(0,1,'o')
grid
```

This results in the output shown in Figure 4.35.

Writing the code to plot the streamlines, the contours on which the imaginary part of Φ is constant is slightly more complicated. This is because $\Phi = \log \dfrac{z+i}{z-i}$ has a branch cut along the imaginary axis between $z = i$ and $z = -i$. To see this, notice that if $x = 0, -1 < y < 1$, then $\dfrac{z+i}{z-i}$ is negative real. Recall that MATLAB uses the principal value of the logarithm. We need be concerned only with the portion of

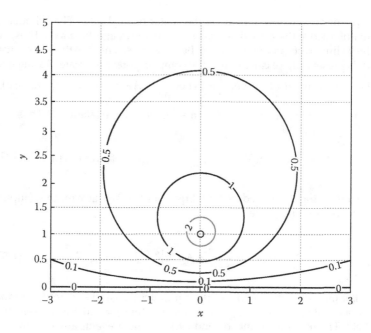

FIGURE 4.35
The equipotentials for line source above a ground plane.

the branch cut in the upper half plane, which is where we are seeking
a solution. Here is the required code:

```
clf;
clear;
x = linspace(-3, 3, 100);
y = linspace( 0, 5, 100);
[x, y] = meshgrid(x, y);
z = x + i*y;
phi = log((z+i)./(z-i));
phim = imag(phi);
del=.05
phim(abs(x) < del & y < 1) = nan; % added this line
values = [-2 -1, -.5,-.3 0,.3, .5, 1 2];
[g,h] = contour(x,y,phim,values,'linewidth',2);
hold on
clabel(g, h);
xlabel('x');
ylabel('y');
axis equal
plot(0, 1, 'o')
grid
```

The line of code phim(abs(x) < del & y < 1) = nan ensures that
MATLAB does not seek to find values of φ, in the **contour** command,

within a strip surrounding the branch cut of width 2∗del. Here 2∗del = 0.1 In the strip, MATLAB is told that φ is simply "not a number," or **nan**. You should try running the code without this statement, and you will see that the resulting streamlines, under the line charge, are erroneous. This is because the **contour** function becomes confused where the function changes discontinuously and seeks to interpolate values between the numerical values on opposite sides of the branch cut.

The line of code values = [-2 -1 -.5 -.3 0 .3 .5 1 2]; was chosen after some experimentation so as to yield an attractive and symmetrical plot. The same is true of the value of del used here, .05. Making this number smaller may result in **contour** seeking to make extrapolations at the discontinuity, while making it too large can cause some of the streamlines not to be plotted. Figure 4.36 is generated by the preceding program.

Because this is a problem in electrostatics and we have assumed that the charge at $z = i$ is positive, we have added arrows to the streamlines as shown. The direction of the streamline is the direction of the electric field and emanates outward from the charge. The infinite conducting plane at $y = 0$ shields the space $y < 0$ from the electric field created by the line charge, and the field in this half space is zero.

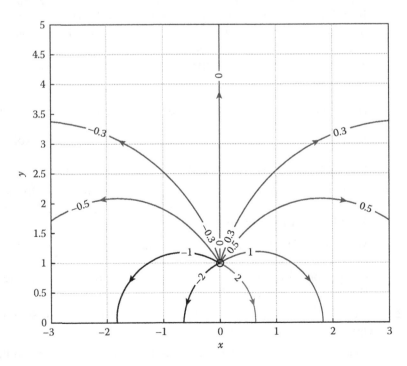

FIGURE 4.36
The streamlines for a line source above a ground plane.

The complex potential can be written as

$$\Phi(z) = A\operatorname{Log}\frac{z+i}{z-i} = A\operatorname{Log}\frac{1}{z-i} - A\operatorname{Log}\frac{1}{z+i}$$

It is customary to interpret this by saying that the potential created above the grounded plane by the line source at $z = i$ is the sum of the potential created by the original line source, of strength A (in the absence of the plane), plus the potential created by an image of that source, located at $z = -i$, and having a strength of the opposite value, namely, $-A$.

Similarly, if the line source were at $z = ih$, we would place its image at $z = -ih$ and use the potential $\Phi(z) = A\operatorname{Log}\frac{z+ih}{z-ih}$.

Exercises

1. a. The space $y > 0$ is filled with a material that conducts heat. Let $\phi(x, y)$ be the temperature in the space $y \geq 0$. The boundary at $y = 0$ is maintained at prescribed temperatures as follows: $-\infty < x < -1$, $\phi(x, 0) = 1; 1 < x < \infty, \phi(x, 0) = -1; -1 < x < 1, \phi(x, 0) = 0$. Following an argument like that used in solving the problem shown in Figures 4.25 and 4.26, find the complex temperature $\Phi(x, y) = \phi(x, y) + i\psi(x, y)$ in the space $y \geq 0$.

   Hint: Try an expression of the form $\Phi(x,y) = A\frac{i}{\pi}\operatorname{Log}(w-1) + B\frac{i}{\pi}\operatorname{Log}(w+1) + C$, where A, B, and C are real constants. Notice that a constant (like C) will satisfy Laplace's equation.

   b. Write a MATLAB program that will use the above result to plot the isotherms on which ϕ assumes the values $\pm$ (0, .1, .3, .5, .7, .9). Try plotting in the space $|x| \leq 3, 0 \leq y \leq 3$.

   c. Write a MATLAB program that will use the same result to plot the streamlines on which ψ assumes the values given above for ϕ, except you will probably want to leave off positive values greater than 0.3. to prevent bunching.

2. A piece of conducting sheet metal is bent into a 60° wedge as shown in Figure 4.37 in the w plane.

   On $y = 0, 0 \leq x < 1$, the electric potential is maintained at $\phi = 1$ volt, on the locus $y = \sqrt{3}x$ the potential on the metal is also at 1 volt for a distance of up to 1 unit from the origin, while on the remainder of

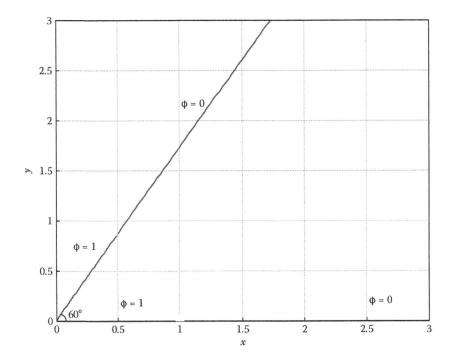

FIGURE 4.37
A piece of conducting sheet metal is bent into a 60° wedge.

the wedge the potential is maintained at zero. Find an expression for the complex potential within the wedge and using MATLAB plot the equipotentials and streamlines on separate plots.

Hint: Study Example 4.7 and use a similar argument, but be sure to use polar coordinates in your code as described at the end of that example. To make your plot more meaningful, include a heavy line showing the line $y = \sqrt{3}x$ for $0 < x < 2$. Some suggested values for the equipotentials are $\phi = 0, .05, .1, .3, .5, .7, .9, 1$ and for the streamlines $\psi = 0, \pm.1, \pm.3, \pm.5$.

3. Fluid flows downward into a corner as shown in Figure 4.38.

   a. At $x = 0+$, $y = 1$, the flow is downward—it must be tangent to the wall—and of magnitude one. As a complex vector this is the velocity $V = -i$. The boundaries of flow on the positive x and y axes must coincide with a streamline. Find the complex velocity potential $\Phi(z)$ in the corner, and write a MATLAB program showing the streamlines of the flow. Show at least five streamlines and place arrows on them to show the direction of flow.

      Which streamlines correspond to the boundary in the given problem?

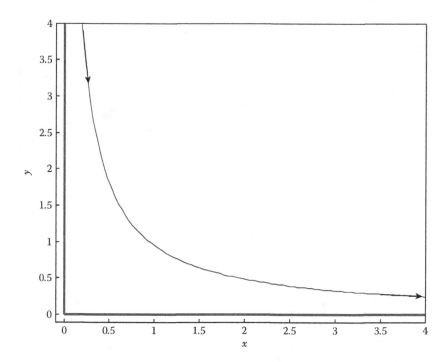

FIGURE 4.38
Fluid flow into a corner.

> *Hint:* Find a simple mapping that will map the corner into the
> real axis in the w plane with the first quadrant of the z plane
> mapped onto the upper half of the w plane. Now solve the prob-
> lem of flow above the streamline $\text{Im}(w) = 0$.

b. Find an expression for the complex velocity vector along the line
   $y = x, x \geq 0$.

4. A Quonset hut in the shape of a semicylindrical conducting metal
   cylinder sits on top of a perfectly conducting earth. The hut is maintained at a potential of 1 volt, and the earth is at ground potential (zero
   volts). The half cylinder has a radius of unity (in some set of units).
   There is a tiny gap between the semicylinder and the ground plane
   so that they can be maintained at different potentials (Figure 4.39).

   a. Find an expression for the complex electric potential, in the
      region $\text{Im}(z) \geq 0$, $|z| \geq 1$, $z \neq \pm 1$ for this problem.

      Hint: Consider the mapping $w = \frac{1}{2}(z + 1/z)$ applied to the given
      configuration. How is the boundary consisting of the semicircular
      roof of the hut and the ground (which is $y = 0$, $|x| > 1$) mapped by
      this function? Note that on the roof we can use $z = e^{i\theta}, 0 < \theta < \pi$. Also,
      notice that $w'(z) = 0$ if $z = \pm 1$. Having applied this transformation,

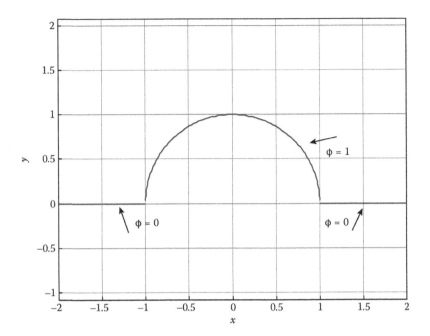

FIGURE 4.39
A metal hut on a grounded plane.

now refer to the problem solved for Figure 4.25. Use the equations derived in that problem to find $\Phi(x, y)$ in the present problem.

b. Write a MATLAB program that will plot the equipotentials for the given problem. They should include $\phi = 1$ and $\phi = 0$.

Hint: Use polar coordinates to set up the grid on which you will evaluate ϕ. Using polar coordinates r and θ, let r range from 1 to around 3. Assume there is no electric field inside the hut, so there is no reason to consider r less than unity. Let θ range from a number a little less than zero to one slightly more than π. If you simply use 0 and π for the ends of the interval, then the command **contour** will not have enough information to plot the contour on which $\phi = 0$.

c. Using a similar program, plot the streamlines, considering both positive and negative values of ψ.

5. A semi-infinite conducting plate is maintained at 1 volt as shown in the z plane in Figure 4.40.

Its cross section occupies the line $y = 1$, $-\infty < x \le 0$. It is one unit above a grounded conducting plate that is kept at zero volt potential. This plane has a cross section occupying the whole x axis. Our goal is to use MATLAB to plot the streamlines and equipotentials created by the conductors.

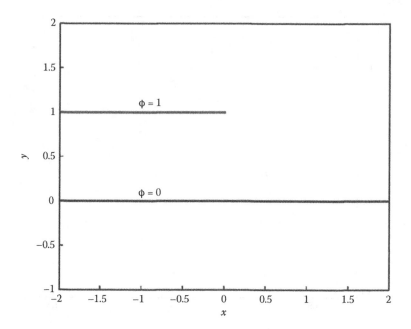

FIGURE 4.40
An infinite plate and a semi-infinite one.

a. Consider the transformation $z = \dfrac{1}{\pi}(w + 1 + \mathrm{Log}(w))$, $w = u + iv$ applied to the line $v = 0$, $0 < u < \infty$ in the w plane. What is the image of this line in the z plane?

b. Under the same transformation, find the image of the line $v = 0$, $-\infty < u < 0$ in the z plane. Be sure to prove that the most positive value achieved by $\mathrm{Re}(z)$ on the image in the z plane is 0.

c. Assigning the values 0 and 1, respectively, to the potential $\phi(u, v)$ on the line segments, in the w plane, in parts (a) and (b), show that the function in the upper half space $\mathrm{Im}(w) \geq 0$ of $\Phi(w) = \dfrac{-i}{\pi}\mathrm{Log}\,w$ will meet the required boundary condition.

d. We cannot solve the equation given in part (a), for w as a function of z, in terms of elementary functions. Thus, we cannot neatly obtain $\Phi(w(z))$ for use in the z plane. However, there is a way to use MATLAB to display both the equipotentials and streamlines without using the **contour** function in MATLAB. Suppose that $\Phi_0 = \phi_0 + i\psi_0$, where ϕ_0 and ψ_0 are specific real values assumed by the potential and stream functions. Show that the value of z where this value of complex potential occurs, is given by

$$z = -\psi_0 + i\phi_0 + \frac{1}{\pi}e^{-\pi\psi_0}\cos(\pi\phi_0) + \frac{i}{\pi}e^{-\pi\psi_0}\sin(\pi\phi_0) + \frac{1}{\pi}.$$

Or,

$$z = i\Phi_0 + \frac{1}{\pi}e^{\pi i \Phi_0} + \frac{1}{\pi}.$$

e. We know, because of the values prescribed on the given boundaries, that all values of the voltage must satisfy $0 \le \phi \le 1$. Suppose we were to choose a value for ϕ_0 satisfying this inequality, say $\phi_0 = 1/2$. From our expression

$$\Phi(w) = \frac{-i}{\pi}\text{Log}\,w = \frac{1}{\pi}\arg(w) - \frac{i}{\pi}\text{Log}|w| = \phi + i\psi$$

we can see that $\frac{-1}{\pi}\text{Log}|w| = \psi$ can assume any real value in the w plane. Thus, fixing the value for ϕ_0 in our expression for z in part (d) and allowing ψ to vary through a set of real values, we can trace out the equipotential on which ϕ_0 has the prescribed value ϕ_0.

Write a MATLAB program that will plot the equipotentials on which $\phi = 0, \frac{1}{4}, \frac{1}{2}, \frac{3}{4}, 1$. Some experimentation might be required to choose the range of values for ψ. For a start, let ψ vary between -1 and 10. Do not make the lower limit on ψ become very negative as the expression for z in part (d) shows that the magnitude of z becomes exponentially unbounded as $\psi \to -\infty$, which creates problems in plotting. Write your program in such a way that each equipotential is labeled with the numerical value of ϕ on that curve.

f. Write a MATLAB program that will generate the streamlines for the given configuration. In electrostatics, these lines are tangent to the electric field. To generate the lines, realize that on each streamline ψ has a fixed real value (positive or negative), while ϕ varies continuously between 0 and 1. Place arrows on the streamlines so that they point in the direction of the electric field (i.e., from higher to lower potential).

Suggestion: It should not be necessary to consider values of ψ more negative than about $-.5$.

6. a. A Neumann problem—a continuation of problem 5 is as follows. The transformation described in problem 5, part (a) can be used to solve a problem arising in fluid mechanics. A fluid is guided between two planes as shown in Figure 4.41.

The upper plane ends at $x = 0$, $y = 1$ allowing the fluid to flow upward as well as from left to right. In the channel, on the far left, we want the flow between the two plates to be uniform and to the right. Thus, we say that the complex velocity between the plates is $V = 1 + i0$ in the limit $x \to -\infty$, $0 < y < 1$. The two planes, the infinite one and the semi-infinite one, must coincide with

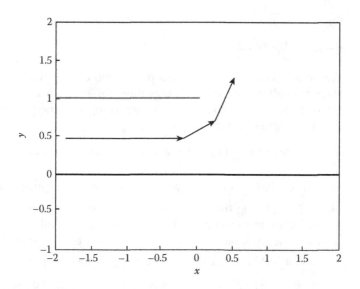

FIGURE 4.41
Fluid flow between two planes.

streamlines of the flow in the z space. If the velocity potential in the image space is given by $\Phi(w)$, then the lines $v = 0$, $u > 0$ and $v = 0$, $u < 0$ (the images of the original z space streamlines) must also be streamlines.

Show that the complex potential $\Phi(w) = A \operatorname{Log} w$ will satisfy the required condition for the streamlines, in the w plane, if A is real. What is the complex velocity vector $V(w)$ for this velocity potential? Answer in terms of A.

b. Show for the transformation given in part (a), in problem 5, that if $|w| \rightarrow 0+$, that $\operatorname{Re}(z) \rightarrow -\infty$.

It is helpful here to switch to the polar representation $w = \rho e^{i\alpha}$. Explain why for the above limit we have $z \rightarrow \dfrac{1}{\pi}\operatorname{Log}(w)$.

c. Show that as $\operatorname{Re}(z) \rightarrow -\infty$, we have for the complex potential in this limit, that $\Phi = A\pi z$. We wish to obtain uniform flow between the plates, as $x \rightarrow -\infty$, which means that $V = 1 + i0$. Prove that this implies that $A = 1/\pi$. Explain why the streamlines in this problem have the same shape as the equipotentials in the previous one, and vice versa.

d. With $A = 1/\pi$, show that $\dfrac{dz}{d\Phi} = 1 + e^{\pi\Phi}$ and that the complex velocity vector is given in general in the z plane by

$$V(z) = \left(\frac{1}{1 + e^{\pi\Phi(w(z))}} \right) = \frac{1}{1 + w(z)}$$

e. Write a MATLAB program that asks you to first supply the coordinates of a point in the z plane and that will then give you the complex velocity vector at that point. Use your program to find the fluid velocity at these coordinates $(0,1/2)$, $(-1,1/2)$, $(0,1/4)$.

Hint: From our transformation $z = \dfrac{1}{\pi}(w+1+\text{Log}(w))$, we can use the **solve** function of MATLAB to obtain w for any given value of z. Now use the result in part (c).

If $z = -5 + i/2$, we are between the guiding planes and far from where the upper plane vanishes. The flow should be toward the right and approximately equal to $V = 1 + i0$.

How well does your result and this asymptotic value agree? Use long format.

7. a. For the following problem, it is convenient to take the results of Example 4.9 and transform them into the w plane. Thus, we have a line charge of strength A placed a distance h above a surface at zero potential whose cross section is the real axis in the w plane. The resulting complex potential in the w plane is

$$\Phi(w) = A\,\text{Log}\,\frac{w+ih}{w-ih} = A\,\text{Log}\left|\frac{w+ih}{w-ih}\right| + iA\,\text{arg}\,\frac{w+ih}{w-ih} \quad \text{for Im}(w) \geq 0$$

Suppose we have two infinite conducting planes separated by a distance π as shown in Figure 4.42. Each plane is at zero electrostatic potential, $\phi = 0$, and the strength of the electric line source, midway between the planes, will be taken as $A = 1$. Show that the transformation $w = e^z$ applied to the configuration in the z plane yields a more familiar (see Example 4.9) geometry in the w plane. Where is the line source now located in the w plane? State the complex potential in the w plane.

b. Now transform your result of part (a) back into the z plane and obtain the complex potential for that problem. Write a MATLAB program, using **contour**, that will plot the equipotentials for that problem with $A = 1$. Choose your equipotentials so as to obtain attractive plots. Note that ϕ cannot be negative because the boundary potentials are zero. For the equipotentials, plot values of ϕ between zero and 1. Verify that the contours for zero coincide with the bounding planes.

c. Repeat the above but plot the streamlines. Place arrows on them to indicate the direction of the electric field, recalling that streamlines leave from a positive charge.

Suggestion: Verify that the function for the complex potential has a branch cut along the line $y = \pi/2$, $x \leq 0$. The imaginary part

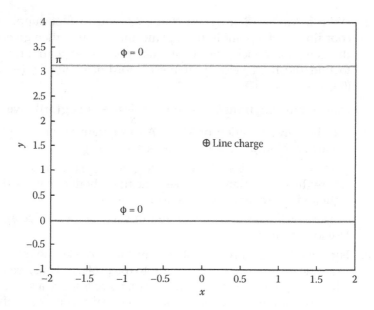

FIGURE 4.42
A line charge between two planes at zero potential.

of this function will be discontinuous as we cross this line. We must avoid values of z in a band containing this line—otherwise, the command **contour** will produce erroneous curves. The line of code

```
phim(abs(y-pi/2) < del & x < 0) = nan
```

can be helpful. Here phim is the imaginary part of the complex potential, namely, ψ, while del is the width of a strip surrounding the branch cut. A good value here is .02. Values of ψ on the streamlines should range between −3 and 3.

8. a. In this problem, we have a line of charge one unit away from the edge of a semi-infinite conducting plane set at zero potential (ϕ = 0) as shown in Figure 4.43.

Note that this half plane can be regarded as the limit as α → 0+ for the configuration shown in Figure 4.44, which employs a wedge of angle 2α.

To begin, let us consider a problem in which there is a grounded plane set at potential ϕ = 0 whose cross section coincides with the entire imaginary axis in the *w* plane.

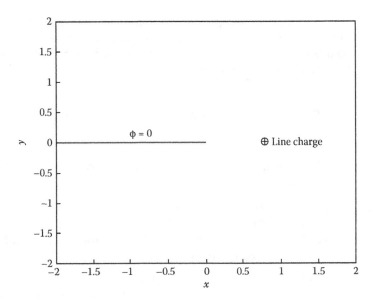

FIGURE 4.43
A line charge and a half plane at zero potential.

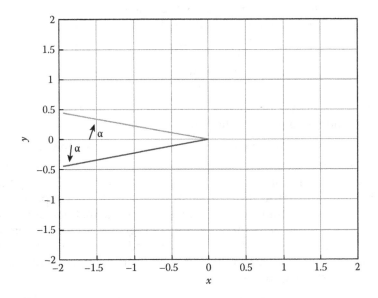

FIGURE 4.44
The half plane is this wedge as $\alpha \to 0+$.

A line charge of strength $A > 0$ is placed a distance h to the right of this plane, and the charge passes through $u = h$, $v = 0$. Use the results of Example 4.9 to argue that the resulting complex potential in the w plane for $\text{Re}(w) \geq 0$ is

$$\Phi(w) = A\text{Log}\frac{w+h}{w-h} = A\text{Log}\left|\frac{w+h}{w-h}\right| + iA\arg\frac{w+h}{w-h}.$$

Hint: Do a simple transformation that will rotate the axes from Example 4.9.

b. Consider the transformation $w = z^{1/2} = e^{\frac{1}{2}\text{Log}\,z}$ applied to the wedge shape (having angle 2α) shown in part (a). How is the wedge mapped by this, and how is the domain outside the wedge mapped by this transformation?

c. Combine the result above and that given in part (a) to show that with $A = 1$ the complex potential for the given problem can be obtained by taking $\alpha \to 0+$ and results in $\Phi(z) = \text{Log}\dfrac{z^{1/2}+1}{z^{1/2}-1}$.

Note for the next two parts that this function has a branch cut along the negative real axis in the z plane because of the function $z^{1/2}$. It will also have a branch cut wherever the argument of the Log, namely, $\dfrac{z^{1/2}+1}{z^{1/2}-1}$, is negative real. This is where $y = 0$, $-1 < x$ < 1, which partly coincides with the first cut. In what follows, use the NaN command for these cuts where needed.

d. Using MATLAB, plot the equipotentials for the line charge/conducting plane configuration.

e. In the same way, plot the streamlines, adding arrows so that they point outward from the line charge.

9. A cylinder of unit radius, whose axis passes through the origin and is perpendicular to the z plane, is maintained at a temperature of $T°$. A similar cylinder of unit radius has an axis that passes through $x = 3$, $y = 0$ and is maintained at $0°$. Our goal is to find the temperature anywhere outside the cylinders and to use MATLAB to plot the isotherms and streamlines in the material. The configuration is presented in Figure 4.45.

a. A bilinear transformation with real coefficients must be found that will map the circle (the cross section of the left-hand first cylinder) into a circle of unit radius, in the w plane, where the points $z = 1$ and $z = -1$ are mapped into $w = 1$ and $w = -1$, respectively. With the same transformation, the second cylinder is mapped into a cylinder in such a way that $z = 4$ is mapped into

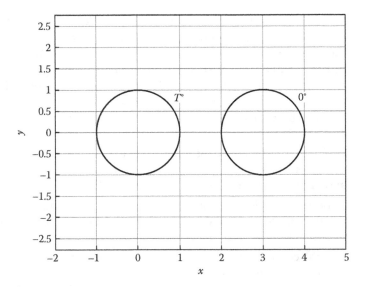

FIGURE 4.45
Two cylinders at different temperatures.

$w = -R$ and $z = 2$ is mapped into $w = R$, where R is real and must
be determined. The two cylinders in the w plane are thus coaxial.
Using the invariance of the cross-ratio (see Section 4.2) and the
solve function of MATLAB, find the numerical value of R and
verify that it is positive. See Example 4.2, Section 4.2.1. Because
R results from the solution of a quadratic equation, MATLAB
should produce two numbers. It is not hard to show that their
product must be 1 and that they are both positive. Verify that
your value of R is positive and real. (You might also want to verify that had you mapped $z = 4$ into $w = R$ and $z = 2$ into $w = -R$
that R would be negative and of no use.)

b. Using the more positive of the two values of R found above, and
the **solve** function of MATLAB, show that the bilinear transformation that maps the two given cylinders (which are in the
z plane) into cylinders of radius 1 and R in the w plane is given
by $-(4*R*z - 11*R + R^2*z + R^2)/(z - 11*R + 4*R*z + 1)$.

Check your result by using this transformation to map each
of the given circles into the w plane and verify that the resulting
curves are the desired circles.

c. To solve our problem in the w plane, where the geometry is
now simpler, observe that the real part of the complex poten-
tial $A\,\mathrm{Log}(R/w)$ is $A\,\mathrm{Log}(R/|w|)$ (here A is real). This function
is harmonic in the w plane and will vanish, as required where

$|w|$ = R. Note this circle is the cross section of the image of the given cylinder at zero potential. Prove that if $A = T/\text{Log}(R)$, the requirements on the potential of the cylinder of unit radius, in the w plane, are satisfied (i.e., it equals T). Now plot the contour lines of the real part of the known complex potential, in the z plane, by using the bilinear expansion that we have derived above. These describe the temperature between the cylinders. Take $T = 1$.

Realize that all the contours must show values satisfying $0 \le T \le 1$, where T is the temperature. This follows from the *maximum* and *minimum* principles in complex variable theory.

Be sure to plot the contours on which $T = 1$ and $T = 0$, and verify that they coincide with the cross sections of the given cylinders in the z plane.

d. Plot the contour lines of the stream function for the temperature assumed above. These lines are tangent to the direction of heat flow. On these contours, ψ can have both positive and negative values, and zero. Try $\psi = \pm(0, .3, .5, .7, 1, 1.5)$. Add arrows to the figure to show heat moving from the hotter to the colder cylinder. Realize that you will be plotting a function of the form $\psi = \text{Im}\left(\text{Log}\dfrac{az+b}{cz+d}\right)$, where the argument is a bilinear transformation. This function will display discontinuities across the branch cut of the function. They can be eliminated from your plot if you include NaN statements for ψ in a narrow strip containing values where $\dfrac{az+b}{cz+d}$ is zero or negative real.

You do not want your plot to show streamlines inside the circles $|z| = 1$ and $|z - 3| = 1$, as they have no physical meaning and will only confuse the picture. These can be eliminated if you include a statement like `psi(abs(z)<1|abs(z-3)<1)=nan;`. Here psi is the imaginary part of the complex potential. To make your plot more meaningful, plot circles indicating the cross section of the cylinders.

5

Polynomials, Roots, the Principle of the Argument, and Nyquist Stability

5.1 Introduction

The question of where an equation like $f(z) = 0$ is satisfied in the complex plane, when $f(z)$ is an analytic function, is a major subject in complex variable theory. This problem often arises in engineering and scientific applications. A value of z for which the equation is satisfied is said to be a "zero" of $f(z)$. Similarly, we must pay attention to the related problem of where $g(z) = 1/f(z)$ possesses pole singularities because of the vanishing of $f(z)$. Some of these questions are directly resolved in MATLAB®, while others require the more sophisticated technique called the *principle of the argument*.

The Nyquist stability criterion, based directly on the principle of the argument, which we discuss, is a favorite technique of engineers; it is accessed directly in a MATLAB toolbox.

5.2 The Fundamental Theorem of Algebra

One of the great pleasures of studying functions of a complex variable is to see how easily one can prove the fundamental theorem of algebra (see W section 4.6 and S problem 5.10). Here is a way to state the theorem: We have a polynomial

$$P(z) = a_n z^n + a_{n-1} z^{n-1} + \ldots a_0 \tag{5.1}$$

in the variable z, where $a_0, a_1, \ldots$ are constants (real and/or complex); n is a positive integer; and $a_n \neq 0$. We say that the polynomial is of *degree n*, although some authors prefer the term *order* instead of *degree*. The theorem states that

$$P(z) = 0 \tag{5.2}$$

has exactly n solutions, or roots, in the complex z plane.* Equivalently, the fundamental theorem can be stated that if $z_1, z_2, \ldots z_n$ are the roots, then we can factor the left side of Equation 5.1 so that

$$P(z) = a_n(z - z_1)(z - z_2) \ldots (z - z_n) \tag{5.3}$$

In applying the fundamental theorem of algebra, we may have to count a root more than once. For example, the equation $z^3 + (2 - i)z^2 + z(1 - 2i) - i = 0$ has roots at $z = i$ and $z = -1$. If you try any other values of z, you will find that the equation is not satisfied. We seem to have two roots, not the expected $n = 3$ from our theorem. But if we factor the left side, we get $(z + 1)^2 (z - i) = z^3 + (2 - i)z^2 + z(1 - 2i) - i = 0$. The root at -1 is of multiplicity two—this follows from the exponent in $(z + 1)^2$.

If we combine terms corresponding to repeated roots in Equation 5.3, any term such as $(z - p)^N$ is said to contribute a *root of order N* or *multiplicity N*. Thus, we assign N roots at p when applying the fundamental theorem.

5.2.1 MATLAB roots

To explore the fundamental theorem in MATLAB, it is easiest to use the function **roots**. The reader may wish to review the documentation for this function using the MATLAB help. Briefly, to find the roots of a polynomial expression like Equation 5.1, you must begin by creating a row vector whose elements are the coefficients of the various powers of z listed in *descending order* of the associated powers. The first element would be the number a_n. If any power of z that is less than n does not appear in the polynomial, then you must enter a zero for that coefficient in the vector. If a polynomial has a root of multiplicity n, the output of **roots** will show that root n times. To find the roots of the polynomial $z^4 + z^3 + 2z + 3$, we proceed as follows in MATLAB:

```
a = [1 1 0 2 3]
>> roots(a)
ans =
   0.7140 + 1.1370i
   0.7140 - 1.1370i
  -1.2140 + 0.4366i
  -1.2140 - 0.4366i
```

Note the presence of the zero, the middle element, in the vector a. It arises because there is no z^2 term in the given polynomial. As the theorem predicted, we obtain four roots. Observe that because the coefficients in the

---

* In some texts, it is stated that the equation has one root. But the existence of the other roots can be proven from this through a procedure of successive division of the polynomial on the left in Equation 5.1 by $(z - z_j)$, where z_j is any of the roots. See W section 4.6, problem 18.

polynomial are real, complex roots must appear in conjugate pairs as is the case here.

We can use **roots** when the coefficients in the polynomial are complex. Let us look at our first case from above: $z^3 + (2 - i)z^2 + z(1 - 2i) - i = 0$. Proceeding as before, we have

```
a = [1 (2 − i) (1 − 2*i) −i]
a = 1.0000   2.0000 − 1.0000i   1.0000 − 2.0000i   0 − 1.0000i
>> roots(a)
ans =
    −0.0000 + 1.0000i
    −1.0000 − 0.0000i
    −1.0000 + 0.0000i
```

Note that −1.0000 − 0.0000i and −1.0000 + 0.0000i are numerically the same, and so we really have only two numerically distinct roots. The root at −1 is of multiplicity 2 and thus is printed twice. However, if you convert our above answer to long format you will get

```
>> format long
>> ans
ans =
    −0.000000000000001 + 1.000000000000000i
    −1.000000016716123 − 0.000000014686955i
    −0.999999983283878 + 0.000000014686955i
```

What we see here is round-off error, leading to the erroneous conclusion that there are three different roots. Ideally the second and third elements in this vector should be identical and equal to $-1 + 0i$. There are sometimes advantages to using the MATLAB short format and, of course, disadvantages.

Comparing the right side of Equation 5.1 with the right side of Equation 5.3, we have that

$$a_n z^n + a_{n-1} z^{n-1} + \ldots a_0 = a_n(z - z_1)(z - z_2) \ldots (z - z_n) \qquad (5.4)$$

On the left side of the preceding, the only term not containing the variable z is a_0, while multiplying out the factors on the right shows that the only term not containing z is $a_n(-z_1)(-z_2) \ldots (-z_n) = a_n(-1)^n z_1 z_2 \ldots z_n$. Thus, $a_0 = (-1)^n a_n z_1 z_2 \ldots z_n$ and so

$$\frac{a_0}{a_n}(-1)^n = z_1 z_2 \ldots z_n \qquad (5.5)$$

This gives us a simple way to check the roots obtained for a polynomial. The product of the roots for the polynomial on the right side of Equation 5.1 is the ratio of the last coefficient to the first, times plus or minus one, depending on the degree of the polynomial. We can say, for example, that the product of the roots of the polynomial $2z^3 + z^2 + i$ is $(-1)^3 i/2 = -i/2$, which the reader should see is the case.

Suppose we were to multiply out all the factors on the right side in Equation 5.3. The root $-z_1$ will multiply z in each of the other terms. There are $n - 1$ of these terms. Thus, $-z_1$ will appear on the right side multiplied by z^{n-1}. Similarly, $-z_2$ will be multiplied by the same factor, as will the other roots. The upshot is that the product of the terms in the parentheses on the right side of Equation 5.3 will contain the expression $(-z_1 -z_2 \ldots -z_n)z^{n-1}$. The variable z^{n-1} does not appear in any other way. The right side of Equation 5.3 when expanded contains $a_n(-z_1 -z_2 \ldots -z_n)z^{n-1}$ as the only term containing z^{n-1}. The left side contains only $a_{n-1}z^{n-1}$ with this same power of z. Equating these expressions gives us another check on the sum of the roots of the polynomial: *the sum of the roots must equal* $\dfrac{-a_{n-1}}{a_n}$. Thus,

$$\left(z_1 + z_2 + \ldots + z_n\right) = -\frac{a_{n-1}}{a_n} \tag{5.6}$$

The reader immediately should see that the sum of the roots of $2z^3 + iz^2 + 17$ is $-i/2$, while the sum of the roots of $z^4 + z + i$ is zero because there is no z^3.

Aside from these above checks, it is interesting to verify that **roots** works correctly by using it to give the roots of a large polynomial for which the roots can be found by some other method. This is illustrated in Example 5.1.

Example 5.1

1. Using **roots**, find all the roots of the polynomial $1 + z + z^2 + \ldots z^9$. Use long format. Notice that there is a negative real root that is available by inspection.
2. Using MATLAB, find the sum of the roots just obtained. Confirm that the result is correct through the use of the coefficients in the polynomial.
3. In a similar way, find the product of the roots obtained in (1), and confirm that the sum is correct by using the coefficients in the polynomial.
4. Check the values of your roots by using a simple formula equal to this finite series and finding where this expression is zero. How well do the results agree with part (1)?

Solution:
Here is the code for parts (1) through (3):

```
%parts 1-3
coeffs=ones(1,10);
('answer to part 1')
```

```
the_roots=roots(coeffs)
('answer to part 2')
sum(the_roots)
('this should equal ratio given in eq(5.6)')
('answer to part 3')
prod(the_roots)
('this should agree with eq(5.5)')
```

Here is our output:

```
ans = answer to part 1
the _ roots =
   0.809016994374947 + 0.587785252292473i
   0.809016994374947 - 0.587785252292473i
   0.309016994374947 + 0.951056516295154i
   0.309016994374947 - 0.951056516295154i
  -0.999999999999999 + 0.000000000000000i
  -0.809016994374948 + 0.587785252292473i
  -0.809016994374948 - 0.587785252292473i
  -0.309016994374948 + 0.951056516295154i
  -0.309016994374948 - 0.951056516295154i
ans = answer to part 2
ans = -1.000000000000001
ans = this should equal ratio given in eq(5.6)
ans = answer to part 3
ans = -1.000000000000000
ans = this should agree with eq(5.5)
```

Comments:
It should be obvious that the given polynomial has a root at –1. Try pairing the first and second terms, the third and fourth, and so on, and you will see that they cancel at –1.

Notice that –0.999999999999999 in the vector giving the roots should ideally have come out as –1, the value of the root.

The sum of the roots should be $-a_8/a_9 = -1$, while our result here is –1.000000000000001, which is close.

The product of the roots should be $(-1)^9 \frac{a_0}{a_1} = -1$, which our program indeed produced.

Recall the finite series and its sum (see W section 5.2 and S problem 2.41):

$$1 + z + z^2 + z^3 + \dots z^n = \frac{1 - z^{n+1}}{1 - z}, \, n \geq 0, z \neq 1 \tag{5.7}$$

The fact that we cannot use the formula on the right for $z = 1$ is of no consequence, because this value is clearly not a root of the polynomial. Equation 5.7 can be verified by our multiplying both sides by $1 - z$.

The roots of the polynomial on the left must coincide with the solutions of $z^{n+1} = 1$ (excluding the solution $z = 1$). We recall (see W section 1.4, and S section 1.13) that roots of the preceding are given by $z = e^{i2\pi k/(n+1)}$ $k = 1, 2, \ldots n$. Notice that we do not have $k = 0$ here, otherwise we would get the unwanted $z = 1$.

To treat the present problem, we take $n = 9$ in Equation 5.7 and in our formula for z.

The numerical values of z are given by the following code and result. We also sum these values.

```
k=[1:9];
format long
new_roots=exp(i*2*k*pi/10).
new_sum_roots=sum(new_roots)
```

The output is

```
for part 4

new _ roots =
    0.809016994374947 + 0.587785252292473i
    0.309016994374947 + 0.951056516295154i
   -0.309016994374947 + 0.951056516295154i
   -0.809016994374947 + 0.587785252292473i
   -1.000000000000000 + 0.000000000000000i
   -0.809016994374947 - 0.587785252292473i
   -0.309016994374948 - 0.951056516295154i
    0.309016994374947 - 0.951056516295154i
    0.809016994374947 - 0.587785252292473i

new _ sum _ roots = -1.000000000000000
```

Notice the excellent—but not perfect—agreement between the values obtained from **roots** and those obtained by our beginning with Equation 5.7. The sum of the roots here is exactly –1, unlike in the previous calculation.

Incidentally, if you have the roots of a polynomial but not the polynomial itself, MATLAB will allow you to create the polynomial by using the command **poly**. This is discussed in problem 6.

5.2.2 Use of Zeros

Suppose you wish to find the roots of a polynomial of a high degree n in which many of the powers of z less than n do not appear (i.e., their coefficients in Equation 5.1 are zero). Rather than key all these zero values into the row vector representing the coefficients of the polynomial, we can begin by assigning all the coefficients the value zero and then correct the few exceptions. The use of **zeros** is given in this example where we solve $(1 - i)z^{10} + (1 + i)z^3 + 2 = 0$. We proceed as follows:

```
a=zeros(1,11);
%above creates a row vector of zeros;
%there are 11 elements in the row vector
```

```
%because of powers 0 thru 10
a(1,1)=(1-i);%coeff of z^10
a(1,11)=2; %coeff of z^0
a(1,8)=(1+i)%coeff of z^3
syms z
check=poly2sym(a,z)
% above checks that we have correct polynomial
rts=roots(a)%gives the roots
```

The above yields the required roots. Notice that if we have a polynomial in which all or many of the coefficients are 1, we can do a similar trick: setting the value of a by using the command **ones** instead of zeros. This was done in Example 5.1.

5.2.3 Roots versus Factor

For some purposes, the MATLAB command **factor** is more useful than **roots** in factoring a polynomial and revealing its roots. For example, to factor $z^4 - 6z^3 + 8z^2 + 6z - 9$ we write

```
>> syms z
>> factor(z^4-6*z^3+8*z^2+6*z-9)
```

```
which yields
ans = (z - 1)*(z + 1)*(z - 3)^2
```

This clearly shows roots at 1, –1, and 3, the latter being a root of multiplicity 2. The terms in the polynomial do not have to be entered into **factor** in descending powers of z.

Had we sought the roots of the polynomial by using **roots**, and if we were working in long format, we would have obtained this less satisfactory result:

```
>> roots([1 -6 8 6 -9])
ans =
   3.000000086034111
   2.999999913965896
  -0.999999999999999
   0.999999999999999
```

Because of rounding error, we do not clearly see that there are roots at 3, 1, and –1. In fact, **roots** seems to show that there are four *different* roots. However, there are a number of reasons *not* to use **factor**. This function, **factor**, will factor polynomials into terms containing $(z - z_k)^{n_k}$, where z_k is a root and n_k its multiplicity only when z_k is rational (which of course means it is real as well). The limitations of **factor** are illustrated in these examples:

```
>> syms z
>> factor(z^2-4)
```

```
ans = (z - 2, z + 2)
% the above is done successfully
>> factor(z^2-5)
ans = z^2 - 5
% the above factorization is not done as it would involve sqrt
(5), an irrational.
>> factor(z^2+4)
ans = z^2 + 4
%the above factorization is not done because it would involve
   i2 which is not real and thus not rational
>> factor(z^3-z^2+4*z-4)
ans = (z - 1, z^2 + 4)
% the above factorization is only partially successful. The
   factors involving +i2 are not present.
```

5.2.4 Derivatives of Polynomials and the Function Roots

It is easy to show that if a polynomial has a zero of multiplicity 1 at z_k, then the derivative of this polynomial cannot have a zero there. A polynomial $P(z)$ with a zero of multiplicity one at z_k can be written as $P(z) = a_n (z - z_k) Q(z)$, where $Q(z)$ is a polynomial of degree $n - 1$ such that $Q(z_k) \neq 0$. This follows from Equation 5.3. The first derivative of this expression is $P'(z) = a_n Q(z) + a_n(z - z_k)Q'(z)$, which does not vanish at z_k.

Suppose the polynomial has a zero of multiplicity 2 at z_k. Then we easily show here that the first derivative of the polynomial will vanish at z_k, while the second will not. The polynomial assumes the form $P(z) = a_n(z - z_k)^2 Q(z)$, where $Q(z)$ is a polynomial of degree $n - 2$ that does not vanish at z_k. We have $P'(z) = 2a_n(z - z_k)Q(z) + a_n(z - z_k)^2 Q'(z)$, which is zero when $z = z_k$. And $P''(z) = 2a_n Q(z) + 4a_n(z - z_k)Q'(z) + a_n(z - z_k)^2 Q''(z)$, which does *not* vanish at z_k. In general, a polynomial with a zero of multiplicity m at some point will vanish at that point as does its derivatives up to $m - 1$. The mth derivative of the polynomial will not vanish at the point. We can investigate this with MATLAB, **roots**, and **diff** as follows.

Before embarking on that, we study here two MATLAB functions **sym2poly** and **poly2sym**, which are useful when we are dealing with polynomials. You may wish to read the documentation on them, although you may recall that we used **poly2sym** in our discussion of row vectors whose elements are the coefficients of a symbolic variable. The numbers appear in descending order and are thus suitable for use in the command **roots**. Here is an example:

```
>> syms w z
>> w=-6*i*z^3+2*z^2+1;
>> sym2poly(w)
ans = 0 - 6.0000i   2.0000   0   1.0000
```

We could easily now apply the command **roots** to the above answer and obtain the roots of the given polynomial. Thus,

```
>> roots(ans)
ans =
   0.0000 + 0.4587i
  -0.4544 - 0.3960i
   0.4544 - 0.3960i
```

The command **poly2sym** takes you from the row vector whose elements are the coefficients of a polynomial (in descending powers) to the polynomial itself. The variable used in the polynomial (e.g., *x* or *z*) is supplied by the user in calling **poly2sym**. Otherwise, *x* is employed by default if you specify no variable. Here, we invoke **poly2sum** to reconstruct the polynomial in *z*, used above, from the numerical values of the coefficients.

```
>> poly2sym([ 0 - 6.0000i 2.0000 0 1.0000], z)
ans = z^3*(-6*i) + 2*z^2 + 1
```

Thus, **sym2poly** and **poly2sym** undo each other.

To take the derivative of a polynomial, we can simply use the function **diff** applied to the polynomial written with its variable (e.g., *z*), or apply the function **polyder** to the row vector whose elements are the coefficients of the polynomial, written in the usual descending order beginning with the coefficient of the highest power.

Example 5.2

Consider the function

$$z^4 - z^3 + (-1 + i)z^2 + (1 - 2i)z + i$$

For the above fourth-degree polynomial, use MATLAB to find not only the roots and their multiplicity but also all the derivatives of this function, up to the fifth; the last should be zero.

Verify that where this function has a root of multiplicity n at a point, the derivatives of the function of all orders up to $n - 1$ are zero there and that the function has a nonzero nth derivative at the same point.

Solution:
```
syms w z
format short
a= [1 (-1) (-1+i) (1-2*i) i]
% note the use of parens in the above to avoid errors
```

```
Roots_of_poly=roots( a)
% above gives roots of given polynomial
k=1;
while k<6
    ('the order of the derivative')
    k
    a=polyder(a)
    rts=roots(a)
    A=poly2sym(a,z)
    % the above is the polynomial in z
    k=k+1;
end
```

This yields

```
a = 1.0000 + 0.0000i  -1.0000 + 0.0000i  -1.0000 +
    1.0000i  1.0000 - 2.0000i  0.0000 + 1.0000i
Roots_of_poly =
 -1.3002 + 0.6248i
  1.0000 - 0.0000i
  1.0000 + 0.0000i
  0.3002 - 0.6248i
ans = the order of the derivative
k = 1


a = 4.0000 + 0.0000i  -3.0000 + 0.0000i  -2.0000 +
    2.0000i  1.0000 - 2.0000i
rts =
 -0.7699 + 0.3877i
  1.0000 + 0.0000i
  0.5199 - 0.3877i
A = z*(- 2 + 2*i) - 3*z^2 + 4*z^3 + 1 - 2*i
ans = the order of the derivative
k = 2


a = 12.0000 + 0.0000i  -6.0000 + 0.0000i  -2.0000 + 2.0000i
rts =
  0.7562 - 0.1646i
 -0.2562 + 0.1646i
A = 12*z^2 - 6*z - 2 + 2*i
ans = the order of the derivative
k = 3
a = 24  -6
rts = 0.2500
A = 24*z - 6
ans = the order of the derivative
k = 4
a = 24
rts = Empty matrix: 0-by-1
A = 24
ans = the order of the derivative
k = 5
a = 0
rts = Empty matrix: 0-by-1
A = 0
```

Comment:
We see that the given polynomial has roots of multiplicity 2 at $z = 1$, a root of multiplicity 1 at $z = -1.3002 + 0.6248i$, and a root of multiplicity 1 at $.3002 - 0.6248i$. After taking one derivative, we obtain a polynomial with one root at 1, but now with roots at $-0.7699 + 0.3877i$ and $0.5199 - 0.3877i$.

Note that the new polynomial obtained by differentiating the given one does have two roots at a location where the first one did not. Now taking the second derivative of the original polynomial, we obtain a polynomial with roots at $0.7562 - 0.1646i$ and $-0.2562 + 0.1646i$.

Taking the third derivative of the original polynomial, we obtain a polynomial that has just one root: at 0.2500. The fourth derivative of our fourth-degree polynomial yields a constant, which has no roots, while the fifth derivative (the first derivative of the constant) is zero.

Exercises

1. Using MATLAB **roots** find all roots of $(1 + i)z^4 + 2z^3 + iz^2 - 3 = 0$, and verify that the number of roots is that predicted by the fundamental theorem of algebra.

2. a. Using MATLAB **roots**, find all the roots of the polynomial $(1 - i)z^{10} + (1 + i)z^9 + z + 2$. Did you find 10 of them, as predicted by the fundamental theorem of algebra?

   b. Find the products of the roots in part (a), and show that they have the value predicted by the coefficients in the given polynomial. Use the MATLAB command **prod**.

   c. Find the sum of the roots found in part (a), and verify that they have the value predicted by the coefficients in the polynomial. Use the MATLAB command **sum**.

3. In using **roots** to find the roots of a polynomial, we can be fooled into accepting a wrong answer if we are not aware of the rounding-off properties of MATLAB.

   a. Do the following in the MATLAB "short" format. Consider this product $w = (z - a)(z - b)$ whose roots are obviously a and b. Let $a = 10^{-6}$ and $b = 10^6$. Using MATLAB, create this expression for w. Employ the command **syms** to make w and z symbolic variables. Now use the command **expand** so that w appears as a polynomial of degree 2 in the variable z. Find the coefficients in the polynomial as a row vector using the command **sym2poly**, and use that result to obtain the roots of the polynomial. Observe that one of the roots of w is obtained incorrectly.

b. Repeat the above using the "long" format, and observe that the correct roots are obtained.

c. Using the long format, repeat part (a) but take $a = 10^{-8}$, $b = 10^8$; observe that one of the roots obtained is incorrect. In general, we must expect serious round-off errors in MATLAB when we are dealing with numbers whose ratios are of the order of 10^{16} or more, which is the case here.

d. MATLAB can do the preceding problem correctly if we ask it to use "variable precision arithmetic." In a MATLAB text or using the Help feature of MATLAB, read up on the commands **vpa** and **digits**. Now repeat part (c) and obtain the correct result.

Hint: 32 digits should be more than sufficient.

4. a. Consider the polynomial $w = 1 + z + z^2 + \ldots z^{23} + z^{25}$. Note that every nonnegative power of z is represented up to the 25th except for the 24th power. Using **roots** and taking advantage of the **ones** command in MATLAB, find all the roots of this expression. You should not have to key in all 26 coefficients into your program. Verify that there are 25 roots. Explain why the sum of the roots is zero. Sum your roots using MATLAB and see what you obtain.

b. Suppose you do have the term z^{24} present in the above. Find the roots now. How different are they from those found in part (a)? Explain why there must be a root at $z = -1$. Do your results confirm this?

c. Consider the polynomial $w = 1 + z + iz^{25}$. There are only three terms. Using **roots** and taking advantage of the **zeros** command in MATLAB, find all the roots of this expression. You should not have to write in all 26 coefficients into your program. Verify that there are 25 roots.

5. a. Consider the polynomial $w(z) = 1 + 2z + 3z^2 + \ldots + nz^{n-1}$, where $n \geq 2$. This is obviously the derivative of the series $W(z) = 1 + z + z^2 + \ldots + z^n$. Earlier, we summed this series into a closed-form expression, valid when $z \neq 1$. Using this sum, show that
$$w(z) = \frac{nz^{n+1} - z^n(n+1) + 1}{(z-1)^2}$$
for $z \neq 1$, $n \geq 2$. If you wish, you can differentiate using the function **diff** in MATLAB.

b. The above result is puzzling as it suggests that our polynomial for $w(z)$, which is of degree $n - 1$, has roots where the numerator in the expression for $w(z)$, above, is zero—that is, where $nz^{n+1} - z^n(n + 1) + 1 = 0$ is satisfied. This would imply— from the Fundamental Theorem of Algebra—that $w(z)$ has $n + 1$ roots. This equation does indeed have $n + 1$ roots, but two of them are at $z = 1$, the very spot where our formula for $w(z)$ is not valid.

To prove the location of the roots, show that the function $A(z) = nz^{n+1} - z^n(n + 1) + 1$ vanishes at $z = 1$, that its first derivative vanishes at the same place but its second does not. Thus, $A(z)$ must be of the form $A(z) = (z - 1)^2 Q(z)$, where $Q(1) \neq 0$. Do these steps in MATLAB using the function **diff**, and to make your evaluation at $z = 1$, the function **subs**. To summarize, we have shown that the roots of $w(z) = 1 + 2z + 3z^2 + \dots nz^{n-1}$ are the solutions of the equation $A(z) = nz^{n+1} - z^n(n + 1) + 1 = 0$ *except* for the two roots in $A(z)$ at $z = 1$.

c. Find the roots of the polynomial $w(z) = 1 + 2z + 3z^2 + \dots 10z^9$ by using the **roots** function in MATLAB. Now compare these roots to those obtained from the equation $nz^{n+1} - z^n(n + 1) + 1 = 0$, solved with MATLAB **roots**, where you must choose the correct value of n and exclude the two roots at $z = 1$.

6. Suppose you have the roots of a polynomial. MATLAB allows you to find the polynomial. The easiest but not the only way is to use the command **poly**, which you might want to read up on. Essentially, you create a row vector, say v, whose elements are the roots of the polynomial, in any order. You must enter a repeated root as many times as it appears in the polynomial. The command **poly**(v) will then yield the polynomial. The output is a row vector whose elements represent the coefficients of the polynomial in descending powers (just as you enter the elements for **roots**). Note that **poly** is designed so that the polynomial produced is such that $a_n = 1$. If **ans** is the row vector obtained for **poly**(v), then the command **poly2sym**(ans, z) will display the actual polynomial having the prescribed roots, with z as the variable in the polynomial, as in Equation 5.1, with $a_n = 1$. However, you will first need the statement **syms** z to establish the symbolic nature of z. Deleting the z in **poly2sym** results in a polynomial in the variable x.

a. Using MATLAB, find the polynomial in the variable z that has the two roots at $1 + i$, one root at $-i$, three roots at the origin, and one root at -4.

b. Apply the command **sym2poly** to the answer to part (a) and use **roots** on that output to check your answer. Is there agreement?

5.3 The Principle of the Argument

The principle of the argument is one of the most useful and elegant theorems in complex variable theory (See W section 6.12 and S problems 5.16 and 5.17). Suppose we have a function $f(z)$ that is analytic on and inside a simple closed contour C. We assume that $f(z) \neq 0$ everywhere *on* C. Let us proceed once around C in the positive direction (counterclockwise if C is just a circle), beginning at any point on C. We call $\Delta_C \arg(f(z))$ the increase in argument, or angle (final value minus the initial value) of $f(z)$ as we negotiate the contour. Let N be the number of zeros of $f(z)$ within C, where the zeros are counted according to their multiplicities.[*] Then $\dfrac{1}{2\pi} \Delta_C \arg f(z) = N$.

If we permit $f(z)$ to have poles inside C, but no other singularities (e.g., branch points, or essential singularities), but keep the other requirements, then we may modify the above and have what is generally referred to as *the principle of the argument*:

$$\frac{1}{2\pi}\Delta_C \arg f(z) = N - P \qquad (5.8)$$

Here P is the total number of poles of $f(z)$ inside C, where the poles are counted according to their order. Thus, for example, if C were a circle $|z| = 4$, and $f(z) = \dfrac{(z+i)(z-5)}{z(z-i)^2(z-12)}$, then $N = 1$ because of the zero of order 1 at $-i$. Also, $P = 3$ because of the pole of order 2 at $z = i$ and the pole of order 1 at $z = 0$. We ignore the zero at $z = 5$ and the pole at $z = 12$ because they are both outside the contour.

In Figure 5.1 we have sketched a hypothetical contour C in the complex z plane.

The arrow shows the direction (positive) in which we negotiate the contour.

In Figure 5.2, we show the locus assumed by $w = f(z)$ as we go once along the closed contour C in the z plane. In this case, the argument of $f(z)$ increased by $2\pi = \Delta_C \arg f(z)$. Notice that we encircled the origin of the w plane exactly once in the positive sense. Thus in Figure 5.2, we show a situation in which $N - P$ on the right in Equation 5.8 is one. This $N - P$ is the difference between the number of zeros and the number of poles of $f(z)$ within C. Observe that if $f(z)$ had a zero *on* C, then the contour in Figure 5.2 would have passed through $w = 0$. And if $f(z)$ had a pole *on* C, then the contour in this same figure would have passed through infinity and could not be drawn on the page.

---

[*] Recall that if an analytic function has a zero of order n at z_0, then it can be expressed as $f(z) = (z - z_0)^n \phi(z)$ in a neighborhood of z_0, where $\phi(z_0) \neq 0$ and $\phi(z)$ is analytic at z_0.

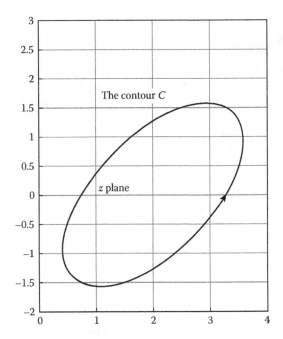

FIGURE 5.1
A hypothetical simple closed contour.

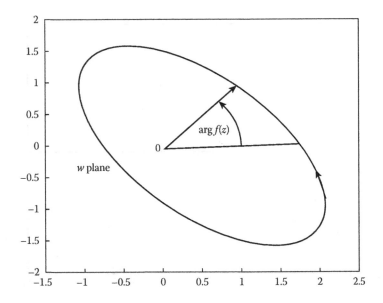

FIGURE 5.2
The image of the preceding contour under the transformation $w = f(z)$.

In Figure 5.3, we show another possibility.

Here, as we negotiate C in the z plane, the locus of $w = f(z)$ takes us twice around the origin in the w plane. Thus, $4\pi = \Delta_C$ arg $f(z)$ and from Equation 5.8 we have $N - P = 2$. In general, as the reader must have seen by now, Δ_C arg $f(z)/(2\pi)$ is the number of encirclements in the positive sense that $w = f(z)$ makes of the origin in the w plane as z negotiates the contour C in the z plane. Sometimes this number of encirclements will be negative as we will see in a specific problem below where we look at $f(z) = \sin(\pi z)/z^3$. In general, we may rewrite Equation 5.8 as

$$E = N - P \tag{5.9}$$

where E is the total number of encirclements of the origin of the w plane (where $w = f(z)$) as z encircles the origin in the positive sense on a contour C. This number, of course, can be negative, positive, or zero.

Example 5.3

Part (a)
We know from the fundamental theorem of algebra that the polynomial $f(z) = z^4 + iz^3 + 4 + i$ has four roots. Using **roots** from MATLAB, we could of course figure out where they are. Use the principle of the argument to see how many roots lie inside the circle $|z| = 3/2$, and check your result by using **roots** and establishing how far the roots are from the origin in the z plane.

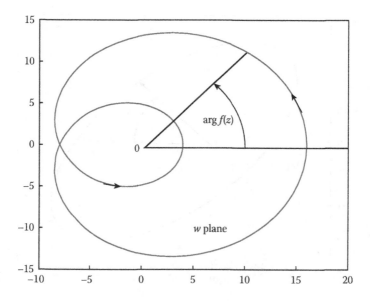

FIGURE 5.3
A contour going around the origin twice.

Solution:

Consider the following code:

```
theta=linspace(0,2*pi,1000);
%above establishes angles on a circle
z=3/2*exp(i*theta);
%above,are the points on a circle, radius 3/2,center origin
w=z.^4+i*z.^3+4+i;
plot(w)
coeffs=[1 i 0 0 (4+i)]
%above are coeffs of the polynomial we are given
the_roots=roots(coeffs)
abs_values=abs(the_roots)
```

Whose output is

coeffs = 1.0000   0 + 1.0000i   0   0   4.0000 + 1.0000i
the_roots =
1.0011 − 1.2337i
−0.8829 − 1.3673i
0.9136 + 0.8635i
−1.0318 + 0.7375i
abs_values =
1.5888
1.6276
1.2571
1.2683

These values are confirmed by the locus as presented in Figure 5.4.

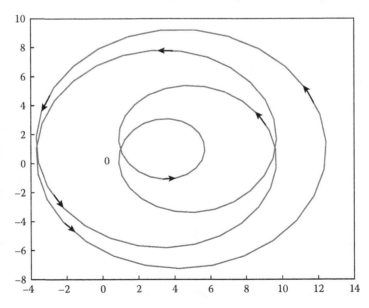

FIGURE 5.4

Locus of $f(z) = z^4 + iz^3 + 4 + i$ as it moves once around the circle $|z| = \dfrac{3}{2}$.

We have placed arrows on the plot in the positive (counterclockwise) direction and have placed a zero at the origin. We know the direction because the given polynomial has no poles in the complex plane. The right side in Equation 5.8 must be positive, and so the encirclement of the origin must be in the positive direction.

The function **roots** produces, as expected, four distinct roots of $z^4 + iz^3 + 4 + i$. Two of them have modulus less than $3/2$ and are 1.2571 and 1.2683. Thus, the locus of $w = f(z)$, as z moves once around $|z| = 3/2$, must twice encircle the origin of the w plane, and it does, as shown in Figure 5.4. Notice that there are two relatively small loops in the locus that do not encircle the origin, and they can be ignored.

Part (b)
Using the principle of the argument, determine how many roots the function in part (a) has in the domain $|z - 1 - i| < 1$.

Solution:
We can use the same code as was employed in part (a) but replace z=3/2*exp(i*theta); with z=1+i+exp(i*theta);. This generates a circle having center at $z = 1 + i$ and radius 1. This is the boundary of the given domain.

The locus generated in the w plane as we negotiate this circle while evaluating $f(z)$ is shown in Figure 5.5.

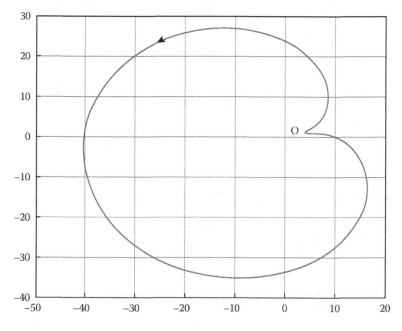

FIGURE 5.5
Locus of $f(z)$ for the contour $|z-1-i| = 1$.

Notice that we have encircled the origin of the w plane exactly once, which indicates that $f(z) = z^4 + iz^3 + 4 + i$ has exactly one root in the domain $|z - 1 - i| < 1$. This is confirmed if we look at the list of roots of the polynomial:

1.0011 − 1.2337i
−0.8829 − 1.3673i
0.9136 + 0.8635i
−1.0318 + 0.7375i

If you were to plot these in the complex plane, you would quickly see that only 0.9136 + 0.8635i lies inside the given domain $|z - 1 - i| < 1$.

Let us now turn to a problem where the function studied has both zeros and poles. We look at $f(z) = \dfrac{\sin \pi z}{z^3}$. Notice that this function has a pole of order 2 at the origin. This is because the denominator has a zero of order 3 at $z = 0$, and the numerator has a zero of order 1 at the same point (see W section 6.2). The order of the pole is the difference of these orders. In general, $\sin \pi z$ has zeros of order 1 when z lies at any integer except zero, which means that $f(z)$ has zeros of order 1 at the same locations. Let us plot the locus of $f(z)$ as z proceeds in the positive sense around the circle $\left| z - \dfrac{1}{2} \right| = 1$. Notice that this circle will enclose the pole of order 2 at the origin and the zero of order 1 at $z = 1$. No other zeros or poles of $f(z)$ are enveloped. Thus, according to Equations 5.8 and 5.9, if we plot $w = f(z)$ as the circle is negotiated, then we should go once around the origin in the w plane in the negative (clockwise) sense. Here is the code and the output that confirms this (Figure 5.6):

```
theta=linspace(0,2*pi,1000);
%above establishes angle on a circle
z=.5+1*exp(i*theta);
%above, the points on a circle, radius 1, center .5
w=sin(pi*z)./z.^3;
plot(w);grid;hold on
w1=w(500)
w2=w(501)
```

The output is

w1 = 7.9987 + 0.1509i
w2 = 7.9987 − 0.1509i

Note that the contour approaches the origin but does not pass through it. The arrow was assigned by our studying the results for w(500) and w(501). The plot moves from having a positive imaginary part to a negative one. We do, as expected, go once around the point $w = 0$ in the clockwise direction.

At this point, the reader may be wondering if the principle of the argument is useful. So far we have used it to confirm that it establishes the zeros

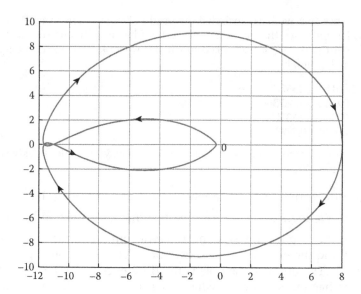

FIGURE 5.6

Image of $\left|z - \dfrac{1}{2}\right| = 1$ under the given transformation.

and poles of functions whose zeros and poles were at known locations. But consider the following situation. Suppose we want to find solutions of the equation $\sin z = e^{-z}$. We might proceed with MATLAB as follows:

```
%solution of exp(-z)=sinz
syms z
solve(exp(-z)==sin(z))
```

The output is

```
ans = 0.58853274398186107743245204570290
```

The above result is entirely plausible. To realize this, we create the plots of $\sin z$ and e^{-z}, where $z = x$ is real (Figure 5.7).

There is no point in making a plot for $x < 0$ as the exponential is greater than one here while $\sin x \le 1$. The intersection of the curves near the value obtained through **solve** (approximately .588) should be evident. Notice too that the curves will have an infinite number of intersections that grow asymptotically closer to $n\pi$, where n is positive integer, increasing without bound. Thus, **solve** has found us just one of an infinite number of real solutions. As for complex solutions, we still do not have a clue.

Our solution using **solve** employed the symbolic math toolbox in MATLAB. This in turn is based on the software called MuPAD, which went into use in September 2008. Before that, MATLAB used the MAPLE software in its toolbox. If you use a version of MATLAB marketed before the above date, you will obtain a different solution to this equation,

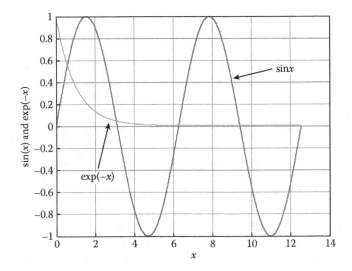

FIGURE 5.7
Plots of $\sin x$ and e^{-x}.

namely, $-2.01+i*2.70$ (approximately).* It turns out that there are other complex solutions; the conjugate of the one just given is also a solution as should
be obvious because $\overline{e^{z}} = e^{\bar{z}}$ and $\overline{\sin z} = \sin \bar{z}$.

Thus, the principle of the argument allows us, in cases such as these, to
explore the complex plane for solutions of a given equation. In the present
instance, we would consider various closed curves and evaluate the change
in argument of $e^{-z} - \sin z$ as we go around the curve. We can refine this process to smaller curves and then seek a solution within a domain bounded by
the curve. In the final attempt to find a root, we again employ **solve** but supply it with a starting value that approximates the root being sought—a value
based on our findings through application of the principle of the argument.
Here is an example. Let us use the principle of the argument to see if there
is a solution to $\sin z - e^{-z} = 0$ near the point $-5+5*i$. Any solution we find will
also have a conjugate that is a solution.

Here is some code that will help with our investigation.

```
z0=input('center of circle=')
r=input('radius of circle=')
theta=linspace(0,2*pi,1000);
z=z0+r*exp(i*theta);
w=sin(z)-exp(-z);
plot(w);grid ;hold on
plot(0,0,'*')
```

---

* See, for example, B. Hunt, R. Lipsman, and J. Rosenberg. *A Guide to MATLAB for Beginners and
Experienced Users*, 2nd ed. (Cambridge, 2007): 21. This book gave me the idea of discussing
this equation.

We supply the program with the center $z0$ and the radius r of a circle. The code generates a plot of the locus of $\sin(z) - \exp(-z)$ as the circle is negotiated. Trying a circle of radius 1 with center at $-5+5*i$, we find that the origin in the w plane is enclosed. Shrinking the radius to one half, we find that the origin is not enclosed. Trying a circle at $-5.5+5.5*i$, with radius 1, in the z plane, we again encircle the origin, and we also encircle the origin if the radius is reduced to .5. You may wish to study this figure using the "zoom in" feature of the MATLAB figure tools to make sure that the origin is indeed encircled. To locate the root, we use the MATLAB command **vpasolve** that allows us to make an initial guess for our solution. In this way, we know that we will get a solution near where we are expecting one. The function **vpasolve** employs the Newton–Raphson method and does occasionally fail—it might produce a root but one that is not the closest one to our initial guess. An advantage that **vpasolve** has over **solve** is that the latter does not permit an initial guess as to the location of the solution, and it might produce a solution that is not near to the one you are hunting for. If you want **vpasolve** to look for complex solutions, begin by providing it with a complex value to begin hunting for solutions. Here is the code for our present problem. The reader may wish to read the MATLAB documentation for **vpasolve**.

```
syms z
vpasolve(exp(-z)==sin(z),z,-5+5*i)
```

Whose output is

```
ans =
-5.1512194593898245955379661256296 +
  5.8443612804879991094759668093818*i
```

This result is not far from the guess made available by the principle of the argument.

For many or perhaps most purposes, it is more convenient to seek roots inside a domain whose boundary is a rectangle with sides parallel to the x–y axes. One reason is that we are seeking solutions in a Cartesian system, so it is natural to use a Cartesian-type contour. The other, equally important, is that the rectangular boundary has 90° corners, and we are interested in seeing how these corners will be mapped. The following code can be used to enter a rectangular contour bounding our domain. We supply the coordinates of the center as well as the width and height of the rectangle. The values of z in the row vector are such that we negotiate the contour in the positive direction. We have chosen to try out our program by plotting the locus of $w = \sin z - \sinh z$ as z negotiates a rectangle, a square in this case, with center at the origin. The rectangle has a width and height of 2. Here are the code and the results:

```
% use of rectangular contour, princ of the argument
cntr=input('real part center of rectangle')
cnti=input('imag part center of rectangle')
ht=input('height of rectangle')
```

```
wdth=input('width of rectangle')
x1=linspace(-wdth/2,wdth/2,1000);
  x2=linspace (wdth/2,-wdth/2,1000);
y1=linspace(-ht/2,ht/2,1000); y2=linspace(ht/2,-ht/2,1000);
%the preceding places 1000 points on the 4 sides of
  rectangle
zbot=x1-i*ht/2;%z on bottom side of rectangle;
ztop=x2+i*ht/2;%z on top of rectangle
zrt= wdth/2+i*y1;%z on right side of rectangle
zlft=-wdth/2+i*y2;%z on left side of rectangle.
z=[zbot zrt ztop zlft]+cntr+i*cnti;%this centers the rectangle
  %where desired
%enter the function w(z) below
w=sin(z)-sinh(z);
plot(w,'linewidth',2);grid;hold on
plot(0,0,'*')
```

The output is Figure 5.8 below.

The arrow has been added to the plot. Figure 5.8 indicates that the function $w(z) = \sin z - \sinh z$ has three zeros inside the given square in the z plane; we encircled the origin three times. If we were to expand $w(z)$ in a Maclaurin

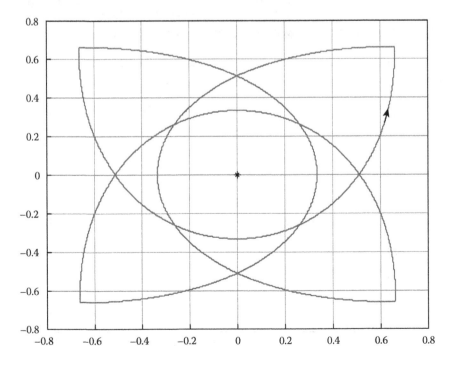

FIGURE 5.8
Locus of $\sin z - \sinh z$ as z negotiates a square of side length z with center at $z = 0$.

series, we would find that the leading term (try this) is $-2z^3/3!$ Thus, there is a zero of order 3 at the origin, and this accounts for the three zeros of $w(z)$ which were found inside the given contour. Of course, this function might have had other zeros in the contour, but our procedure shows that there are no others. Notice here the four cusps in the curve where the locus suddenly changes direction by 90°. These arise from the four corners of the square that we are mapping using this analytic function. The angles of the square are preserved because this is a conformal transformation.

Suppose we apply the above program to the function $w(z) = (\sin z - \sinh z)^2$, using the same square in the z plane. We would expect the locus of $f(z)$ to encircle the origin of the w plane six times. This is because the leading term in the Maclaurin expansion of the new $w(z)$ would be $(-2z^3/3!)^2 = 4z^6/(3!)^2$.

However, we obtain the plot as shown in Figure 5.9.

There is something puzzling about this result: We expected to find six encirclements of the origin, but the figure shows only three. Another puzzling feature, which is a clue to explaining our result, is that the plot exhibits only two corners of 90° and not the expected four.

Notice that $w = f(z) = (\sin z - \sinh z)^2$ is an even function of z (it is the square of an odd function). For every point on the square contour (in the z plane) we employed, there is another point on the contour that is its negative. Both these points are mapped into the same point in the w plane by the given function. Thus, if we proceed just halfway around the given square, we generate the entire curve shown in Figure 5.9. Proceeding the rest of the way

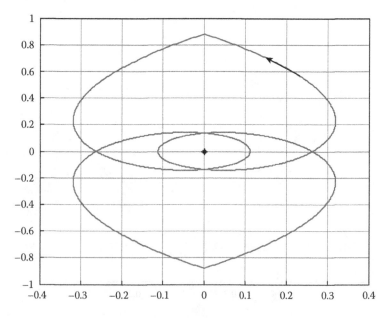

FIGURE 5.9
The locus of $(\sin z - \sinh z)^2$ as z negotiates the square being discussed.

around causes us to revisit these same points in the w plane. This is why we see only two right angles in Figure 5.9. We trace each of them out twice. We go along the above curve twice and encircle the origin six times as we let z move once around our given rectangle in the z plane.

Thus, we must be cautious in interpreting a contour. The matter can be resolved if we generate a table of numerical values of the values of z that we employ and list next to them the corresponding values of w. Of course, in the problem just presented, we would not need to list all 4,000 values used for z as well as the corresponding values of w. We could have used, for example, just 16 values in our list by adding on the following additional lines of code:

```
x1=linspace(-wdth/2,wdth/2,4);x2=linspace(wdth/2,-wdth/2,4);
y1=linspace(-ht/2,ht/2,4); y2=linspace(ht/2,-ht/2,4);
zbot=x1-i*ht/2;%z on bottom side of rectangle;
ztop=x2+i*ht/2;%z on top of rectangle
zrt= wdth/2+i*y1;%z on right side of rectangle
zlft=-wdth/2+i*y2;%z on left side of rectangle.
z=[zbot zrt ztop zlft]+cntr+i*cnti;

w=(sin(z)-sinh(z)).^2;
('z and w')
disp([z.' w.'])
```

Adding this code shows in the program's output that every value of w that we generate is encountered twice. Thus, we negotiate the curve shown in the preceding figure twice.

Suppose we had to investigate the locus of $(\sin z - \sinh z)^2$ as z proceeds around a rectangle whose center is displaced from the origin. Then it would no longer be true that for every point on this rectangle, there is a point on the rectangle that is its negative. Thus, the locus traced in the w plane would not, unlike the preceding case, be covered twice. This matter is investigated in problem 3.

5.3.1 Rouché's Theorem

This theorem is a nice result, easily proven (see W section 6.12 and S problem 5.18), and is useful in determining the location of zeros of analytic functions. The theorem refers to two functions $f(z)$ and $g(z)$, both of which are analytic in and on a simple closed curve. We assume that on C we have everywhere $|f(z)| > |g(z)|$. The theorem states that with these conditions $\Delta_C(\arg(f(z) + g(z))) = \Delta_C \arg(f(z))$ from which it of course follows that the number of zeros (or roots) of $f(z) + g(z)$ inside C is equal to the number of zeros of $f(z)$ inside C. Again we count zeros according to their multiplicities. Note that the theorem does *not* state that the locations of the zeros of $f(z) + g(z)$ and $f(z)$ are identical.

The theorem can provide a concise proof of the Fundamental Theorem of Algebra. Here is a simple example solved with Rouché's theorem.

Example 5.4

Show that the function $w(z) = z^4 + z^3 + i$ has four zeros inside $|z| = 1.5$ and no zeros inside $|z| = .8$.

Solution:
Let us first take $f(z) = z^4$ and $g(z) = z^3 + i$. We take the contour C as the circle $|z| = 1.5$. On this contour, $|f(z)| = (1.5)^4 = 5.0625$. Also, $|g(z)| = |z^3 + i| \leq |z^3| + |i| = |(1.5)^3| + |i| = 3.375 + 1 = 4.375$, from which we see that indeed $|f(z)| > |g(z)|$ on the contour. It follows that the number of zeros of z^4 inside the contour must be the same as the number of zeros of $z^4 + z^3 + i$. It is obvious that z^4 has four zeros, all at the origin. From this, it follows that $z^4 + z^3 + i$ has four zeros inside the contour $|z| = 1.5$. Of course, the locations of the zeros are different in these two cases.

The preceding should give an indication of how to prove the Fundamental Theorem of Algebra. We must argue that the polynomial $a_n z^n + a_{n-1} z^{n-1} + \ldots a_0$ has n roots in the complex plane. We take $f(z) = a_n z^n$ and $g(z) = a_{n-1} z^{n-1} + \ldots a_0$. We then choose a circle, $|z| = r$ centered at the origin, whose radius is sufficiently large that $|a_n z^n| > |a_{n-1} z^{n-1} + \ldots a_0|$. One can always find such a circle—the reader should convince herself of this. The number of roots $f(z)$ inside the circle is obviously n. Invoking Rouché's theorem, we proved the theorem.

To return to the problem, we now take $g(z) = z^4 + z^3$ and $f(z) = i$. On the contour $|z| = .8$ we have $|g(z)| \leq |z|^3 + |z|^4 = .8^3 + .8^4 = 0.9216$. On the same contour, we have $|f(z)| = |i| = 1$. Thus, it is clear that on the contour we have $|g(z)| < |f(z)|$. This tells us that $f(z) + g(z) = z^4 + z^3 + i$ has the same number of zeros inside $|z| = .8$ as does the number i. Thus, there are no zeros. We can confirm all of the above by applying the function roots to the polynomial $z^4 + z^3 + i$. We have the following:

```
>> roots([1 1 0 0 i])
ans =
  -1.2865 + 0.3197i
  -0.6382 - 0.8183i
   0.1834 + 0.8614i
   0.7413 - 0.3628i
>> abs(ans)
ans =
   1.3256
   1.0378
   0.8807
   0.8253
```

The absolute values of all four roots exceed .8 and are less than 1.5, as predicted.

Exercises

1. a. Consider $w(z) = \dfrac{z^3 + 1}{z}$. Identify the zeros and poles of this function, and give the orders of each.

   b. Using MATLAB plot the locus of $w(z)$ as z negotiates the circle $|z| = .9$ once in the counterclockwise direction. Explain how your result confirms the principle of the argument by referring to part (a). Put an arrow on your locus to show the direction in which it is negotiated.

   c. Repeat part (b) but use as your locus $|z| = 1.2$.

2. a. Consider the function $w(z) = z^3$, which obviously has a zero of order 3 at the origin. Using MATLAB plot the locus assumed by $w(z)$ as z goes once around the circle $|z| = 1$. Notice that this locus apparently encircles the origin in the w plane just once, but the principle of the argument would suggest three encirclements. Explain this apparent contradiction.

   b. Repeat part (a) but use the circle $|z - 1| = 3/2$. Observe that your locus does indeed encircle the origin three times as expected. Explain why this differs from the result in part (a).

   c. Repeat part (c) but use as your locus in the z plane the square centered at the origin with sides parallel to the x and y axes. Each side has length one. Did you get three encirclements?

3. Plot the locus of $f(z) = (\sin z - \sinh z)^2$ as z goes, in the positive sense, along the boundary of a domain bounded by a square. The length of each side is 2. The sides are parallel to the x–y axes. The center of the square is at $x = .1, y = .1$. How does your result confirm the conclusion obtained from a Maclaurin expansion that $f(z)$ has a zero of order 6 at $z = 0$, and how does it show that there are no other zeros for this function in this domain? How many right angles are in your locus? Explain why?

4. a. Using the principle of the argument and a MATLAB plot, show that the equation $\sin(e^z) = e^z$ has no solution in the z plane inside the domain $|x| < 1/2, |y| < 1/2$.

   b. Repeat the above problem and show that there is no solution inside the domain $|x| < 1, |y| < 1$. You will have to use the **zoom in** feature of MATLAB to see that your locus does not encircle the origin.

   c. Use **vpasolve** to try to find a solution near the upper-right-hand corner of the square domain (i.e., in the vicinity of $x = 1, y = 1$). Is your solution outside the square, as it should be?

5. a.   Where are the zeros and poles of $w = f(z) = \dfrac{\text{Log}(1+z)}{z^2}$? Give the
location and the order. Note that this is the principal branch of
the logarithm.

   b.   Using MATLAB, plot the locus of this function as z goes around
the circle $|z| = .9$ in the positive sense. Be sure to place an arrow
on the plot showing the direction in which it is negotiated.
Explain how the number of encirclements of the origin confirms
the principle of the argument.

   c.   The function $w = f(z) = \dfrac{\text{Log}(1+z)}{z}$ has a removable singularity at
$z = 0$. How should this function be defined at this point if the
function is to be analytic in a neighborhood of the origin?

   d.   Repeat part (b) for the function given in (c). Discuss the number
of encirclements of the origin.

   e.   The function $w = f(z) = \text{Log } z$ has a zero of order 1 at $z = 1$. Yet a
plot generated by evaluating $f(z)$ as z proceeds once around the
circle $|z| = 2$ will show no encirclements of the origin in the w
plane. Try this. Why is there no encirclement?

6. a.   Using the principle of the argument, show that the equation
$\sin(i \cos(z)) = e^z$ has just one root inside the square domain
$-1 < x < 1, -1 < y < 1$.

   b.   Does this root lie in the left half of this domain, the right half,
or on the imaginary axis? Answer using the principle of the
argument.

   c.   Using **solve** or **vpasolve**, find the location of *this* root. You may
have to experiment to find the root we are seeking. Did you find
other roots?

7. a.   Show that on the circle $|z| = 1.5 \ |z^4| > |e^z|$.

   b.   From the above, state how many solutions the equation $z^4 + e^z$
has inside this circle. How many roots does $z^4 - e^z$ have inside the
same circle?

   c.   Plot the locus of $z^4 + e^z$ as z goes around the given circle.
Also plot $z^4 - e^z$. How do your results confirm your findings of
part b?

   d.   A plot of z^4 as z goes around the circle once seems to show a
single encirclement of the origin. Does this contradict what you
have found in part (b)?

   e.   Using **solve**, find four solutions of $z^4 + e^z = 0$ inside the given circle. Note that **solve** yields solutions in terms of the lambert function. To get numerical (decimal) expressions from these results,
apply the MATLAB command **eval** to the results containing the

lambert function. Work in long format to make comparisons with part (f) easier.

f. Using **vpasolve**, find a solution of the above equation. Begin with the starting (trial) value of $z = 1 + i$ and see whether you obtain one of the values obtained in part (e). Note that you have not proved in this problem that there are only four solutions to the given equation. You have shown only that there are exactly four solutions inside the given circle.

8. The purpose of this problem is to show that $5 \sin z + e^z$ and $5 \sin z$ have an identical number of zeros inside a certain square domain centered at the origin.

   a. Recall that (see W section 3.2) $|\sin z| = \sqrt{\sinh^2 y + \sin^2 x}$. Consider the square-shaped domain, centered at the origin described by $|x| < \pi/2$, $|y| < \pi/2$.

      Show that on the square boundary of this domain, which we call C, that $|e^z| < 5|\sin z|$. Thus, from Rouché's theorem, we can conclude that on C we have $\Delta_C \arg(5 \sin z + e^z) = \Delta_C \arg(5 \sin z)$. Verify this by plotting the locus of $5 \sin z + e^z$ and $5 \sin z$ as z goes in the positive direction around the given C. Verify that the increase in argument is identical for both plots. How many roots does $5 \sin z + e^z = 0$ have inside the given square?

   b. Study this equation and conclude with the aid of sketches of $5 \sin z$ and $-e^z$ that it has an infinite number of real roots in the complex plane.

   c. Solve the equation using the **solve** function. Is the solution inside the square of part (a)?

9. a. Show mathematically that on the circle $|z - 2| = 1$, we have $|e^{1/z}| < 3|z|$.

   b. Explain why this establishes that in the domain bounded by this circle we have that the equation $e^{1/z} = 3z$ has no solution.

   c. Plot the locus of $e^{1/z} - 3z$ as z progresses around this circle in the positive sense, and verify that the origin is never encircled, as confirmed by Rouché's theorem.

5.4 Nyquist Plots, the Location of Roots, and the Pole-Zero Map

A useful special case of the principle of the argument deals with the counting of the number of zeros and poles of an analytic function in a half plane (e.g., the right half of the z plane, the left half of the same plane, or the top or bottom halves). Let us focus on the right half of the z plane, as this is of most interest in engineering applications.

Suppose we have a polynomial of the form

$$w(z) = a_n z^n + a_{n-1} z^{n-1} + \ldots a_0 \tag{5.10}$$

Refer now to Figure 5.10. Imagine that we have a semicircle whose diameter is the imaginary axis satisfying $-R \leq y \leq R$. We move downward along the imaginary axis from $z = iR$ to $z = -iR$ and then close the contour by moving upward (counterclockwise) along the arc $|z|$ and return to the starting point. While on this trip, we compute the values assumed by $w(z)$ and plot them in the w plane so as to obtain the locus of w. We assume that the value of R is sufficiently large that no roots of $w(z)$ in the right half plane lie outside this circle. Now assuming that $w(z)$ has no roots on the imaginary axis, we can say, from the principle of the argument, that the number of encirclements of the origin in the w plane is equal to the number of roots of $w(z)$, where roots are counted according to their multiplicities. If we were to allow $w(z)$ to have roots on the imaginary axis,

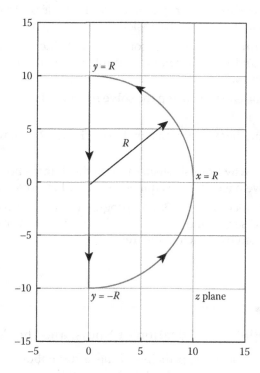

FIGURE 5.10

Semicircle with diameter on the imaginary axis and center at the origin.

then when we pass through these roots, the locus would go through the origin in the w plane. The above, with some modification shown later, is the essence of the Nyquist method, a technique for finding the number of roots of a function in the right half plane (RHP). We could, of course, use a similar technique to find the number of roots in the left half plane or the upper or lower half planes.

It is an interesting little exercise to see how large we must take R in order for there to be no roots of $w(z)$ in the RHP outside the semicircle. Suppose we write

$$w(z) = P(z) + Q(z) \tag{5.11}$$

where $P(z) = a_n z^n$ and $Q(z) = a_{n-1} z^{n-1} + a_{n-2} z^{n-2} + \ldots a_0$. Note that where $w(z) = 0$, then

$$P(z) = -Q(z) \tag{5.12}$$

Notice that $|P(z)| = |a_n||z^n|$ and $|Q(z)| = |a_{n-1} z^{n-1} + a_{n-2} z^{n-2} + \ldots a_0| \leq |a_{n-1}||z^{n-1}| + |a_{n-2}||z^{n-2}| + \ldots |a_0|$, where we use a generalization of the triangle inequality.

Now we assume that $|z| \geq 1$. Thus, $|Q(z)| \leq |a_{n-1}||z^{n-1}| + |a_{n-2}||z^{n-1}| + \ldots |a_0||z^{n-1}| = |z^{n-1}|(|a_{n-1}| + |a_{n-2}| + \ldots |a_0|)$. Let A be the largest of the n numbers $|a_{n-1}|, |a_{n-2}|, \ldots |a_0|$. Using this in the preceding inequality, we have $|Q(z)| \leq |z^{n-1}|nA$. Suppose we require that $|P(z)| > |Q(z)|$. This is ensured if $|a_n||z^n| > |z^{n-1}|nA$. Or $|z| > \dfrac{nA}{|a_n|}$. Thus, if we take

$$R > \frac{nA}{|a_n|} \tag{5.13}$$

$|z| \geq R$, and $|z| \geq 1$, then $|P(z)| > |Q(z)|$, and Equation (5.12) is not satisfied, which means that $w(z) = a_n z^n + a_{n-1} z^{n-1} + \ldots a_0 \neq 0$ for $|z| \geq R$.

If the right side of Equation 5.13 is less than or equal to one, it is not hard to show that $w(z)$ has no roots outside $|z| = 1$.

Let us verify the above by establishing the number of roots of $w(z) = 9z^5 + z^4 + 2$ in the right half of the z plane. Of course, the MATLAB function **roots** will do this immediately, but this little exercise will lead us into the kind of thinking involved in the Nyquist plot. From Equation 5.13, we see that the radius of the circle used must exceed $(5 \times 2)/9 = 1.111 \ldots$. We choose $R = 1.2$. There are advantages to making R not much bigger than we have to. If R greatly exceeds its minimum value, then from Equation 5.10, we have (approximately) on the semicircular portion of the contour, which is of radius R, that $w(z) \approx a_n R^n e^{in\theta}$, where $\theta = \arg(z)$. Thus, the mapping of this semicircle into the *w* plane will result in an arc of large radius $a_n R^n$, which might cause portions of the rest of the locus in the *w* plane to appear very close to *w* = 0. The shape of this locus might not be discernible.

The following code generates the locus describing our polynomial *w(z)* as *z* negotiates the closed semicircle of Figure 5.10 in the positive (counterclockwise) sense.

```
R=1.2;
y=linspace(R,-R,1000);
theta=linspace(-pi/2,pi/2,1000);
z1=i*y;%gives points on the diameter
z2=R*exp(i*theta);%gives points on the arc
w1=9*z1.^5+z1.^4+2;% the function at points on the diameter
w2=9*z2.^5+z2.^4+2;%the function at points on the arc
u1=real(w1);v1=imag(w1);u2=real(w2);v2=imag(w2);
plot(u1,v1);hold;
plot(u2,v2); grid;axis equal
plot(0,0,'o');%puts a o at the origin
roots([9 1 0 0 0 2])
%the above tells us where the roots are
```

The output is shown in Figure 5.11 and the roots are:

```
ans =
  -0.7639 + 0.0000i
  -0.2513 + 0.7027i
  -0.2513 - 0.7027i
   0.5777 + 0.4344i
   0.5777 - 0.4344i
```

We have added arrows to the plot to show the direction in which it is negotiated as we move along the contour of Figure 5.10. It is easy to establish their direction. Because the function *w(z)* has only zeros, and no poles, the locus can encircle the origin in the *w* plane, only in the positive sense.

It is clear that in this case the origin is enclosed twice, indicating two roots in the RHP. This result is confirmed if we use **roots** to find the zeros of the polynomial. The results are above.

We see that the last two roots are in the RHP, and there are no others there. There are no roots on the imaginary axis. Had there been, the locus in Figure 5.11 would have passed through the origin.

The Nyquist technique, which has been much used by control system engineers, is a graphical method designed to answer whether a certain class of functions has zeros in the right half of the complex plane. The method was invented in 1932 by Harry Nyquist of Bell Telephone Laboratories. A system described by a function having one or more zeros in the RHP is said to

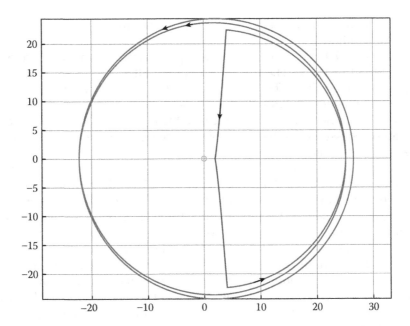

FIGURE 5.11
Locus of $w(z) = 9z^5 + z^4 + 2$ as the semicircle in the z plane is negotiated in the z plane.

be unstable. The technique has to a considerable extent been superseded by other functions in the MATLAB toolboxes. Employing the function **nyquist**, which is in the Control Systems Toolbox of MATLAB, gives us a figure called the *Nyquist diagram*.

In the case of **nyquist**, the Toolbox does not employ the variable z but instead the complex variable $s = \sigma + i\omega$, where σ and ω are real. The technique is applied to an expression like

$$F(s) = 1 + T(s) \tag{5.14}$$

where $T(s)$ is a rational function of s, with real coefficients, and is often multiplied by an exponential function of s. In an elementary form, it is expressible as

$$T(s) = \frac{a_n s^n + a_{n-1} s^{n-1} + \dots + a_0}{b_m s^m + b_{m-1} s^{m-1} + \dots + b_0} e^{-\tau s} \tag{5.15}$$

In MATLAB's **nyquist**, $T(s)$ is called the "transfer function" of an electrical or mechanical system. In some texts (e.g., W pages 494–498), the term *transfer function* is applied to an expression containing Equation 5.14 in its denominator and a polynomial in s in its numerator. We use the language of MATLAB here.

To use **nyquist**, the quantity τ in the exponential in Equation 5.15 must be real and not negative. The function **nyquist** will accept more complicated expressions than those shown in Equation (5.15), provided they are combinations of polynomials in s (with real coefficients) and exponentials of the form $e^{-s\tau}$, $\tau \geq 0$. Some examples appear in the problems below. The function $T(s)$ above is assumed to have no poles in the right half of the s plane or on the imaginary axis.

To determine the number of zeros of $F(s) = 1 + T(s)$ in the right half of the s plane, we move around a semicircular boundary like that in Figure 5.10, but one created in the complex s plane. Then, computing the values assumed by $F(s)$, we plot them in a complex plane. If the locus of $1 + T(s)$ encircles the origin in the complex F plane, then we know that there are zeros in the RHP. The number of encirclements is equal to the number of zeros that $1 + T(s)$ has in the right half of the s plane. Rather than plot $1 + T(s)$, we can simply plot $T(s)$, which is the previous plot but moved one unit to the left. Rather than see if the origin is enclosed, we see if the point $-1+i0$ is enclosed. This is the procedure followed by the function **nyquist** in MATLAB when it generates the Nyquist plot. The point $-1+i0$ is placed directly on the plot so that we can see if it is enclosed.

Another feature of the function **nyquist** is that the plot mentioned above is generated by our moving in the clockwise direction, rather than in the counterclockwise direction we employed in Figure 5.10. We must reverse the arrow on that plot. Thus, in the s plane, we begin at $\sigma = 0$, $\omega = -R$, proceed upward to $\sigma = 0$, $\omega = R$, and then move downward (clockwise) along the arc of radius R. The function **nyquist** places an arrow on the locus of $T(s)e^{-s\tau}$, and we now look for the number of times that this locus encloses $-1+i0$ in the clockwise sense.

In physically realizable situations, as we let $R \rightarrow \infty$ in Figure 5.10, so that our contour encloses the entire right-hand plane, all the values assumed by $T(s)$ on the curved portion of the contour become in this limit, a constant. The function **nyquist** automatically chooses R sufficiently large so that we can assume that $T(s)$ is essentially constant, on the semicircular portion of the contour in the s plane, for purposes of drawing the plot. Thus, in studying the plot generated by **nyquist**, you will not see a curve corresponding to the image of the curved portion of the contour in Figure 5.10.

At this point you should read the documentation in MATLAB help for **nyquist**. We use only a small part of the information provided there. Here is an example of how to use Nyquist. In every case, you must enter the transfer function that is created from the MATLAB function **tf**; you might wish to look this up in the help feature.

Example 5.5

Obtain a Nyquist plot to determine the number of roots in the RHP of

$$F(s) = 1 + \frac{-6(2s^2 + s + 1)}{s^3 + 5s^2 + 9s + 5} e^{-s\tau}$$

where $\tau = .01$.

Solution:
We use the following code:

```
s = tf('s');%establishes that s is to be used to create a
%transfer function.
tau=.01;
transfer_func=-6*(2*s^2+s +1)/(s^3+5*s^2+9*s+5)*exp(-tau*s)
nyquist(transfer_func)
%the above will create the Nyquist plot
```

Notice the especially important line of code s = tf('s'). This establishes that a transfer function is to be established whose variable is s. Having stated this, you do not have a choice of variables here; you must use s.

The output of this program is Figure 5.12 and

$$\text{transfer\_func} = \exp(-0.01 * s) * \frac{-12s^2 - 6s - 6)}{s^3 + 5s^2 + 9s + 5}$$

We have added the notation "minus one" to Figure 5.12. The Nyquist plot always shows you the location of –1 via a plus sign. Since minus one is encircled, clockwise, three times by the locus of

$$F(s) = \frac{-6(2s^2 + s + 1)}{s^3 + 5s^2 + 9s + 5} e^{-s\tau}, \tau = .01$$

we can conclude that

$$F(s) = 1 - 6\frac{2s^2 + s + 1}{s^3 + 5s^2 + 9s + 5} e^{-s\tau}, \tau = .01$$

has three zeros in the right half of the s plane. Engineers would say that this transfer function describes an unstable system. This expression for F(s) as may be verified has no poles in the RHP. We apply the MATLAB function **roots** to the denominator of the preceding fraction to establish this. In this example, the resulting Nyquist plot does not pass through –1, confirming that there are no zeros of F(s) on the imaginary axis in the s plane.

If the preceding problem is changed to one where

$$F(s) = 1 + 6\frac{2s^2 + s + 1}{s^3 + 5s^2 + 9s + 5} e^{-s\tau}$$

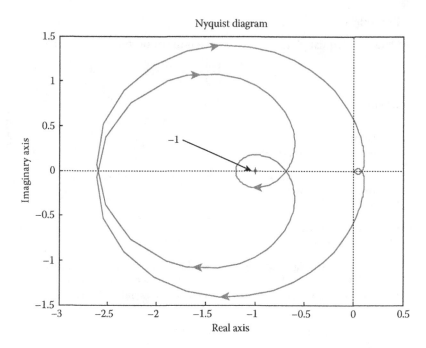

FIGURE 5.12
Nyquist plot for Example 5.5.

with the same value of τ as before, it is found that the resulting Nyquist plot does not encircle the origin. There are no zeros of $F(s)$ in the right half of the s plane. The reader should verify this.

5.4.1 The Use of pzmap

Another approach to finding the roots of an expression like the one given in Equations 5.14 and 5.15 is to use the MATLAB function **pzmap**, which is in the Signal Processing Toolbox. We use this in much the same way as we used **nyquist**, except we regard the transfer function as the function $F(s)$ in Equation 5.14. We can ask **pzmap** to plot the zeros and poles of this function. Again s is used as the variable. The plot obtained uses an x to show where the poles of a function are and a 0 to show the locations of zeros. Here is Example 5.5 solved with **pzmap**.

The following is the code and the resulting figure:

```
clear; clf
s = tf('s');
tau=.01;
transfer_func=1-6*(2*s^2+s +1)/(s^3+5*s^2+9*s+5)*exp(-tau*s)
pzmap(transfer_func)
```

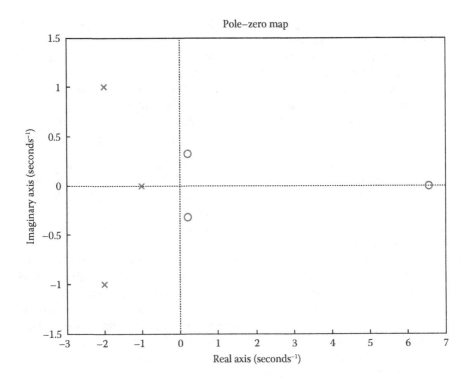

Note that, as predicted from the example, the function

$$F(s) = 1 - 6\frac{2s^2 + s + 1}{s^3 + 5s^2 + 9s + 5}e^{-s\tau}$$

with the given τ has three zeros in the RHP. All three poles are in the left half plane.

5.4.2 A Caveat on the Use of pzmap

The function whose poles and zeros are being investigated in **pzmap** has certain restrictions that arise from communications theory.

Rational functions and exponentials can be used; these are functions that are analytic except for poles. You cannot use functions that contain a branch cut, like Log s or sqrt(s), and you cannot employ functions with an infinite number of zeros like $\frac{\sin s}{s^2 + 1}$, or poles, like 1/sin s. When a function is not suitable for the **pzmap**, you may get an error message and/or simply a wrong answer.

At this juncture, the attentive reader might wonder what point there is in using **nyquist** since **pzmap** shows you the actual location of poles and zeros, while the former function merely establishes the number of poles in the RHP of a narrower class of functions.

To answer this question, we must see how Nyquist diagrams are used in an electrical laboratory. Recall that the variable s is $\sigma + i\omega$. We can sometimes make a series of electrical measurements on an electrical system by measuring $T(s)$ (see Equations 5.14 and 5.15) while varying ω from 0 to such a large positive value that $T(s)$ becomes essentially constant. This constant is $\lim_{|s|\to\infty} T(s)$ in the RHP. While making these measurements, we keep $\sigma = 0$. The results are typically complex numbers. Because the coefficients in the polynomials in Equation 5.15 are real numbers, it follows that $\overline{T}(s) = T(\overline{s})$. With $s = i\omega$, this means $\overline{T}(i\omega) = T(-i\omega)$. Thus, if we measure the complex values of T for positive ω and take their conjugates, it follows that we have the values for negative ω. For example, if T measured at $\omega = 1000$ is $3+i4$, then T measured at $\omega = -1000$ would be $3-i4$.

Shown below is a hypothetical set of measurements for $\omega \geq 0$. We have displayed the data points, with a small o, and connected them with a smooth curve. We have also plotted with a broken curve the conjugate of the curve for positive ω. Thus, we have a hypothetical Nyquist plot. In this case, we see that our plot encloses the point $-1+i0$ two times as we proceed along the ω axis from $-\infty$ to ∞. There are two zeros of $1 + T(s)$ in the right half of the complex plane.

Notice that Nyquist plots made from experimental data do not use the function **nyquist** but just the **plot** function.

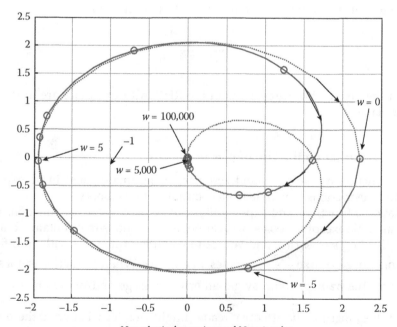

Hypothetical experimental Nyquist plot

Exercises

1. a. Explain why all roots of the polynomial $3z^4 + z^3 + 1$ must lie inside any circle in the complex plane, centered at the origin, whose radius R exceeds $4/3$.

   b. How many of the roots lie in the RHP? Answer this not by finding the roots but by moving along a contour, like that in Figure 5.11, and plotting the results. Make a suitable choice of radius for the semicircle. How does your contour show that the polynomial has no roots on the imaginary axis?

   c. Check your answer by using the MATLAB function **roots**.

2. a. Use the command **Nyquist** to determine if

   $$F(s) = 1 + 20 \frac{(s^2 + 1)e^{-s/10}}{s^5 + 10s^4 + 40s^3 + 80s^2 + 80s + 33}$$

   has any zeros in the right half of the s plane or on the imaginary axis.

   b. Make a pole–zero map for the above function, and see whether it confirms your answer.

3. a. Consider the function

   $$F(s) = 1 + 20 \frac{\beta(s^2 + 1)e^{-s/10}}{s^5 + 10s^4 + 40s^3 + 80s^2 + 80s + 33}$$

   where β is a real number that can be positive or negative. Study the Nyquist plot obtained in the preceding problem (which is this problem except that in problem 2 we have $\beta = 1$), and use that result to estimate (approximately) the smallest value for $|\beta|$ that will result in the function having at least one zero in the right half of the s plane. Prove your result by generating a Nyquist plot with this value of β, which you should have found is <0.

   b. Confirm the result of part (a) by means of a pole–zero map of the given function where you employ the value of β found above.

   c. For the value of β that you selected, use the function **fzero** to find the numerical value of s in the right half of the s plane that causes the function $F(s)$ to have a zero in that plane. The **pzmap** can help you to choose where to search in the plane.

4. a. Suppose in Equation 5.14 that we have

   $$T(s) = \frac{-\beta}{(s+1)(1 + e^{-1}e^{-s})}$$

   where β is a positive real number.

How can you be certain that this expression has no poles in the right half of the s plane?

b. Taking $\beta = 1$, obtain a Nyquist plot for $F(s)$.

c. By studying the above plot, figure out approximately how large β can be made so that $F(s)$ will not have any zeros in the right half of the s plane. Confirm this with a Nyquist plot using such a value of β.

6

Transforms: Laplace, Fourier, Z, and Hilbert

6.1 Introduction

Anyone working in engineering or the physical sciences will know some of the transformations named in this chapter's title. Laplace and Fourier transforms are much used in electric circuit theory, electromagnetic theory (including optics), and signal processing. The Z transformation has in the past two generations proved a major tool in the field of digital signal processing, while the Hilbert transform is useful in subjects as diverse as the modulation of radio signals as well as the analysis of brain waves. Because of the similarity of Laplace and Fourier transforms, we treat them first and begin with the easier of the two, Laplace.

6.2 The Laplace Transform

The Laplace transform of $f(t)$, a function of the real variable t, is generally written as

$$Lf(t) = \int_0^\infty f(t)e^{-st}\,dt \tag{6.1}$$

The function produced by this integral is written as $F(s)$.

The preceding equation will be defined as follows, where some special attention is paid to the limits

$$Lf(t) = F(s) = \lim L \to \infty \lim \varepsilon \to 0 - \int_\varepsilon^L f(t)e^{-st}\,dt \tag{6.2}$$

The order of taking the two limits on the right should be immaterial if the definition is to be valid. Note the lower limit is achieved by our letting ε shrink toward zero through *negative* values. For functions that are continuous at $t = 0$, this subtlety can be ignored. Otherwise it should be observed. Note that some textbooks use a lower limit in which ε shrinks toward zero through *positive* values, and this can cause different results when we are transforming functions that are discontinuous at the origin. The convention we use here is consistent with MATLAB®. In general, we should think of *s* as a complex variable:

$$s = \sigma + i\omega \quad \sigma \text{ and } \omega \text{ real} \tag{6.3}$$

Thus, the Laplace transformation takes a function (which might be complex, e.g., e^{it}) of a real variable and creates a function of a complex variable. In most cases, there are restrictions on *s* if the integral is to exist. For example, the reader should verify that if $f(t) = 1$ and if *s* is purely imaginary or has a negative real part, then the integral fails to exist. But, if $f(t) = 1/t$, then the integral will not exist for any value of *s* owing to the singularity of this function at $t = 0$.

Typically, when $F(s)$ does exist, it is an analytic function in some domain in the complex *s* plane. At this point, the reader should peruse the MATLAB documentation on the function **laplace**, which creates Laplace transforms. He or she might also wish to review a textbook showing how Laplace transforms are useful in solving linear differential and sometimes integral equations. Note that because the defining integral does not require our knowing or using values assumed by $f(t)$ for $t < 0$, many authors and practitioners simply take $f(t) = 0$ for $t < 0$, and we will follow that here even though it means that $f(t)$ may now be discontinuous at $t = 0$. In using MATLAB to find Laplace transforms, it is usually best to supply a function of the variable *t*, as is used in the above definition. This is the "default variable." If you wish to use some other variable like *x*, you should read the documentation for **laplace** to see how this is accomplished.

Here is a simple example of using MATLAB to find

$$Le^{-bt} = \int_0^\infty e^{-st} e^{-bt}\, dt$$

```
>> syms b t
>> laplace (exp(-b*t))
ans = 1/(b + s)
```

Notice the answer gives no indication of what restrictions must be placed on *s* for this result to be valid. Clearly, $s \neq -b$. Let us perform this simple integration to see what limitations might be required. We assume that *b* is complex so that $b = \alpha + i\beta$. Thus,

$$Le^{-at} = \lim_{L\to\infty}\lim_{\varepsilon\to 0-}\int_\varepsilon^L e^{-bt}e^{-st}\, dt = \lim_{L\to\infty}\lim_{\varepsilon\to 0-}\frac{e^{-(s+b)L}-e^{-(s+b)\varepsilon}}{(-b-s)} = \lim_{L\to\infty}\frac{e^{-(\alpha+\sigma)L}e^{-i(\beta+\omega)L}-1}{-(b+s)}$$

Studying the expression on the above right, we see that the limit exists only if $\alpha + \sigma > 0$ or $\text{Re}(s) > -\text{Re}(b)$. Thus, the result obtained by MATLAB is valid only to the right of a vertical line in the complex s plane. Here, as is typically the case, the Laplace transform is an analytic function of s. Thus, one must be cautious in using the **laplace** operation in MATLAB.

The function $1/(b + s)$ is the Laplace transform of e^{-bt} only in the domain of the s plane where $\text{Re}(s) > -\text{Re}(b)$. However, we see that this function, $1/(b + s)$, is defined and analytic in any domain in the s plane from which the point $s = -b$ is excluded. The function $1/(b + s)$ evaluated anywhere in such a domain where $\text{Re}(s) > -\text{Re}(b)$ is not satisfied is said to be the *analytic continuation* of the Laplace transform of e^{-bt}. Usually we are not aware of this subtlety and say simply that $1/(b + s)$ is the Laplace transform of e^{-bt} and use it, except at $s = -b$, in the s plane. Such analytic continuations will occur for other Laplace transforms and are especially useful when we are inverting Laplace transforms (recovering $f(t)$) by means of residues as described below. See W section 5.7 and S section 10.1 for more information on analytic continuation.

There is a sufficient condition that can guarantee that $F(s) = Lf(t)$ is analytic in a half plane to the right of a vertical line in the complex s plane. Briefly stated, assuming that the integral in Equation 6.2 exists for some values of s and that there exist real constants k, p, and T such that

$$|f(t)| < ke^{pt} \quad \text{for } t \geq T \tag{6.4}$$

then it can be shown* that the function $F(s)$ is analytic in the half space $\text{Re}(s) > p$.

Let us use MATLAB to find the Laplace transform of $\cos(bt)$, where $b = \alpha + i\beta$ is complex. We proceed as follows:

```
>> syms b t
>> laplace(cos(b*t))
ans = s/(b^2 + s^2)
```

Where in the s plane is this expression for $F(s)$ valid?

Notice that

$$\cos(bt) = \frac{1}{2}\left(e^{i(\alpha+i\beta)t} + e^{-i(\alpha+i\beta)t}\right) = \frac{1}{2}\left(e^{-\beta t}e^{i\alpha t} + e^{\beta t}e^{-i\alpha t}\right)$$

$$|\cos(bt)| \leq \frac{1}{2}\left|e^{-\beta t}e^{i\alpha t}\right| + \frac{1}{2}\left|e^{\beta t}e^{-i\alpha t}\right| = \frac{1}{2}e^{-\beta t}\left|e^{i\alpha t}\right| + \frac{1}{2}e^{\beta t}\left|e^{-i\alpha t}\right| = \frac{1}{2}e^{-\beta t} + \frac{1}{2}e^{\beta t} \leq e^{|\beta|t}$$

Thus, referring now to Equation 6.4, we see that by taking $T = 0, p > |\beta| = |\text{Im}(b)|$, the inequality is satisfied, and we are guaranteed that the Laplace

---

* See, for example, R.V. Churchill. *Operational Mathematics*, 3rd ed. (McGraw-Hill, New York, 1971).

transform obtained by MATLAB is analytic in the half space Re(s) > |Im(b)|. A simple additional calculation can show you that in this half space the denominator in s/(b^2 + s^2) is nonvanishing, which of course means that F(s) is analytic in this domain, as predicted.

In general, such calculations are not necessary each time you employ MATLAB to find a Laplace transform. To find the half space in which the result holds, you should look for the singularities of F(s) and assume that the result is valid in the infinite half space that is to the right of all these singularities.

In using MATLAB to obtain Laplace transforms, you should be aware that the result might be in terms of a perhaps unfamiliar transcendental function, as in the following example:

```
>> syms n s t
>> laplace( t^(1/n))
ans = piecewise([−1 < real(1/n), gamma(1/n + 1)/s^(1/n + 1)])
```

The gamma in the preceding refers to the Γ function that you may wish to review (see W section 6.11 and S section 10.8). Recall that it is defined by

$$\Gamma(w) = \int_0^\infty e^{-t} t^{w-1}\, dt$$

where w is a complex variable whose real part is positive.

Thus, we learn that the Laplace transform of the function $t^{1/n}$ is $\dfrac{\Gamma(1/n+1)}{s^{(1/n+1)}}$. There is an additional constraint, $-1 < \text{real}(1/n)$, which ensures that the argument of the gamma function in the preceding has a positive real part; this is a requirement of the definition of the gamma function via an integral.

MATLAB in Release 2015a is still unable to obtain the Laplace transform of some relatively simple functions as shown by the following:

```
>> laplace(cosh(sqrt(t)))
ans = laplace(cosh(t^(1/2)), t, s)
```

We are trying to get the Laplace transform of $\cosh(t^{1/2})$. MATLAB merely returns our question to us. In fact, the transform is to be found in any good set of tables of Laplace transforms, but it does involve the error function, which is defined further in this section.

Sometimes you must be careful in employing MATLAB to obtain a Laplace transform of a function. The statement **laplace** takes as the default variable of integration the variable t. If this is not your intention, you must so instruct MATLAB. Suppose you wish to obtain the Laplace transform of the function $f(x) = \cos(x + 1/t)$, where t is a parameter and x is the variable of integration.

Consider

```
>> syms s x t
>> laplace (cos(x+1/t))
ans = laplace(cos(x + 1/t), t, s)
```

Notice that MATLAB returns the question to us. It is unable to do the integration on the variable t. We wanted the integration done on the variable x. We proceed as follows:

```
>> syms s x t
>> laplace (cos(x+1/t),x,s)
ans = -(sin(1/t) - s*cos(1/t))/(s^2 + 1)
```

MATLAB obtains the relatively simple answer. The characters x,s in the **laplace** command instruct MATLAB to do the integration in Equation 6.1 on the variable x, not t, and to produce a function of s.

6.2.1 Branch Points and Branch Cuts in the Laplace Transform

Sometimes in using MATLAB to find a Laplace transform you will receive a function that must be defined by means of a branch cut. This would be true in the above case where we transformed $t^{1/n}$. Here is an example where we get the transform of $\dfrac{e^t}{\sqrt{t}}$ and must also deal with branch cuts:

```
>> syms s t
>> laplace(t^(-1/2)*exp(t))
ans = pi^(1/2)/(s - 1)^(1/2)
```

Our result is

$$F(s) = \frac{\sqrt{\pi}}{(s-1)^{1/2}}$$

How do we define this function in the complex s plane? Notice that the given function of t, namely, $e^t/\sqrt{t}$, can be made to satisfy Equation 6.4 if we take $p = 1$ and $k > 1$ and $T = 1$. Thus, we are guaranteed that this Laplace transform is analytic in the half plane $\mathrm{Re}(s) > 1$. We may use any branch cut of $(s - 1)^{1/2}$ in the complex s plane that originates at $s = 1$, extends to infinity, and does not pass into the domain $\mathrm{Re}(s) > 1$. We might have guessed this, because we know that if the Laplace transform exists, it must be analytic in a domain to the right of a vertical line in the complex s plane. We can choose any vertical line to the right of $s = 1$ and achieve this.

To establish a branch of a multivalued function, we require not only a branch cut (or cuts) but also the numerical value of the function at one point.

Suppose $s = 2$ or any real value greater than 1. Using $s = 2$ in Equation 6.1, with our given $f(t) = e^t/\sqrt{t}$ must result in a positive real value, and this completes the specification of the branch of $F(s) = \dfrac{\sqrt{\pi}}{(s-1)^{1/2}}$, because $F(2) = \sqrt{\pi}$.

It is entirely consistent with the preceding, and convenient, for us to take the principal branch of $(s - 1)^{1/2}$ in our result for $F(s)$. It is usually safe to take the principal branch of $F(s)$, but to be absolutely sure, one should verify the choice as was done here.

6.2.2 Heaviside and Dirac Delta Functions

Both the Heaviside and Dirac delta functions are often used in conjunction with Laplace transforms, especially in electric circuit theory and mechanics. Following the conventions of MATLAB, we use the notation heaviside(t) to mean a function that is equal to 1 for $t > 0$ and equal to zero for $t < 0$. In MATLAB, the function is defined as .5 for $t = 0$, but this is by no means universal, and many authors simply leave it undefined there. The reader should verify by pencil-and-paper integration that the Laplace transform of this function is given by $\mathbf{L}$(heaviside(t)) = $1/s$ for Re(s) > 0. The Heaviside function is also often known as the *unit step function*, and we use the common notation $u(t)$ to mean the Heaviside function as one soon tires of writing out the word *Heaviside*. Thus,

$$u(t) = \text{heaviside}(t)$$

The plot shown in Figure 6.1 should make clear where the language "unit step" comes from. The function $u(t - \tau)$ or heaviside $(t - \tau)$ (here τ is real) would just be a step like the one shown in Figure 6.1 but with the discontinuous rise displaced τ units to the right if $\tau > 0$. Here we obtain the Laplace transform of the Heaviside function with MATLAB:

```
>> syms s t
>> laplace (heaviside(t))
ans = 1/s
```

The Heaviside function is defined only as a function of a real variable. Thus, to take the Laplace transform of the function $f(t)$ = heaviside($t - t1$), we proceed as follows:

```
>> syms s t t1
>> assume(t1>0)
>> laplace (heaviside(t–t1))
ans = exp(–s*t1)/s
```

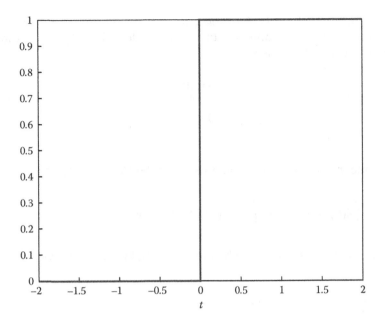

FIGURE 6.1
Heaviside function for $-2 < t < 2$.

Notice that the line of code **assume(t1>0)** establishes that t1 is positive and that it is real. Without such an **assume** statement, MATLAB cannot evaluate the Laplace transform if t1 is a symbol whose numerical value has not been specified. If you used **assume(t1<0)**, the answer would be 1/s, and t1 would not appear. You should explain to yourself why this is true. In the preceding work, you may use any real number instead of t1. But MATLAB will not evaluate the Laplace transform of heaviside($t-i$) because of the presence of the non-real i. Verify this.

Studying Figure 6.1, we would expect that the derivative of the function heaviside(t) might be thought of as infinite at $t = 0$ and zero for all $t \neq 0$. A rigorous treatment, using the usual definition of the derivative as the limit of a difference quotient, would show that this function has no derivative at $t = 0$.* However, we nevertheless use the following:

$$\mathrm{dirac}(t) = \frac{d(\mathrm{heaviside}(t))}{dt}$$

---

* A branch of advanced calculus called *generalized functions* permits the repeated differentiation of the Heaviside function.

The function on the left, the *Dirac delta function*, is often written as $\delta(t)$ and is nonzero only at $t = 0$ and is to be thought of loosely as "positive infinity" at $t = 0$. Thus, we have also

$$\delta(t) = \text{dirac}(t)$$

$$\delta(t) = \frac{du(t)}{dt} = \frac{d}{dt}\text{heaviside}(t)$$

The integral of this expression must yield the Heaviside function. Thus,

$$\int_L^t \delta(t')\,dt' = u(t) \text{ or } \int_L^t \text{dirac}(t')\,dt' = \text{heaviside}(t) \quad L < 0 \text{ and } t > L$$

We can verify the preceding by means of MATLAB as follows:

```
>> syms b t
>> diff(heaviside(t-b))
ans = dirac(b - t)
>> int(dirac(t))
ans = heaviside(t)
```

Note that dirac(b − t)=dirac(t−b) as the dirac function is an *even* function of its argument. Observe also that the operator **diff** treats t as the variable for which differentiation is to take place. MATLAB assumes that you do not wish to differentiate with respect to b and takes t as the default variable for differentiation.

A useful result is called the *sampling property* of the dirac delta function:

$$\int_{-\infty}^{\infty} \text{dirac}(t - \tau)f(t)\,dt = f(\tau) \tag{6.5}$$

The integral returns the value (or sample) of the function $f(t)$ at $t = \tau$. This is sometimes known also as the *sifting property* as it sifts through possible values of the function and supplies one. We assume that $f(t)$ is continuous at $t = \tau$. Because the dirac function is zero except where its argument is zero, the preceding equation is still valid if $-\infty$ is replaced by the symbol b, and ∞ is replaced by the symbol c, provided $b < \tau < c$.

Because of the sampling property, we have that

$$L\,\text{dirac}(t) = \lim L \to \infty \lim \varepsilon \to 0 - \int_{\varepsilon}^{L} \text{dirac}(t)e^{-st}\,dt = 1$$

which shows the significance of having ε approach zero through *negative* values.

Similarly, if τ is a nonnegative real number, we have

$$L \, dirac(t - \tau) = \lim L \to \infty \lim \varepsilon \to 0 - \int_{\varepsilon}^{L} dirac(t - \tau)e^{-st}\, dt = e^{-s\tau}$$

You should see that if τ is negative real, then the Laplace transform is zero. Here is the same result from MATLAB where we use t1 instead of τ.

```
>> syms s t t1
>> laplace (dirac (t-t1))
ans = piecewise([t1 < 0, 0], [0 <= t1, exp(-s*t1)])
```

Notice the additional information in this "piecewise" answer: that if t1 is negative, the Laplace transform is zero. Observe also that the form of the answer shows that MATLAB assumes that t1 is real.

The Heaviside function is useful in describing pulse-type functions. Suppose a function $f(t)$ is zero for $t < t1$, and also for $t > t2$, where $t1 < t2$. Also suppose that $f(t) = 1$ for $t1 < t < t2$. The reader should verify that this is conveniently written as $f(t) = heaviside(t - t1) - heaviside(t - t2)$. Here MATLAB finds the Laplace transform of $f(t)$, where t1 = 1, t2 = 4. The function is shown in Figure 6.2.

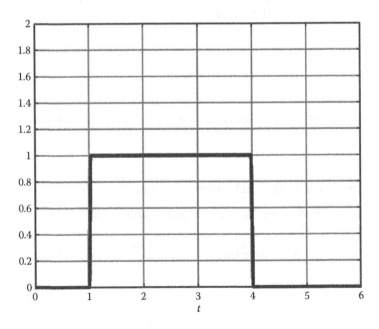

FIGURE 6.2
Plot of $f(t) = heaviside(t - 1) - heaviside(t - 4)$.

And here is its transform:

```
>> syms s t t1 t2
>> t1=1;t2=4;
>> laplace(heaviside(t-t1) - heaviside(t-t2))
ans = exp(-s)/s - exp(-4*s)/s
```

A function that is similar to the Heaviside function is the **sign** function. It is defined only for real arguments. Thus, $sign(x) = 1$, $x > 0$ and $sign(x) = -1$, $x < 0$, while $sign(0) = 0$. Thus, $sign(x)$ assumes the values 1 or –1 according to whether its argument is positive or negative.

The reader who is at all familiar with Laplace transforms knows that much of their utility resides in how easily one can find the derivatives of the functions $f'(t)$, $f''(t)$, $f'''(t)$, and so on, once the Laplace transform of $f(t)$ is known. With $Lf(t) = F(s)$, we have

$$Lf'(t) = sF(s) - f'(0-) \tag{6.6}$$

$$Lf''(t) = s^2F(s) - sf(0-) - f''(0-) \tag{6.7}$$

The first of these formulas is obtained by our integrating Equation 6.1 by parts, using $f'(t)$ in place of $f(t)$ in that equation. A comparable procedure yields the second of these. We can also obtain a formula for the Laplace transform of the integral of a function in terms of the Laplace transform of that function:

$$L\int_0^t f(t')dt' = \frac{F(s)}{s} \tag{6.8}$$

This is also obtainable if we do an integration by parts of Equation 6.1 but use $\int_0^t f(t')dt'$ in place of $f(t)$.

If we know the Laplace transform of the function $f(t) = -\dfrac{\cos at}{a}$, then it should be easy to find the Laplace transform of the function which is its first derivative: $f'(t) = \sin(at)$.

From Equation 6.6, we have $L\sin(at) = sL\left(\dfrac{-\cos at}{a}\right) - (-1/a)$. From MATLAB, we obtain

```
>> syms s a t
>> s*laplace (-cos(a*t)/a) - (- 1/a)
ans = 1/a - s^2/(a*(a^2 + s^2))
>> simplify (ans)
```

ans = a/(a^2 + s^2)
>> laplace(sin(a*t))
ans = a/(a^2 + s^2)

Note the agreement between the two ways of getting the Laplace transform of sin(*at*).

The derivative of the Heaviside function is the Dirac function. Thus, the Laplace transform of the Dirac function should be obtained if we use MATLAB to obtain the Laplace transform of the Heaviside function, multiply this result by *s*, and from this subtract the value of the Heaviside function at 0− (this value is zero). We verify this as follows:

>> syms s t
>> s*laplace (heaviside(t))
ans = 1

which is indeed the Laplace transform of dirac(*t*).

The derivative of the sign function is twice the Dirac function. This is because the discontinuity of heaviside(t) at $t = 0$ is half the discontinuity of sign(t) at $t = 0$. The size of the first jump is unity, and the size of the second is 2. A sketch of the two functions illustrates this.

Notice that from MATLAB we have

>> syms t
>> diff(heaviside(t))
ans = dirac(t)
>> diff(sign(t))
ans = 2*dirac(t)

The Dirac function has, if we permit ourselves the use of *generalized functions*, a first derivative. Generalized functions include the Dirac function and/or any of its derivatives. In the language of MATLAB, we have that $\frac{d}{dt}$dirac(*t*) = dirac(*t*,1), where the notation dirac(*t*,1) means simply the first derivative of the Dirac function. MATLAB can handle higher-order derivatives of the Dirac function and uses dirac(*t*,*n*) to mean the *n*th derivative of dirac(t). If the number *n* is left unspecified, it is taken as zero—no derivative is taken. Just as the function dirac(*t* − *τ*), where *τ* is real, samples the value of a function that is continuous at *τ* when we do an integration (see Equation 6.5), the function dirac(t−τ, 1) will sample the *negative* of the first derivative of a function with a continuous first derivative. Thus,

$$\int_{-\infty}^{\infty} dirac(t-\tau,1)f(t)dt = -f'(\tau) \qquad (6.9)$$

The preceding equation is still valid if $-\infty$ is replaced by the symbol b and ∞ is replaced by the symbol c, provided $b < \tau < c$. The above equation can be derived by our performing this integration

$$\int_{-\infty}^{\infty} f(t) \frac{d}{dt} \text{dirac}(t - \tau) dt$$

by parts and employing Equation 6.5. A generalization of the above is

$$\int_{-\infty}^{\infty} \text{dirac}(t - \tau, n) f(t) dt = (-1)^n f^{(n)}(\tau) \tag{6.10}$$

Thus, the nth derivative of the Dirac function, $\delta^{(n)}(t - \tau)$, samples the nth derivative of the function $f(t)$ and multiplies it by either 1 or –1.

6.2.3 The Inverse Laplace Transform

If $f(t)$ has Laplace transform $F(s)$, then we say that $F(s)$ has *inverse Laplace transform $f(t)$*.

Using complex variable theory, we can obtain the inverse Laplace transform for a wide class of functions $F(s)$ (see W section 7.1).

We assume that the function $F(s)$ is analytic in a right half portion of the complex s plane described by $\text{Re} s \geq a$. We also assume that there exist positive constants m, R_0, and k such that

$$|F(s)| \leq m/|s|^k \tag{6.11}$$

for all $|s| > R_0$ in this half plane.* Under these conditions, the function $f(t)$, for $t > 0$, can be obtained by a contour integral along an infinite vertical line in the complex s plane as follows:

$$f(t) = \mathbf{L}^{-1} F(s) = \frac{1}{2\pi i} \int_{a-i\infty}^{a+i\infty} F(s) e^{st} \, ds \tag{6.12}$$

The integration is done along the line $\text{Re}(s) = a$ or along any other contour into which this line can legitimately be deformed without affecting the value of the integral. The preceding is known as the *Bromwich integral* and is performed along a straight infinite vertical line like that shown in Figure 6.3.

The line is called the *Bromwich contour*. If we perform the integration for $t < 0$, it can be shown that the function of t obtained is zero.

---

* The reader might wish to verify that the function $F(s) = \dfrac{1}{s+1}$ will satisfy the preceding if we were to take $m = 1$, $k = 1$, $a = 0$, and R_0 as any positive number.

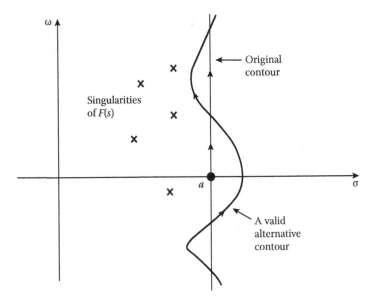

FIGURE 6.3
A Bromwich contour and its valid deformation.

If the function $f(t)$ has a step-type discontinuity at $t = t_0 > 0$, and if we obtain $F(s)$ and then seek to get back $f(t)$ by means of Equation 6.12, there is a subtlety. We obtain a function whose value at t_0 is $\frac{1}{2}(f(t_0+) + f(t_0-))$, which is the average of the values on either side of the step.

Suppose the function $F(s)$ satisfies the requirements for using Equation 6.12, and suppose also that its singularities are all poles and that they are finite in number. They of course lie to the left of the line in Figure 6.3. Assume also that there exist positive constants m, R_0, and k such that $F(s) \leq m/|s|^k$ when $|s| > R_0$ is in the half plane lying to the left of the vertical line described above. Then the integral can be evaluated using the method of residues (see W section 7.1) with this result:

$$f(t) = \text{the sum of the residues of } F(s)e^{st} \quad \text{at all poles of } F(s) \qquad (6.13)$$

If there are branch cut singularities of $F(s)$ lying to the left of the vertical line, the method cannot be used.

Let us see how this applies to the function $F(s) = \dfrac{1}{s^2(s+1)}$. Notice that this function is analytic in a half plane $\text{Re}(s) \geq a$, where $a > 0$. Also, in this half plane $|s| < |s+1|$, which means that in this same space $\left|\dfrac{1}{s^2(s+1)}\right| \leq \left|\dfrac{1}{s^2|s|}\right| = \dfrac{1}{|s|^3}$.

Thus, we may employ the Bromwich integral to find the inverse transform.

We may evaluate this integral using the residues described in Equation 6.13. Note that if $|s| > 1$, we have that $|s+1| > |s| - 1 > 0$. Thus, $|s+1| > |s| - 1 = |s/2| + |s/2| - 1 > |s/2| > 0$ provided $|s| > 2$. Furthermore, in this same domain,

$$\left|\frac{1}{s^2(s+1)}\right| = \left|\frac{1}{s^2||s+1|}\right| \le \frac{1}{|s|^2|s|/2} = \frac{2}{|s|^3} \text{ if } |s| > 2.$$

The conditions for using Equation 6.13 are satisfied, and we assert that

$$\mathbf{L}^{-1}\frac{1}{s^2(s+1)} = \text{Res}\frac{e^{st}}{s^2(s+1)} @s=0 + \text{Res}\frac{e^{st}}{s^2(s+1)} @s=-1$$

The first pole is of order 2 and has residue

$$\lim_{s\to 0}\left(\frac{d}{ds}\frac{s^2e^{st}}{s^2(s+1)}\right) = t-1.$$

The second pole is of order 1 and has residue

$$\lim_{s\to -1}\left(\frac{(s+1)e^{st}}{s^2(s+1)}\right) = e^{-t}.$$

Summing residues, we have finally that $f(t) = t - 1 + e^{-t}$.

All of the preceding can very easily be performed with MATLAB by means of the function **ilaplace** in the symbolic math toolbox. The reader should read the documentation for this function. Here we perform the above problem:

```
>> syms s t
>> ilaplace (1/(s^2*(s+1)))
ans = t + exp(-t) - 1
```

which agrees with what we obtained using the calculus of residues.

The use of **ilaplace** did not seem particularly helpful when we did the above problem, as an answer using residues also works. But there are many functions for which Equation 6.12 or Equation 6.13, or both cannot be applied. This is especially true when $f(t)$ is composed of one or more Dirac functions or their derivatives. The function $f(t) = \delta(t-1)$ has Laplace transform $F(s) = e^{-s}$. Yet, we cannot apply the Bromwich integral, Equation 6.12, to $F(s)$ and get back $f(t)$ because $F(s) = e^{-s}$ cannot be made to satisfy Equation 6.11, as the reader should verify.

On the other hand, the function $F(s) = e^{-s}/s$ can be made to satisfy this same equation, but we cannot apply Equation 6.13 to find its inverse transform, because there do not exist constants such that $F(s) \le m/|s|^k$, $|s| > R_0$, is satisfied to the *left* of the vertical line used in the Bromwich integral.

The function $F(s) = \dfrac{1}{(s-b)^{1/2}}$, where b is a constant and the function is defined by the principal branch, cannot be found from Equation 6.13 because of the presence of branch point singularities to the left of the Bromwich contour. Yet in all three cases just cited, MATLAB will yield the inverse Laplace transform. Here is one example:

```
>> syms s b t
>> ilaplace (1/(s–b)^(1/2))
ans = exp(b*t)/(pi^(1/2)*t^(1/2))
```

The result is that the inverse Laplace transform of $F(s) = \dfrac{1}{(s-b)^{1/2}}$ is $f(t) = \dfrac{e^{bt}}{\sqrt{\pi}t^{1/2}}$. Notice that MATLAB recognizes that s is the variable and b the constant in the expression that it is asked to invert. It always takes s as the default variable in these situations unless instructed otherwise. The branch of $(s - b)^{1/2}$ is assumed to be the principal one, and in general **ilaplace** uses the principal branch of $F(s)$. The square root of t in the preceding answer should be interpreted as the principal value.

The preceding inverse transform *can* be found by analytic means (see W section 7.1). Some idea of the method is outlined in problem 6 where $b = -1$.

In the present release of MATLAB (R2015a) that the author is using, there are some quirks that perhaps will be ironed out in future versions. For example, consider this operation:

```
>> syms s t
>> ilaplace (1)
```

whose output is
Undefined function 'ilaplace' for input arguments of type 'double'.

MATLAB refuses to provide the desired inverse Laplace transform because the **ilaplace** function is anticipating a function of s. The number *one* is not such a function. The matter can be dealt with as follows:

```
>> ilaplace(sym('1'))
ans = dirac(t)
```

The number 1 is here treated as a symbol. But an easier ploy is this:

```
>> syms s t
>> ilaplace (s/s)
ans = dirac(t)
```

where the number 1 is disguised here as a function of s. This subtlety is also considered in problem 3.

Sometimes when we use the functions **laplace** or **ilaplace**, we may receive from MATLAB a result that might not be familiar. Consider the following, where we want the Laplace transform of $f(t) = e^{-at^2}$, where a is a positive real number.

```
>> syms s t a
>> assume(a>0)
>> laplace (exp(-a*t^2))
ans = (pi^(1/2)*exp(s^2/(4*a))*erfc(s/(2*a^(1/2))))/(2*a^(1/2))
```

Our result is

$$\frac{\sqrt{\pi}}{2\sqrt{a}} e^{s^2/(4a)} \operatorname{erfc}\left(\frac{s}{2\sqrt{a}}\right)$$

The function erfc is known as the *complimentary error function*. It is defined as

$$\operatorname{erfc}(x) = \frac{2}{\sqrt{\pi}} \int_x^{\infty} e^{-t^2/2} \, dt$$

and appears often both in Laplace transformations and their inverses. A related function is the *error function* which is

$$\operatorname{erf}(x) = \frac{2}{\sqrt{\pi}} \int_0^x e^{-t^2/2} \, dt.$$

The sum of these functions is one.

Incidentally, had we left off the line of code, assume($a>0$), we would still have obtained the same result from MATLAB. However, the result is not valid if $a < 0$, a fact not stated by the output from MATLAB.

To remind the reader of the power of the Laplace transform in solving differential equations or integro-differential equations, we look at the following problem:

Example 6.1

Solve the equation.

$$\frac{d^2 f}{dt^2} + \int_0^t f(\tau) d\tau = e^t \sin t \tag{6.14}$$

Assume $f'(0) = 0$ and $f(0) = 0$.

Solution:

Recalling that the Laplace transform is a linear operator, we take the Laplace transform of the left-hand side of the above equation by adding together the Laplace transform of each of the two terms on the left.

We have with the aid of Equation 6.7 and 6.8 that the Laplace transform of the left side of Equation 6.14 is $s^2 F(s) + F(s)/s$. The Laplace transform of the right side is best obtained from MATLAB, as follows:

```
>> syms s t
>> laplace (exp(t)*sin(t))
ans = 1/((s – 1)^2 + 1)
```

Thus, equating the Laplace transform of the right- and left-hand sides of Equation 6.14, we have

$$s^2 F(s) + F(s)/s = \frac{1}{(s-1)^2 + 1}$$

which yields

$$F(s) = \frac{1}{(s-1)^2 + 1} \frac{1}{s^2 + 1/s}$$

We find $f(t)$ by applying **ilaplace** as follows:

```
>> syms s t
>> ilaplace (1/((s–1)^2+1)*1/(s^2+1/s))
ans = (2*exp(t/2)*cos((3^(1/2)*t)/2))/3 – exp(–t)/15 – (3*exp(t)*(cos(t)
     – sin(t)/3))/5
```

This is more comfortably displayed as

$$f(t) = \frac{2}{3} e^{t/2} \cos\left(\frac{\sqrt{3}}{2} t\right) - \frac{e^{-t}}{15} - \frac{3}{5} e^t \left(\cos t - \frac{1}{3}\sin t\right)$$

Of course, there are other more direct ways to solve equations like the one just treated by using MATLAB. Differentiating both sides of Equation 6.14 results in a differential equation that can be solved with the MATLAB function **dsolve**, but this topic takes us far from complex variable theory.

Exercises

1. By evaluating the Bromwich integral with the method of residues, find the inverse Laplace transform of the following functions and compare your results with those obtained by MATLAB.

   a. $F(s) = \dfrac{1}{(s^2 - 1)}$

   b. $F(s) = \dfrac{1}{(s-1)^2 + 1}$

   c. $F(s) = \dfrac{1}{s^2 + s + 1} + \dfrac{1}{s+1}$

2. a. Suppose a function $f(t)$ has Laplace transform $F(s)$. Show that the Laplace transform of the function $f(t - \tau)u(t - \tau)$ is $e^{-s\tau}F(s)$, where τ is a nonnegative real.

   Hint: Write the Laplace transform of $f(t - \tau)u(t - \tau)$ as an integral, and make a simple change of variable. Our result is known as a *translation theorem*. The inverse transform of $e^{-s\tau}F(s)$ is thus the inverse transform of $F(s)$ shifted τ units to the right on the t axis and multiplied by $u(t - \tau)$. The latter ensures that the resulting function is zero for $t - \tau$. Thus, $\mathbf{L}^{-1}e^{-s\tau}F(s) = u(t - \tau)f(t - \tau)$, where $f(t) = \mathbf{L}^{-1}F(s)$.

   b. Find the residue of the function $\dfrac{e^{st}}{(s-1)^2}$ at its pole and use that result together with Equation 6.13 as well as the result derived in part (a) to find the inverse Laplace transform of $\dfrac{e^{-2s}}{(s-1)^2}$.

   c. Check the result by using MATLAB to find the inverse transform of $F(s) = \dfrac{e^{-2s}}{(s-1)^2}$.

3. This problem explores some peculiarities of MATLAB in its handling of inverse Laplace transforms.

   a. Using **ilaplace**, find the inverse Laplace transform of the function $\dfrac{s+1}{s}$.

   b. Verify that you get the same result by applying **ilaplace** to the function $\left(1 + \dfrac{1}{s}\right)$.

   c. Since the inverse Laplace transformation is a linear operation, you should get the same result by using **ilaplace**(1) + **ilaplace**$\left(\dfrac{1}{s}\right)$. Try this. You will see that there is a difficulty. This arises because the operation **ilaplace** expects a symbolic function in its argument.

   d. Show that you can resolve this difficulty by first stating one=sym('1') in your code. This establishes that one is now a *symbol* equal to unity. Now try taking the sum **ilaplace**(one) + **ilaplace**$\left(\dfrac{1}{s}\right)$ and show that you get an inverse Laplace transform of $\dfrac{s+1}{s}$.

   Also try taking the sum **ilaplace**$\left(\dfrac{s}{s}\right)$ + **ilaplace**$\left(\dfrac{1}{s}\right)$ and verify that this works, too.

4. a. Consider the Laplace transform $F(s) = \dfrac{1}{(s^2+1)^{1/2}}$. The principal branch of the function is used. Where in the s plane is $F(s)$ analytic (i.e., where is the branch cut?)? Explain why $F(s)$ satisfies Equation 6.11 for suitable m and k.

   Use MATLAB to find the inverse Laplace transform. Notice that the answer is given in terms of the Bessel function of the first kind, order zero: $J_0(t)$. You can learn more about this function by typing **help besselj** in MATLAB.

   b. Repeat the above using $F(s) = \dfrac{1}{(s^2-1)^{1/2}}$. The principal branch of the function is used. Notice that the answer is again given in terms of the Bessel function of the first kind, order zero: $J_0(t)$, but the argument of the function is purely imaginary. Some books use the function I_0 but with a real argument to mean the same thing.

5. Let $f(t)$ and $g(t)$ be two functions having Laplace transforms $F(s)$ and $G(s)$. The *convolution* of $f(t)$ and $g(t)$, written $f(t) \otimes g(t)$, is defined as $f(t) \otimes g(t) = \displaystyle\int_0^t f(\tau)g(t-\tau)d\tau$. It is a simple matter to show that $f(t) \otimes g(t) = g(t) \otimes f(t)$. This means that the order of the convolution is immaterial. In standard texts on complex variable theory or Laplace transforms, it is shown that $L[f(t) \otimes g(t)] = F(s)G(s)$ (i.e., the Laplace transform of the convolution of two functions is the product of the Laplace transform of each). In this problem, we use MATLAB to verify this property in a specific case. A more general definition of the convolution of two functions is given in the following section of this book.

   a. Use the function **int** in MATLAB to find the function of t that is the convolution of the functions $f(t) = \sin(t + 1)$ and $g(t) = \cos 2t$. Compute both $f(t) \otimes g(t)$ and $g(t) \otimes f(t)$, and verify that you get the same answer. You may wish to read the documentation for **int** in the help window.

   b. Using MATLAB, find the Laplace transforms of $f(t)$ and $g(t)$.

   c. Find the Laplace transform of the function found in part (a). Verify that it is indeed the product of the Laplace transforms of $f(t)$ and $g(t)$.

   d. Show that the general relationship $L[f(t) \otimes g(t)] = F(s)G(s)$ can be used to derive Equation 6.8 if we make a suitable choice of $g(t)$ and know the Laplace transform of the unit step function.

6. a. Using MATLAB, find the inverse Laplace transform of the function

$$F(s) = \frac{1}{(s+1)^{1/2}}$$

where the principal branch of the function is used.

We check this result by using contour integration. Follow the steps below.

b. Explain why the inverse Laplace transform can be obtained by your doing the integration in Equation 6.12, when $a = 0$.

c. Explain why if we integrate

$$\frac{1}{2\pi i} \oint \frac{e^{st}}{(s+1)^{1/2}} ds$$

along the closed contour, in the s plane, shown with arrows in Figure 6.4, we get zero. Note that we make a detour around $s = -1$ by using a nearly circular contour of radius ε. The horizontal portions of the contour are just above and below the branch cut for $(s + 1)^{1/2}$. This cut lies along the line $\text{Im}(s) = 0$, $-\infty < \text{Re}(s) \le -1$.

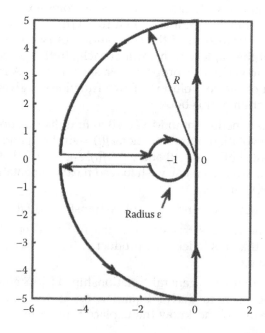

FIGURE 6.4
Contour for Part (c).

d. In the next two steps we prove that as $R \to \infty$ the portion of the integral along the arc in the second quadrant above goes to zero.

Hint: Make the polar substitution $s = Re^{i\theta}$, where $\pi/2 \le \theta \le \pi$. Recall the inequality

$$\left| \int_p^q u(\theta) d\theta \right| \le \int_p^q |u(\theta)| d\theta \quad \text{if } q > p$$

We may also substitute a function $g(\theta)$ for $|u(\theta)|$ on the right provided that $g(\theta) \ge |u(\theta)|$ everywhere along the interval of integration $p \le \theta \le q$. Show that the magnitude of the integral along the arc satisfies the inequality

$$\left| \frac{1}{2\pi i} \oint \frac{e^{st}}{(s-1)^{1/2}} ds \right| \le \frac{1}{2} \frac{R}{\pi \sqrt{R-1}} \int_{\pi/2}^{\pi} e^{Rt \cos\theta} d\theta$$

Hint: Make the change of variable $s = Re^{i\theta}$ for the integral on this arc, where $\pi/2 \le \theta \le \pi$. Find $\left| e^{t Re^{i\theta}} \right|$.

e. Observe that, by making a sketch, for $\pi/2 \le \theta \le \pi$, we have the inequality $1 - 2\theta/\pi \ge \cos\theta$. Thus, show that

$$\left| \frac{1}{2\pi i} \oint \frac{e^{st}}{(s-1)^{1/2}} ds \right| \le \frac{1}{2} \frac{R}{\pi \sqrt{R-1}} \int_{\pi/2}^{\pi} e^{Rt(1-2\theta/\pi)} d\theta$$

along the arc in the second quadrant.

Now perform the right-hand integration to show that

$$\left| \frac{1}{2\pi i} \oint \frac{e^{st}}{(s-1)^{1/2}} ds \right| \le \frac{1}{4t\sqrt{R-1}} \left[e^{-Rt} - e^{-2Rt} \right]$$

Argue that for $t > 0$, we must have that the integral on the left goes to zero as R tends to infinity.

If we perform this integration along the arc in the third quadrant, we will again get zero, in this same limit, because the integrand assumes the conjugate of the values it assumed in the second quadrant at conjugate points on the two arcs.

f. It is a similar but easier exercise to show that the integral

$$\frac{1}{2\pi i}\oint \frac{e^{st}}{(s+1)^{1/2}}ds$$

around the small circle of radius ε, on the contour, goes to zero as $\varepsilon \to 0+$. Take $s = \varepsilon e^{i\theta}$. Carry out the steps using the "ML inequality."

g. Having passed to the two limits $\varepsilon \to 0+$, $R \to \infty$, show that the inverse Laplace transform of $1/(s+1)^{1/2}$ consists of the sum of the integrals:

$$\frac{1}{2\pi i}\int_{-1}^{-\infty} \frac{e^{xt}}{|x+1|^{1/2}i}dx + \frac{1}{2\pi i}\int_{-\infty}^{-1} \frac{e^{xt}}{|x+1|^{1/2}(-i)}dx$$

which arise, respectively, from integrations along the top and bottom of the branch cut in the above figure in the limits. Combine these integrals and make the change of variable $x' = -x$ to show that

$$\mathbf{L}^{-1}\frac{1}{(s+1)^{1/2}} = \frac{1}{\pi}\int_{1}^{\infty} \frac{e^{-x't}}{\sqrt{x'-1}}dx'$$

You can evaluate this integral by using the well-known result

$$\int_{0}^{\infty} e^{-au^2}du = \frac{1}{2}\sqrt{\pi/a} \quad a>0.$$

Let $u = \sqrt{x'-1}$. Thus, finally verify that the inverse Laplace transform of $\dfrac{1}{(s+1)^{1/2}}$ is $\dfrac{e^{-t}}{\sqrt{\pi t}}$.

7. The present version of MATLAB (R2015a) will not give the Laplace transform of the error function defined as $\operatorname{erf}(t) = \dfrac{2}{\sqrt{\pi}}\int_{0}^{t} e^{-\tau^2} d\tau$.

Investigate this for your version. Now show how to get the Laplace transform of this function by using MATLAB to obtain the Laplace transform of the function $\dfrac{2}{\sqrt{\pi}}e^{-t^2}$ and then employing Equation 6.8.

It is helpful in this problem to treat π as a symbol using the statement syms pi. In this way, the program will not reduce the square root of π to a number. Notice that your answers will be in terms of the complimentary error function: $\operatorname{erfc}(t) = 1-\operatorname{erf}(t)$.

6.3 The Fourier Transform

The Fourier transform (or transformation) is a tool that is much used by engineers and scientists in such disciplines as signal analysis, system theory, electromagnetics, optics, and mechanics. Like the Laplace transform, it can be used to solve both linear differential and integral equations. Suppose we are given a function $f(t)$ of a real variable t. Then its Fourier transformation, $\mathbb{F}f(t)$, is obtained by converting $f(t)$ to a function $f(w)$ defined by the integral

$$\mathbb{F}f(t) = F(w) = \int_{-\infty}^{\infty} f(t)e^{-iwt}\,dt \tag{6.15}$$

Here w is to be regarded as a real variable, although we often employ the analytic continuation of the function defined above and continue to call it $F(\omega)$. A sufficient condition for the existence of the above integral is that the integral $\int_{-\infty}^{\infty}|f(t)|\,dt$ exist and that discontinuities of the function $f(t)$, if any, be limited to a finite number of jump discontinuities in any finite interval on the t axis. In reality, the engineer or scientist very often deals with functions that do not satisfy this necessary condition but whose transforms do in fact exist. A simple function $\sin t$ is such an example.

The function $f(t)$ can be recovered from $F(\omega)$, and we write $\mathbb{F}^{-1}F(w) = f(t)$. The procedure requires an integral:

$$f(t) = \mathbb{F}^{-1}F(w) = \frac{1}{2\pi}\int_{-\infty}^{\infty} F(w)e^{iwt}\,dw \tag{6.16}$$

The integration is done along the real axis in the complex w plane, and we employ the analytic continuation of $F(\omega)$ as mentioned above. Like the inversion formula for the Laplace transform, this formula comes with a *caveat*: If $f(t)$ has a jump discontinuity at a point, say t_0, then the integral on the right yields $f(t_0) = \frac{1}{2}[f(t_0+) + f(t_0-)]$, the average of the values assumed by $f(t)$ on each side of the jump.

The reader should note that there are a variety of ways in which to consistently define the Fourier transform and its inverse. We have used the one employed by MATLAB. In other definitions, $\frac{1}{2\pi}$ is swapped from the right side of Equation 6.16 to the right side of Equation 6.15. This is the procedure in W (pages 405–411). Sometimes the minus sign is swapped from the exponent in the integral Equation 6.15 to the exponent in the integral in Equation 6.16. Note that in the Wikipedia tables of Fourier transforms, three different options are offered for defining the transform, see http://en.wikipedia.org/wiki/Fourier_transform#Distributions.

The integrals in Equations 6.15 and 6.16 are both to be interpreted as Cauchy principal values—that is,

$$\int_{-\infty}^{\infty} ...(dt \text{ or } dw) = \lim_{R \to \infty} \int_{-R}^{R} ...(dt \text{ or } dw).$$

The limits move out from the origin symmetrically. If $f(t)$ has a singularity at the real value t_0, then we must perform our integration as follows:

$$\lim_{R \to \infty} \lim_{\varepsilon \to 0+} \left[\int_{-R}^{t_0 - \varepsilon} f(t)e^{-iwt}dt + \int_{t_0 + \varepsilon}^{R} f(t)e^{-iwt}dt \right]$$

The limits move toward t_0 symmetrically, and this is also known as a Cauchy principal value. In all practical cases, the result does not depend on the order in which we take the limits on R and ε. The method is readily extended to cases where there are more than one such singularity for real t. If the limit as ε shrinks to zero does not exist, then the function $f(t)$ does not have an integrable singularity and the integration cannot be done. However, as we see, both MATLAB and the community that uses Fourier transforms take liberties with this fact. If $F(w)$ in Equation 6.16 has a singularity as some real value, say, ω_0, it is handled in an analogous way.

Unlike the Laplace transformation, the Fourier transformation does not directly take into account conditions, present at $t = 0-$ in an electrical or mechanical system. However, the Laplace transformation will not treat the behavior of a system that has been receiving excitation for $t < 0-$.

Much of the utility of the Fourier transform, like that of the Laplace transform, resides in our ability to easily find the Fourier transform of derivatives of any order of a function $f(t)$ in terms of $F(w)$. We can also find the Fourier transform of the integral of $f(t)$ in terms of $F(w)$. These results can be found in any standard text on Fourier transforms.

$$\mathbb{F}f(t) = F(w),\ \mathbb{F}\frac{df}{dt} = iwF(w),\ \mathbb{F}\frac{d^2 f}{dt^2} = (iw)^2 F(w)\ ...\ \mathbb{F}\frac{d^n f}{dt^n} = (iw)^n F(w) \quad (6.17)$$

$$\mathbb{F}\int_{-\infty}^{t} f(\tau)d\tau = \frac{F(w)}{iw} + F(w)\delta(w) \quad (6.18)$$

The first group, Equation 6.17, can be derived formally if we permit ourselves to differentiate both sides of Equation 6.16 with respect to the variable t and bring the derivative under the integral sign.

Equation 6.18 can be derived from the properties of the convolution integral described below. It is especially helpful if we are seeking the Fourier transform of a function such that we can readily transform its derivative. A similar procedure was used for Laplace transforms.

These formulas in Equations 6.17 and 6.18 are useful for converting differential/integral equations in the variable t to algebraic equations in the variable w. These equations involve the unknown $F(w)$, and the technique is similar to that employed for Laplace transforms.

It is instructive for us to perform a Fourier transformation using the calculus of residues and verify these results with MATLAB. Here is an example.

Example 6.2

Using complex variable theory, find $\mathbb{F}\dfrac{1}{t^2+1}$ and check your result by using MATLAB.

Solution:
From Equation 6.15, we must evaluate

$$F(w) = \int\limits_{-\infty}^{\infty} \frac{1}{t^2+1} e^{-iwt} dt$$

Let us think of t here as a complex variable and integrate around the semicircular contour shown in Figure 6.5 in the complex t plane.

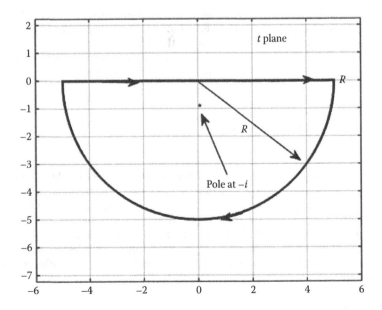

FIGURE 6.5
Contour for Example 1.

We assume at first that w is a nonnegative real quantity. Note that our contour is in the lower half of the t plane so that we can argue using Jordan's lemma (see W section 6.6 and S problem 7.7) that the integral along the arc of radius R vanishes as $R \rightarrow \infty$. The function $\dfrac{e^{-iwt}}{t^2+1}$ has a simple pole in the lower half of the complex plane, at $-i$, with residue $\dfrac{e^{-iwt}}{2t}$, which, evaluated at the pole, yields $\dfrac{e^{-w}}{-2i}$. Thus, passing to the limit in R, we are left with

$$F(w) = \int_{-\infty}^{\infty} \frac{1}{t^2+1} e^{-iwt} dt = -2\pi i \frac{1}{-2i} e^{-w} = \pi e^{-w} \tag{6.19}$$

Note the presence of the minus sign in front of $2\pi i$, which arises because the pole is enclosed in the negative (clockwise) direction. The preceding result holds when w is a nonnegative real. We can write

$$F(w) = \int_{-\infty}^{\infty} \frac{1}{t^2+1} e^{-iwt} dt = \int_{-\infty}^{\infty} \frac{1}{t^2+1} \cos wt \, dt - i \int_{-\infty}^{\infty} \frac{1}{t^2+1} \sin wt \, dt$$

where we have employed Euler's identity for the exponential. The second of these integrals is zero as it represents the integral of an odd function between symmetrical (even) limits. The first integral on the right, $\int_{-\infty}^{\infty} \dfrac{1}{t^2+1} \cos wt \, dt$, where w is real, is unaffected by a change in the sign of w because $\cos wt$ is an even function of w. Our result for $F(w)$, Equation 6.19, which was derived from residue calculus, required our assuming that $w \geq 0$. We now realize that we can adapt our answer to the case $w \leq 0$ by simply placing absolute values around w wherever it may occur. This renders our result, from residues, applicable for $w \leq 0$. Thus, to summarize,

$$\mathbb{F}\frac{1}{t^2+1} = F(w) = \int_{-\infty}^{\infty} \frac{1}{t^2+1} e^{-iwt} dt = \int_{-\infty}^{\infty} \frac{1}{t^2+1} \cos wt \, dt = \pi e^{-|w|}$$

where w is real.

Let us verify the preceding result with the use of MATLAB. To take a Fourier transform of a function of t, we use the command **fourier**. This yields, unless MATLAB is instructed otherwise, a function of the variable w. To recover the function of t from the function of w we use the command **ifourier**. MATLAB will return, unless told otherwise, a function of the variable x. Because we are often solving problems involving time, or t, we may wish to have a function of t. This is accomplished as shown below.

```
>> syms w t
>> fourier (1/(t^2+1))
ans = pi*exp(−w)*heaviside(w) + pi*heaviside(−w)*exp(w)
```

```
>> ifourier (ans, t)
ans = -(pi/(t*i - 1) - pi/(t*i + 1))/(2*pi)
>> simplify(ans)
ans = 1/(t^2 + 1)
```

The answer, pi*exp(-w)*heaviside(w) + pi*heaviside(-w)*exp(w), for the Fourier transform, requires some study. If $w > 0$, then heaviside(-w) is zero, while heaviside(w) = 1, and we obtain πe^{-w}. Similarly, if $w < 0$, then heaviside(w) is zero, while heaviside(-w) = 1. Notice that now exp(w) = exp(-|w|), and we obtain $\pi e^{-|w|}$ for $w < 0$. If $w = 0$, then MATLAB uses heaviside (0) = 1/2, and the answer pi*exp(-w)*heaviside(w) + pi*heaviside(-w)*exp(w) is equal to π, which will agree with $\pi e^{-|w|}$ at $w = 0$. Thus, MATLAB's answer for the Fourier transform agrees with our result of $\pi e^{-|w|}$.

Consider the line of code **ifourier** (ans, t). We have included the variable t so as to obtain a function of t after the inversion. Had we not included t, then MATLAB would have given us a function of x. This is the default variable if none is specified.

We could have performed the inverse transformation using elementary calculus as follows:

$$\mathbb{F}^{-1}\pi e^{-|w|} = \frac{1}{2\pi}\int_{-\infty}^{\infty}\pi e^{-|w|}e^{iwt}\,dw = \frac{1}{2\pi}\int_{-\infty}^{\infty}\pi e^{-|w|}(\cos wt + i\sin wt)\,dw$$

$$= \frac{1}{2\pi}\int_{-\infty}^{\infty}\pi e^{-|w|}(\cos wt)\,dw$$

The integral involving $\sin wt$ vanishes because it contains on odd function. Now, we evaluate the final integral by noticing that $e^{-|w|}\cos wt$ is an even function of w. We also employ Euler's identity:

$$\mathbb{F}^{-1}\pi e^{-|w|} = \frac{1}{2}\int_{-\infty}^{\infty}e^{-|w|}(\cos wt)\,dw = \int_{0}^{\infty}e^{-w}(\cos wt)\,dw = \mathrm{Re}\int_{0}^{\infty}e^{-w}(e^{iwt})\,dw$$

$$= \mathrm{Re}\int_{0}^{\infty}e^{-w(1-it)}\,dw$$

The last integral on the right is elementary, and we have

$$\mathbb{F}^{-1}\pi e^{-|w|} = \mathrm{Re}\left.\frac{e^{-w(1-it)}}{-1+it}\right]_{0}^{\infty} = \mathrm{Re}\frac{1}{1-it} = \mathrm{Re}\frac{1+it}{1+t^2} = \frac{1}{1+t^2}$$

which is the desired result.

Here is a *caveat* in using the MATLAB command **fourier**. If you supply MATLAB with a function that contains the variable x, it will perform the Fourier integration by integrating on the variable x, even if, say for example, the variable t is present. Suppose we wish to take the Fourier transform of the function $f(t) = \dfrac{1}{t^2 + x}$ by using the definition in Equation 6.2. We might carelessly proceed as follows:

```
>> syms t x
>> fourier(1/(t^2+x))
ans = -pi*exp(t^2*w*i)*i - pi*exp(t^2*w*i)*sign(imag(t^2))*i + pi*heaviside(-w)
*exp(t^2*w*i)*2*i
```

This is incorrect, as MATLAB treated t as a parameter and integrated on x. The correct method is this:

```
>> syms t x
>> fourier(1/(t^2+x),t,w)
(pi*exp(-w*x^(1/2))*heaviside(w))/x^(1/2) + (pi*heaviside(-w)*exp(w*x^(1/2)))/
x^(1/2)
```

In **fourier**(1/(t^2+x),t,w) the t,w instructs MATLAB to integrate on t and produce a Fourier transform that is a function of w.

Practitioners of applied mathematics frequently use Fourier transforms in situations where a mathematician, accustomed to rigor, might be appalled because he or she would quickly realize that one or both of the integrals in Equations 6.15 and 6.16 do not exist. Here is a simple example. Suppose $f(t) = \delta(t)$, the Dirac delta function. Then using the sampling property of the delta function, we have from Equation 6.15,

$$\mathbb{F}\delta(t) = \int_{-\infty}^{\infty} \delta(t)e^{-i\omega t}\,dt = 1 = F(\omega) \tag{6.20}$$

Now suppose we wish to recover our function of t, namely, $\delta(t)$, from $F(\omega)$. Employing Equation 6.16, we have

$$f(t) = \mathbb{F}^{-1}(1) = \frac{1}{2\pi}\int_{-\infty}^{\infty} e^{i\omega t}\,dw = \frac{1}{2\pi}\int_{-\infty}^{\infty}\cos wt\,dw + \frac{i}{2\pi}\int_{-\infty}^{\infty}\sin wt\,dw$$

Evaluating these as Cauchy principal values, we have

$$f(t) = \lim_{R\to\infty}\left[\frac{1}{2\pi}\int_{-R}^{R}\cos wt\,dw + \frac{i}{2\pi}\int_{-R}^{R}\sin wt\,dw\right] = \frac{2}{2\pi}\lim_{R\to\infty}\left(\frac{\sin Rt}{t}\right)$$

We have used the fact that the second integral above, $\dfrac{i}{2\pi} \displaystyle\int_{-R}^{R} \sin wt\, dw$, which

involves an odd function $\sin wt$, vanishes because of the symmetric limits.

Notice that the limit on the above far right, $\dfrac{2}{2\pi} \lim_{R \to \infty} \left(\dfrac{\sin Rt}{t} \right)$, does not exist.

Thus, we might argue that the inverse Fourier transform of 1 does not exist. However, MATLAB has other ideas as the following shows:

```
>> syms t
>> ifourier(1,t)
ans = dirac(t)
```

The delta function has been recovered. To deal with this contradiction, we might simply postulate that because the Fourier transform of $\delta(t)$ is 1, then *by definition* the inverse Fourier transform of 1 must be $\delta(t)$. In other words, we dispense with the integral definition of the inverse transform. Sometimes we will be forced to do this—a widespread practice in engineering but one generally abhorrent to mathematicians.[*]

Here is a further example: The reader should verify with Equation 6.15 that the Fourier transform of $\delta(t - \tau)$ is $e^{-iw\tau}$, where τ is a real number. The reader should also verify that the inverse transformation of $e^{-iw\tau}$ cannot be obtained through the use of Equation 6.16 as the integral does not exist. We have, however, from MATLAB:

```
>> syms t tau
>> assume(tau,'real')
>> fourier(dirac(t–tau))
ans = exp(–tau*w*i)
>> ifourier(ans,t)
ans = dirac(t – tau)
```

The **assume** statement above tells MATLAB that tau is real. This is a precaution. In the present version that we are using (R2015a), MATLAB assumes as a default that the argument of the Dirac function is real.

The reader might find it helpful to memorize not only that $\mathbb{F}\delta(t - t_0) = e^{-iw t_0}$ (which includes that $\mathbb{F}\delta(t) = 1$), but also $\mathbb{F}^{-1}\delta(w - w_0) = \dfrac{e^{iw_0 t}}{2\pi}$ which means that $\mathbb{F}^{-1}\delta(w) = \dfrac{1}{2\pi}$.

Sometimes we can be surprised, in the subject of Fourier transforms, when comparing results obtained from calculus with the results produced

---

[*] A presentation of a rigorous treatment of such practices can be found in the book by Hans Bremermann, *Distributions, Complex Variables, and Fourier Transforms*. (Reading, MA: Addison Wesley, 1965).

by MATLAB. For example, let us look at the transform of $f(t) = 1/(t-1)^2$. The definition of Fourier transforms requires that we do this integration:

$$\int_{-\infty}^{\infty} \frac{e^{-iwt}}{(t-1)^2}dt$$

Let us look at the real part:

$$\int_{-\infty}^{\infty} \frac{\cos wt}{(t-1)^2}dt$$

and consider an integration along the subinterval $1-\tau < 1 < 1+\tau$. We consider

$$\int_{1-\tau}^{1+\tau} \frac{\cos wt}{(t-1)^2}dt$$

and assume that $\cos wt > 0$ in the interval of integration. If $\cos wt > m > 0$, we have

$$\int_{1-\tau}^{1+\tau} \frac{\cos wt}{(t-1)^2}dt > \int_{1-\tau}^{1+\tau} \frac{m}{(t-1)^2}dt$$

If we attempt the second integral, we find the result is infinite. Thus, the integral on the left of the inequality cannot exist and, therefore, neither does the Fourier transform we are seeking. Note that it does not exist as a Cauchy principal value either. However, consider this from MATLAB:

```
>> syms t
>> fourier(1/(t−1)^2)
ans = −pi*w*exp(−w*i) + 2*pi*w*heaviside(−w)*exp(−w*i)
>> simplify (ans)
ans = −pi*w*exp(−w*i)*(2*heaviside(w) − 1)
```

According to MATLAB, the function $\dfrac{1}{(t-1)^2}$ has the Fourier transform

$$\mathbb{F}\left(\frac{1}{(t-1)^2}\right) = -\pi w e^{-iw}\left(2u(w)-1\right)$$

How can we explain this? Notice that the Fourier transform of the function $g(t) = \dfrac{1}{(t-1)}$ does exist as a Cauchy principal value. The expression $\dfrac{e^{-iwt}}{(t-1)}$ has an integrable singularity at $t = 1$.

We can find the Fourier transform

$$\int\limits_{-\infty}^{\infty} \frac{e^{-iwt}}{(t-1)} dt$$

by means of residue calculus or MATLAB.

Now recall (see Equation 6.17) that if we know the Fourier transform of a function, then the transform of its derivative can be found simply by multiplying this known transform by iw. The derivative of $1/(t-1)$ is $-1/(t-1)^2$. Thus, $-iw$ times the Fourier transform of $1/(t-1)$ is the Fourier transform of $1/(t-1)^2$. We use residues and verify that this is the procedure consistent with MATLAB.

Let us assume that $w > 0$. Noting the presence of the simple pole on the real axis at $t = 1$ in the complex t plane and invoking Jordan's lemma for the lower half of the t plane, we have

$$\mathbb{F}\left(\frac{1}{t-1}\right) = -\pi i \operatorname{Res} \frac{e^{-iwt}}{t-1} @ t = 1 = -\pi i e^{-iw}$$

Figure 6.6 illustrates the contour of integration. We ultimately pass to a limit where $R \to \infty$ and $\varepsilon \to 0+$.

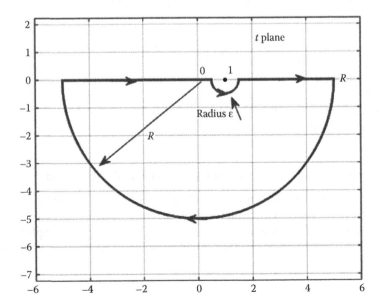

FIGURE 6.6
An indented contour for this problem.

Similarly, if $w < 0$, we use the above contour reflected about the real axis, and Jordan's lemma for the upper half plane, and obtain $\mathbb{F}\left(\dfrac{1}{t-1}\right) = \pi i \operatorname{Res} \dfrac{e^{-iwt}}{t-1} @ t = 1 = \pi i e^{-iw}$. To summarize

$$\mathbb{F}\left(\frac{1}{t-1}\right) = -\pi i e^{-iw}(2u(w) - 1)$$

We have used the Heaviside function to bring about the required change in sign. Multiplying the preceding by $-iw$, we obtain $-\pi w e^{-iw}(2u(w)-1)$, which is indeed the result that will be produced by MATLAB for $\mathbb{F}\dfrac{1}{(t-1)^2}$. The preceding is also verified with MATLAB as follows:

```
>> syms w t
>> a1=fourier(1/(t−1))
a1 = pi*exp(−w*i)*(2*heaviside(−w) − 1)*i
% the above is the Fourier transform of the given function.
>> a2=−i*w*a1
a2 = pi*w*exp(−w*i)*(2*heaviside(−w) − 1)
% the above is −iw times times the transform of 1/(t−1)
>> a3=fourier(1/(t−1)^2)
a3 = −pi*w*exp(−w*i) + 2*pi*w*heaviside(−w)*exp(−w*i)
% the above is the transform of 1/(t−1)^2
>> a2−a3
ans = pi*w*exp(−w*i) + pi*w*exp(−w*i)*(2*heaviside(−w) − 1)
    − 2*pi*w*heaviside(−w)*exp(−w*i)
>> simplify(ans)
ans = 0
% the preceding shows that the Fourier transform of 1/(t−1)^2 is identical to −iw
%times the transform of 1/(t−1)
```

6.3.1 Changing Variables: Fourier to Laplace and Vice Versa

A technique that requires shifting between Laplace and Fourier transforms by means of a change of variables must be applied with some caution (see W page 380). For example, we can use the MATLAB function **laplace** to find the Laplace transform of the Heaviside function. The Laplace transform of heaviside(t) is given by

$$\mathrm{L}\,\text{heaviside}(t) = \int_0^\infty e^{-st}\,dt = \frac{1}{s}$$

while from MATLAB we have that

$$\mathbb{F}\,\text{heaviside}(t) = \int_{0}^{\infty} e^{-iwt}\,dt = \pi\delta(w) - \frac{i}{w}$$

Here is the code and output:

```
>> syms t
>> laplace(heaviside(t))
ans = 1/s
>> fourier(heaviside(t))
ans = pi*dirac(w) - i/w
```

If we replace s in the first result with iw, we do not get the second result—the Fourier transform. Simply put, the function obtained from the first integral is not defined at $s = 0$, so the variable change $w = -is$, when applied to this function, is not valid at $w = 0$, which is exactly where the two transforms disagree. In general, if a result obtained from a Laplace transform is used to obtain a Fourier transform by the substitution $s = iw$, then we should, if possible, perform the inverse transform on the Fourier result, as a check, using the command **ifourier**. The same applies when going from Fourier to Laplace transforms.

6.3.2 Convolution of Functions

If you did problem 5 in the previous section, you would have been exposed to a somewhat restricted definition of the convolution of two functions. The more general definition is this. The *convolution* of functions $f(t)$ and $g(t)$, written $f(t) \otimes g(t)$, is given by

$$f(t) \otimes g(t) = \int_{-\infty}^{\infty} f(t - \tau)g(\tau)\,d\tau \qquad (6.21)$$

This definition is consistent with the one used in the previous section. This is because if we agree to employ functions $f(t)$ and $g(t)$ that are 0 for $t < 0$, as we did in dealing with Laplace transforms, then the lower and upper integration limits in Equation 6.21 can be replaced by 0 and t, respectively. It can be shown that the *Fourier transform of the convolution of two functions is the product of the Fourier transform of each*. Conversely (and less frequently used), it can be shown that if $H(w)$ is a Fourier transform obtained by the convolution of the functions $F(w)$ and $G(w)$, then the inverse Fourier transform of $H(w)$ is $f(t)g(t)$, where $f(t)$ and $g(t)$ are the inverse transforms

of *F*(*w*) and *G*(*w*), respectively. Let us try out Equation 6.21 with MATLAB.
We take $f(t) = \dfrac{1}{t^2+1}$ and $g(t) = \dfrac{1}{t^2+9}$. We use MATLAB to form the convolution of these functions, and take the Fourier transform of this result, using
MATLAB. Then we use MATLAB to find the Fourier transform of *f*(*t*) and
g(*t*) and show that the product of these results is the Fourier transform of
the convolution of *f*(*t*) with *g*(*t*).

Here is the MATLAB code:

```
syms t tau
 assume(t,'real')
 assume(tau,'real')
 %f(t)=1/(t^2+1)and g(t)=1/(t^2+9)our funcs of t.
 ftau=1/((tau)^2+1);
gtau=1/((t-tau)^2+9);
h=ftau*gtau;
the_conv=int(h,tau,-inf,inf)
% the above is the convolution of two functions
syms w
assume(w,'real')
Fourier_of_conv=fourier(the_conv,t,w)
%the above is the Fourier transform the convolution of the two
  %functions
F=fourier(ftau,tau,w)
G=fourier(1/(t^2+9),t,w)
prod_transforms=simplify(F*G)
%the above is the product of the Fourier transforms of each
  %function
check_me=prod_transforms-Fourier_of_conv
% the above should work out to zero, it checks to see that the
  %product
%of the transforms is the transform of their convolution.
```

whose output is

```
the_conv = (4*pi)/(3*(t^2 + 16))
Fourier_of_conv = (pi^2*exp(-4*abs(w)))/3
F = pi*exp(-abs(w))
G = (pi*exp(-3*abs(w)))/3
prod_transforms = (pi^2*exp(-4*abs(w)))/3
check_me = 0
```

The line **check_me** should produce zero, and it does.

Thus, our conjecture that the Fourier transform of a convolution is equal to
the product of the Fourier transform of each function is valid in this problem.

Exercises

1. a. Using MATLAB, find the Fourier transform of the function $f(t) = \dfrac{t}{t^2 + a^2}$, where a is assumed to be a positive real. Be sure to use the statement **assume** $(a > 0)$ in your code. Use the **simplify** statement to get your final result. Now check your work by using the statement **ifourier** on the transform, and verify that you recover the given $f(t)$. Explain why the Fourier transform of your $f(t)$ must be a pure imaginary odd function.

   b. Find the Fourier transform requested above by using residue calculus. Consider w positive and also negative.

2. a. Repeat problem 1, parts (a) and (b), but take as $f(t) = \dfrac{\sin(at)}{t}$ and a is again real. Explain why the Fourier transform of $f(t)$ must be an even function of w.

   b. Taking $a = 1$, use MATLAB to plot $F(w)$ for $-5 < w < 5$.

3. a. Using the definition of the inverse Fourier transform (Equation 6.16), find the inverse transform of $\delta(w - w_0)$, where w_0 is a real ≥ 0. Give your result as a function of t, not x (the default). Repeat this for the function $\delta(w + w_0)$. Check these results using MATLAB and the function **ifourier**.

   b. Using the above results for the inverse Fourier transform, state the Fourier transform of the function $\sin(w_0 t)$.

   Hint: Write this as the sum of two exponentials. Now check this result by applying the operation **fourier** to this function.

   c. Find the Fourier transform of $\cos(w_0 t)$ from the transform for $\sin(w_0 t)$, which you found in part (a), by using the second entry in Equation 6.17. Check your result by using MATLAB to find the Fourier transform of $\cos(w_0 t)$.

4. Assuming that the convolution property for Fourier transforms, following Equation 6.21, is true, use it to prove Equation 6.18.

   Hint: Let $f(t - \tau) = $ heaviside$(t - \tau)$. Simplify the integral in Equation 6.21 for the Heaviside function in the integrand and find the Fourier transform of heaviside$(t - \tau)$.

5. a. The present release of MATLAB, R2015 a, will not produce the Fourier transform of the function arctan(t). Verify that this is so in the version that you are using. If indeed your MATLAB will compute the transform of this function, take note of the result as it will confirm the calculations to be made in what follows.

b.   There is a way in which MATLAB can easily compute the Fourier transform of this function. Note that $\dfrac{d\arctan(t)}{dt} = \dfrac{1}{1+t^2}$ so that $\arctan(t) = \displaystyle\int_{-\infty}^{t} \dfrac{d\tau}{1+\tau^2} - \pi/2$. Confirm this expression for $\arctan(t)$ by doing a symbolic integration in MATLAB.

c.   Use this result to get the Fourier transform of $\arctan(t)$ by employing Equation 6.18 and using MATLAB to first get the Fourier transform of $\dfrac{1}{t^2+1}$. You will also need the Fourier transform of $\dfrac{\pi}{2}$ which you should be able to figure out, or get from MATLAB.

6. a.   This problem explores how the **assume** command in MATLAB can simplify the results obtained from MATLAB. Consider the problem of getting the Fourier transform of $f(t) = \dfrac{1}{t^2+2t+a}$. Let us assume that a is real. Explain why the question of whether $a > 1$ or $a < 1$ affects the nature of the answer.

b.   Using MATLAB, compute the Fourier transform of the function $f(t) = \dfrac{1}{t^2+2t+a}$, where a is real and $a < 1$. Use the **assume** command on a. Use **simplify** to simplify your answer.

c.   Using MATLAB, compute the Fourier transform of $f(t) = \dfrac{1}{t^2+2t+b}$, where no restrictions are placed on b. It can be any complex number. Was it advantageous to use **assume** in part (b)?

7. Use Fourier transforms to solve the following integro-differential equation:

$$\int_{-\infty}^{\infty} f(\tau)e^{-|t-\tau|}d\tau = \frac{d^2 f}{dt^2} + f(t) + \cos(t)$$

Proceed as follows:

a.   Take the Fourier transform of both sides, recognizing that the left side is a convolution, and obtain an algebraic equation in $F(w)$.

b.   Solve this equation for $F(w)$.

c.   Using MATLAB, obtain the inverse Fourier transform of $F(w)$, thus solving the problem. Simplify your answer and show that it is real, as it must be, as the equation that it satisfies involves real functions of a real variable.

6.4 The Z Transform

The Z transform is useful in solving linear difference equations; such equations arise in the field of digital signal processing. Not surprisingly, the history of the Z transform is relatively recent and has its origins in the twentieth century.

A difference equation uses differences (and sums) of the values assumed by a function for specified finite uniform intervals of its independent real variable. This is in contrast to differential equations that employ derivatives of functions that are more or less continuous functions of an independent variable. Let $n \geq 0$ be an integer, and let $f(n)$ be a function that is defined for all the nonnegative values of n. If we should need $f(n)$ for a negative value of n, which sometimes happens, we use $f(n) = 0$, $n < 0$. This is analogous to $f(t) = 0$, $t < 0$ in our work in Laplace transforms.

Here are some examples of difference equations:

$$f(n + 1) - f(n) = 2 \tag{6.22}$$

$$f(n + 1) - 2f(n) = 0 \tag{6.23}$$

$$f(n + 2) - f(n + 1) = f(n) \tag{6.24}$$

$$f(n + 1) = (n + 1)f(n) \tag{6.25}$$

Equation 6.22 says, with slight rearrangement, that the value of $f(n + 1)$ is 2 added to $f(n)$. Thus, if we take as a "boundary condition" that $f(0) = 1$, then it is apparent that $f(1) = 3$, $f(2) = 5$ ... or $f(n) = 2n + 1$.

Equation 6.23 says that each term in the sequence $f(0)$, $f(1)$, ... is twice the one before it. Taking the zeroth ($n = 0$) term as 1, you quickly see that the nth term is $f(n) = 2^n$. Equation 6.24 says that each term is the sum of the two preceding ones—that is, $f(n + 2) = f(n + 1) + f(n)$; this is the basis of the famous Fibonacci sequence that often appears in biology and population growth. By definition, Fibonacci numbers assume that $f(0) = 0$, $f(1) = 1$, which would mean that $f(2) = 1$, $f(3) = 2$, $f(4) = 3$, $f(5) = 5$, ...

In Equation 6.25, we have that the $n + 1$ element, $f(n + 1)$ is $n + 1$ times the preceding element. Taking $f(0) = 1$, you quickly discover that $f(n) = n!$.

Except for Equation 6.24, the problem of finding the nth term was quite simple. The method of Z transforms provides us with a systematic way of solving equations such as these, especially, as in Equation 6.24, where the answer is not readily apparent. Notice that in all these equations $f(n)$, $f(n + 1)$, ... appear only to the first power and that these functions are not multiplied together as, for example, $f(n + 1) f(n - 1)$. This is why we say that we are dealing with linear difference equations.

In the sequence $f(0), f(1), f(2), \ldots,$ the first element is of course $f(0)$, and the second is $f(1)$, and so on. But we also refer to the zeroth element, the first element, and so on, in correspondence with the number n. The usage should be clear from the context.

Sometimes one is concerned not with $f(n)$ but with $f(nT)$, where T is a positive real number that represents a time interval from which samples of a time-varying function $f(t)$ are collected.

Expressions involving $f(nT)$ instead of $f(n)$ frequently appear in the field of digital signal processing. Such an expression might be $f((n+2)T) = \frac{1}{2} f((n+1)T) + \frac{1}{2} f(nT),$ which would indicate that each new value of the sample is the average of the preceding two samples.

Figure 6.7 illustrates a signal from which values $f(T), f(2T), f(3T), f(4T),$ have been sampled.

MATLAB is set up so as to assume that $T = 1$, and we follow that convention.

A Z transformation creates an analytic function of the complex variable z by first forming a Laurent series in the variable z whose coefficients are found from $f(0), f(1), \ldots f(n), \ldots$. The series contains no values of z raised to positive powers. The closed-form expression for the sum of this series, an analytic function, is said to be the Z transform of the sequence. (See W section 5.8.)

Here is the form of the series:

$$\frac{f(0)}{z^0} + \frac{f(1)}{z} + \frac{f(2)}{z^2} + \ldots \qquad (6.26)$$

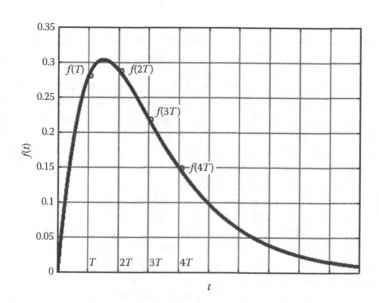

FIGURE 6.7
Sampling a signal or function at intervals of length T.

As we know from complex variable theory, when such a Laurent series converges, it will do so in a ring-shaped domain, centered at the origin, of the form, $a < |z| \leq \infty$. It is possible that a can be as small as zero. We also know from complex variable theory that if the series converges, it does so to an analytic function. If we determine a closed-form expression for the sum of the series in Equation 6.26 and if we call this analytic function $F(z)$, we say that

$$Z[f(n)] = F(z) \tag{6.27}$$

that is, "the Z transform of $f(n)$ is $F(z)$." Equivalently,

$$F(z) = \frac{f(0)}{z^0} + \frac{f(1)}{z} + \frac{f(2)}{z^2} + \ldots \tag{6.28}$$

We also say that "the inverse Z transform of $F(z)$ is $f(n)$." Mathematically,

$$Z^{-1}[F(z)] = f(n) \tag{6.29}$$

Even if we are unable to determine a closed form for the series in Equation 6.26, we can refer to the sum of this series—an analytic function in a domain—as the Z transform of $f(n)$ and, likewise, that the inverse Z transform of the sum of this series is $f(n)$.

A property of the Z transform (and its inverse) that is nearly self-evident and that can be useful is this: the Z transformation is a linear operation, and so the Z transformation of the sum of two or more functions is the sum of the Z transform of each. The inverse Z transformation has the same property.

Here is another useful result: Suppose in Equation 6.26 we make the change of variable $w = 1/z$. We then have

$$F(1/w) = f(0) + f(1)w + f(2)w^2 + \ldots$$

which is valid in a disc-shaped domain centered at the origin. This is a Maclaurin series. If we let $w \to 0$ in the preceding, we obtain on the right $f(0)$. Thus,

$$f(0) = \lim_{w \to 0} F(1/w),$$

or equivalently,

$$f(0) = \lim_{z \to \infty} F(z) \tag{6.30}$$

If $F(z)$ has no limit as $z \to \infty$, for example, it becomes unbounded, then it cannot be a Z transformation of $f(n)$.

Sometimes a Z transformation is obvious by inspection. An easy example is if $f(n) = 1$ for all $n \geq 0$. Then, our Laurent series in Equation 6.26 is $1 + \dfrac{1}{z} + \dfrac{1}{z^2} + \dfrac{1}{z^3} + \dots$. Recalling the Maclaurin series,

$$\frac{1}{1-w} = 1 + w + w^2 + w^3 + \dots \quad |w| < 1$$

we replace w with $1/z$ and find that

$$\frac{1}{1-\dfrac{1}{z}} = \frac{z}{z-1} = 1 + \frac{1}{z} + \frac{1}{z^2} + \frac{1}{z^3} + \dots \quad |z| > 1$$

Thus, $Z(1) = \dfrac{z}{z-1}$ is the Z transformation of our given function. The inverse Z transform of $\dfrac{z}{z-1}$ is simply 1. In other words, $Z^{-1}\left(\dfrac{z}{z-1}\right) = 1$.

Suppose $f(n) = 1/n!$, $n = 0, 1, 2, \dots$. Thus, our Laurent series in Equation 6.19 is

$$\frac{1}{z^0} + \frac{1}{z} + \frac{1}{2!z^2} + \frac{1}{3!z^3} + \dots + \frac{1}{n!z^n} + \dots$$

The alert reader will recognize this as the Laurent series for $e^{1/z}$. We say that $Z\left[\dfrac{1}{n!}\right] = e^{1/z}$. Of course, the inverse Z transform of $e^{1/z}$ is $\dfrac{1}{n!}$.

It should quickly become apparent that we can choose an expression for $f(n)$ such that we cannot find its Z transform by inspection. Here is where MATLAB comes in handy. The reader should now read the documentation for the MATLAB function **ztrans**.

Suppose we want to obtain the Z transform of the function $f(n) = \sin(n)$. We proceed as follows:

```
>> syms n
>> ztrans(sin(n))
ans = (z*sin(1))/(z^2 - 2*cos(1)*z + 1)
>> pretty(ans)

    z sin(1)
-----------------
z^2 - cos(1)z^2 + 1
```

Thus, the Z transform of $\sin n$ is $\dfrac{z\sin 1}{z^2 - 2z\cos 1 + 1}$, which is the sum of the Laurent series

$$\frac{\sin 1}{z} + \frac{\sin 2}{z^2} + \frac{\sin 3}{z^3} + \dots$$

Suppose you want to use MATLAB to find the Z transform of $f(n) = 1$. If you try **ztrans(1)**, you get an error message because the function **ztrans** is expecting a symbolic function, not a number, in its argument. However, **ztrans(sym(1))** does work because 1 is now interpreted as a symbolic function. Try this on your computer, and verify that your result agrees with what we got by a mathematical calculation.

You can use the Heaviside function to take the Z transform of a constant in MATLAB. Suppose you want the Z transform of $f(n) = 1$. We proceed as follows:

```
>> syms n
>> ztrans(heaviside (n))
ans = 1/(z – 1) + ½
```

The preceding is not the Z transform of unity. From earlier in this section, we know the correct answer is $\frac{z}{z-1}$. The problem is that when $n = 0$, the Heaviside function yields one half. However, we can do the following trick:

```
>> ztrans(heaviside (n+1))
ans = z*(1/(z – 1) + 1/2) – z/2
>> simplify(ans)
ans = z/(z – 1)
```

This does yield the desired result because **heaviside (n+1)** is unity for $n = 0, 1, 2, \dots$.

The question of the inverse Z transform (see Equation 6.29) is potentially vexing. The reader should at this point read the help documentation for the MATLAB function **iztrans**, which yields the inverse Z transform. If we have a function that is analytic in the domain $|z| > a > 0$, then this function has a unique Laurent expansion in integer powers of $1/z$—that is, the coefficients $f(0), f(1), \dots$ in Equation 6.26 are determined. The Z transform is unique. However, there can often be more than one closed-form expression that will generate these coefficients. For example, suppose these numbers are $1, \frac{1}{2}, \frac{1}{3}, \dots$. These could have been derived from $f(n) = \frac{1}{n+1}$ or from

$$f(n) = \frac{\sin^2 ((2n+1)\pi/2)}{n+1}$$

or from a limitless number of similar expressions. In taking the inverse Z transform of an analytic expression or its Laurent expansion (which is also analytic), the best we can say is that the ordered set of numbers that are the coefficients in the Laurent expansion are uniquely determined.

In general, MATLAB deals with this issue by giving us the formula $f(n)$ that is obtained from the method of residues in complex variable theory, as follows. Consider $\oint_C \frac{1}{z^m} dz$, where m is *any* integer, positive, negative, or zero, and C is any simple closed contour whose interior contains $z = 0$. If $m = 1$, the value of this integral is $2\pi i$, otherwise its value is zero. (See W section 4.3 and S section 5.1.) Thus,

$$\oint_C z^{-m} dz = 2\pi i, \ m = 1, \quad \oint_C z^{-m} dz = 0, \ m \neq 1 \tag{6.31}$$

Now we have, taking a Z transform,

$$Zf(n) = F(z) = \frac{f(0)}{z^0} + \frac{f(1)}{z^1} + \frac{f(2)}{z^2} + \ldots + \frac{f(n)}{z^n}$$

We assume that this series converges for $|z| > a > 0$.

Suppose we multiply both sides of the above by z^{n-1}, where $n \geq 0$ is an integer. We have

$$z^{n-1} F(z) = \frac{z^{n-1} f(0)}{z^0} + \frac{z^{n-1} f(1)}{z^1} + \frac{z^{n-1} f(2)}{z^2} + \ldots + \frac{z^{n-1} f(n)}{z^n}$$

and now suppose we integrate both sides around a simple closed curve C that encloses the origin and lies entirely in a domain $|z| > a$.

$$\oint_C z^{n-1} F(z)\, dz = \oint_C z^{n-1} f(0)\, dz + \oint_C z^{n-2} f(1)\, dz + \ldots + \oint_C z^{-1} f(n)\, dz$$

$$+ \oint_C z^{-2} f(n+1)\, dz + \ldots$$

From Equation 6.31, we see that every integral on the right has the value zero except the one containing z^{-1}. Thus, from Equation 6.31,

$$\oint_C z^{n-1} F(z)\, dz = 2\pi i f(n)$$

and we have a contour integration that will provide the inverse Z transformation:

$$f(n) = \frac{1}{2\pi i} \oint_C z^{n-1} F(z) dz \quad n = 0, 1, 2, \dots \tag{6.32}$$

The preceding integration is most often done with residues.

Here is a very simple example that we first perform with **iztrans** and later verify with residues.

Example 6.3

What is the inverse Z transform of $F(z) = \dfrac{z}{1+z}$?

Solution:

```
>> syms z
>> iztrans(z/(1+z))
ans = (–1)^n
```

Thus, the answer is $f(n) = (-1)^n$.

Using residues we see from Equation 6.32 that we must evaluate

$$f(n) = \frac{1}{2\pi i} \oint_C z^n \frac{1}{1+z} dz$$

The contour C must enclose all poles of the integrand. There is but one pole—it is simple—and it lies at $z = -1$. Using the calculus of residues, we have that

$$f(n) = \frac{2\pi i}{2\pi i} \operatorname{Res} \frac{z^n}{1+z} @ z = -1$$

The preceding is z^n evaluated at $z = -1$, which yields $(-1)^n$.

Here is a more sophisticated example.

We learned from MATLAB that the Z transform of $f(n) = \sin n$ is

$$F(z) = \frac{z \sin 1}{z^2 - 2z \cos 1 + 1}$$

Let us use residues to take the inverse Z transform of $F(z)$ and try to obtain our $f(n)$. We must find

$$\frac{1}{2\pi i} \oint_C z^{n-1} \frac{z \sin 1}{z^2 - 2z \cos 1 + 1} dz \quad n = 0, 1, 2, \dots \tag{6.33}$$

We choose as a contour a circle of radius r centered at $z = 0$. The radius is chosen large enough so as to include all singularities of the integrand. Otherwise, we would not be justified in using a Laurent expansion to represent $F(z)$ outside this circle. Notice that all the singularities of

$$\frac{z^{n-1}z\sin 1}{z^2 - 2z\cos 1 + 1} \quad n = 0, 1, 2, \ldots$$

are poles and lie where $z^2 - 2z\cos 1 + 1 = 0$. Thus

$$z = \frac{2\cos 1 \pm \sqrt{4\cos^2 1 - 4}}{2} = e^{\pm i}$$

Hence, our inverse Z transform is the sum of the residues of

$$z^{n-1}\frac{z\sin 1}{z^2 - 2z\cos 1 + 1}dz @ z = e^{\pm i}$$

Notice the canceling of the $2\pi i$ factors when we apply residue calculus to the expression in Equation 6.33. The expression

$$z^{n-1}\frac{z\sin 1}{z^2 - 2z\cos 1 + 1}$$

has only simple poles at its two singularities. The residues are the numerator divided by the first derivative of the denominator, evaluated at the poles (see W section 6.3 and S section 7.2).

The residues are

$$z^{n-1}\frac{z\sin 1}{2z - 2\cos 1}$$

evaluated at e^i and e^{-i}. Because these numbers are conjugates of each other, it should be seen that the residues themselves are also (try plugging these numbers into this expression for the residue). The sum of the residues is twice the real part of the residue at e^i. You may use the other pole if you prefer. You should convince yourself that the residue at e^i is

$$\frac{e^{in}}{2\sin 1}\left(e^{-i} - \cos 1\right)$$

Twice the real part of this expression is $\sin n$, which you should confirm. This is the desired answer.

The preceding example was not entirely simple, and it would have been much easier to obtain the desired inverse Z transform with MATLAB. Here is the procedure:

```
>> syms a b z
>> a=sym('sin(1)');
>> b=sym('cos(1)');
>> iztrans(a*z/(z^2-2*z*b+1))
ans = sin(n)
```

There is an important subtlety here: We used *symbols* to represent sin 1 and cos 1. Had we simply placed sin 1 and cos 1 in the argument of **iztrans**, then MATLAB would have sought to replace these irrational numbers by the quotient of rationals. The final answer would then contain an elaborate collection of numbers, and the result sin(n) would in no sense be apparent. As an exercise, try running this piece of code where you use sin1 and cos1 in the argument and they are regarded as numbers, not symbols. Notice the confusing answer, which I will not supply here as it takes up a lot of space.

```
>> syms z
>> iztrans(z*sin(1)/(z^2 – 2*cos(1)*z + 1))
```

In obtaining Z transforms from the function **iztrans**, we might obtain an unfamiliar expression.
Here is an example:

```
>> syms z
>> iztrans(1/(z^2–1))
ans = (–1)^n/2 – kroneckerDelta(n, 0) + ½
```

The function **kroneckerDelta**(a, b) contains two real arguments. If they agree ($a = b$), the function assumes the value 1. If they disagree ($a \neq b$), the function has value zero. Some versions of MATLAB prior to R2015a may not contain this function.

Let us study the answer just obtained. Note that $(-1)\texttt{\^{}}n/2 = \frac{1}{2}(-1)^n$...
Thus, $f(n) = \frac{1}{2}(-1)^n - \text{kroneckerDelta}(n,0) + \frac{1}{2}$. With $n = 0$, this is equal to
$\frac{1}{2} - 1 + \frac{1}{2} = 0$.
When $n = 1$ it is again zero. Some study will show you that when $n > 0$ is even, we have that $f(n) = 1$; otherwise, $f(n) = 0$. In Laurent series form, we have

$$F(z) = \frac{1}{z^2} + \frac{1}{z^4} + \frac{1}{z^6} + \dots$$

This result is confirmed with residue calculus in problem 4.
Sometimes MATLAB will not allow us to perform an inverse Z transformation simply because the inverse Z transform does not exist. In other words, the given function $F(z)$ cannot be the sum of a series like that in Equation 6.26. Consider the following:

```
>> syms z
>> iztrans(exp(1/z))
ans = 1/factorial(n)
>> iztrans(exp(z))
ans = iztrans(exp(z), z, n)
```

Notice that MATLAB gave us the inverse Z transform of $e^{1/z}$ without any trouble but refused to give us the inverse transform of e^z. This is because (refer to Equation 6.30) $\lim\limits_{z\to\infty} e^z$ does not exist. In fact, this function has an essential singularity at infinity.

How can we use Z transforms to solve difference equations? Note the following.

Suppose the function $f(n)$ has the Z transform shown on the right of Equation 6.28. What is the Z transform of $f(n-1)$? Recall that $f(n-1)$, by definition, will be zero if $n-1 < 0$, or equivalently if $n < 1$. Thus,

$$Zf(n-1) = \frac{f(0)}{z} + \frac{f(1)}{z^2} + \frac{f(2)}{z^3} + \ldots$$

and in similar fashion, we have

$$Zf(n-2) = \frac{f(0)}{z^2} + \frac{f(1)}{z^3} + \frac{f(2)}{z^4} + \ldots$$

In general,

$$Zf(n-k) = \frac{f(0)}{z^k} + \frac{f(1)}{z^{k+1}} + \frac{f(2)}{z^{k+2}} + \ldots$$

And if we refer back to Equation 6.28, we have

$$Zf(n-k) = \frac{F(z)}{z^k} \qquad (6.34)$$

where $k \geq 0$ is an integer. This is the first of our shifting or translation formulas. Now suppose we want $Zf(n+1)$. We have from our definition of $Zf(n)$ in Equation 6.28 that

$$Zf(n+1) = \frac{f(1)}{z^0} + \frac{f(2)}{z^1} + \frac{f(3)}{z^2} + \ldots$$

Now refer to Equation 6.28; combining it with the above, we have

$$Zf(n+1) = \frac{f(1)}{z^0} + \frac{f(2)}{z^1} + \frac{f(3)}{z^2} + \ldots = zF(z) - zf(0) \qquad (6.35)$$

Now in a similar fashion, we find

$$Zf(n+2) = \frac{f(2)}{z^0} + \frac{f(3)}{z^1} + \frac{f(4)}{z^2} + \ldots = z^2 F(z) - z^2 f(0) - zf(1) \qquad (6.36)$$

And in general we have the second translation formula:

$$Zf(n+k) = \frac{f(k)}{z^0} + \frac{f(k+1)}{z^1} + \frac{f(k+2)}{z^2} + \ldots$$

$$= z^k F(z) - z^k f(0) - z^{k-1} f(1) - z^{k-2} f(2) \ldots - zf(k-1)$$

(6.37)

where $k \geq 0$ is an integer.

With the preceding formulas (Equations 6.34 through 6.37), we can solve some simple difference equations.

Example 6.4

Suppose each term in a sequence is simply the average of the two preceding terms, and that the initial term in the sequence is 1 and the subsequent term is 2. Calling these the zeroth and first terms, what is the general (nth) term?

We must solve this difference equation:

$$f(n+2) = \frac{1}{2} f(n+1) + \frac{1}{2} f(n).$$

We have $f(0) = 1$, $f(1) = 2$. The Z transform of the left side is according to Equation 6.36: $Zf(n + 2) = z^2 F(z) - z^2 f(0) - zf(1) = z^2 F(z) - z^2 - 2z$. The Z transform of the right side is, from the linearity property, the sum of the Z transforms of each term. The Z transform of $\frac{1}{2} f(n+1)$ is according to Equation 6.36,

$$\frac{1}{2}(zF(z) - zf(0)) = \frac{z}{2} F(z) - \frac{z}{2},$$

while obviously the Z transform of $\frac{1}{2} f(n)$ is $\frac{1}{2} F(z)$. We are left with this transformed equation:

$$z^2 F(z) - z^2 - 2z = \frac{z}{2} F(z) - \frac{z}{2} + \frac{1}{2} F(z)$$

whose solution is

$$F(z) = \frac{2z^2 + 3z}{2z^2 - z - 1}.$$

We need the inverse Z transform of this. The easy way is with MATLAB. We have

```
>> syms z
>> iztrans((2*z^2+3*z)/(2*z^2-z-1))
ans = 5/3 - (2*(-1/2)^n)/3
```

```
>> pretty(ans)
```

$$\frac{5}{3} - \frac{2}{3}\left(\frac{-1}{2}\right)^n$$

Thus, we see that our answer $f(n)$ is $\frac{5}{3} - \frac{2}{3}\left(\frac{-1}{2}\right)^n$. As $n \to \infty$, this has a limit of $\frac{5}{3}$. We should check our answer by trying the values $n = 0$ and 1 and seeing that we do indeed obtain the initial values of 1 and 2, respectively. It is interesting to plot this function against n to see how rapidly the sequence converges to its limit.

Here is the code we used for Figure 6.8 on the next page:

```
n=[0:10];
y=5/3-2/3*(-1/2).^n;
plot(n,y,'k*:');grid
```

We might check our answer with residue calculus. According to Equation 6.32,

$$f(n) = \sum \text{residues } z^{n-1} \frac{2z^2 + 3z}{2z^2 - z - 1} \, @ \text{ all poles.}$$

This is the same as finding the sum of the residues of

$$z^n \frac{z + 3/2}{z^2 - z/2 - 1/2} \, @ \text{ its poles.}$$

The zeros of the denominator are readily found to be at 1 and –1/2. Thus, the poles are simple. The residue at 1 is 5/3, while the residue at –1/2 is $-\frac{2}{3}(-1/2)^n$. The sum of these residues is exactly what MATLAB gave us above.

The preceding problem led to a sequence that has a limit as $n \to \infty$. Here is a more interesting sequence, whose limit is unbounded, and which leads to the famous Fibonacci sequence of numbers.[*] Each element is the sum of the previous two. Thus,

$$f(n + 2) = f(n + 1) + f(n)$$

This is the same equation as the one in the previous problem, but there is no one half on the right side. The convention is to take $f(0) = 0, f(1) = 1$, which would of course mean that $f(2) = 1, f(3) = 2, f(4) = 3, f(5) = 5$. We want the

---

[*] The Wikipedia entry on this subject is short and good. See http://en.wikipedia.org/wiki/Fibonacci_number

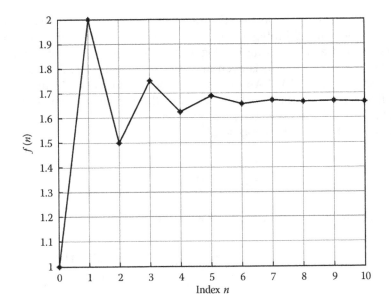

FIGURE 6.8

Plot of $\dfrac{5}{3} - \dfrac{2}{3}\left(\dfrac{-1}{2}\right)^{n}$.

general, nth term. We take Z transforms of both sides of the above equation and apply the conditions on $f(0)$ and $f(1)$. The problem is very similar to the previous one. We have, after transforming,

$$z^2 F(z) - z = z F(z) + F(z)$$

The solution is $F(z) = \dfrac{z}{z^2 - z - 1}$. Let us try to get the inverse transform of this function with MATLAB.

```
>> syms z
>> iztrans(z/(z^2–z–1))
ans=(2*(–1)^n*cos(n*(pi/2+asinh(1/2)*i)))/i^n–(2*(–1)^(1–n)*(–1)^n*5^(1/2)*(1/2
    –5^(1/2)/2)^(n–1))/5+(2*(–1)^(1–n)*(–1)^n*5^(1/2)*(5^(1/2)/2+1/2)^(n–1))/5
>> simplify(ans,100)
ans = 2*(–1)^(n/2)*cos(n*(pi/2 + asinh(1/2)*i)) + 1/2^n*(5^(1/2)/5 – 1)*(5^(1/2) +
    1)^n – 1/2^n*(5^(1/2)/5 + 1)*(1 – 5^(1/2))^n
```

The above is certainly confusing. We rewrite it more clearly as

$$f(n) = 2i^n \cos\left(\frac{n\pi}{2} + ni\,\mathrm{arcsinh}(1/2)\right) + \frac{1}{2^n}\left(\frac{\sqrt{5}}{5} - 1\right)\left(\sqrt{5} + 1\right)^n - \frac{1}{2^n}\left(\frac{\sqrt{5}}{5} + 1\right)\left(1 - \sqrt{5}\right)^n$$

The first term on the left is still daunting. One could simplify it by first assuming n is an odd integer, applying the formula $\cos(a + b) = \cos a \cos b - \sin a \sin b$, and also by using the logarithmic representation of the arcsinh. Later one assumes that n is even and repeats the process. A more fruitful approach is to use MATLAB to again compute the inverse Z transform, but one first simplifies the problem by using the method of partial fractions (see W section 5.5, or any standard calculus book).

Notice that $z^2 - z - 1 = 0$ has solutions

$$z_1 = \frac{1}{2} + \frac{\sqrt{5}}{2} \text{ and } z_2 = \frac{1}{2} - \frac{\sqrt{5}}{2},$$

which means that we have the representation

$$F(z) = \frac{z}{z^2 - z - 1} = \frac{1}{\sqrt{5}} \frac{\left(\sqrt{5}+1\right)/2}{z - \frac{1}{2}\left(1+\sqrt{5}\right)} + \frac{1}{\sqrt{5}} \frac{\left(\sqrt{5}-1\right)/2}{z - \frac{1}{2}\left(1-\sqrt{5}\right)}$$

The reader may wish to review the method of partial fractions to see how this result was obtained. We then find the inverse Z transform of $F(z)$ by taking the inverse transform of each of these fractions separately and adding the results. To make our answer useful, we should enter the square root of 5 as a *symbol* in MATLAB, otherwise it will be converted to a decimal expression. Here is our code for finding $f(n)$:

```
syms a z b c f_n
a=sym('sqrt(5)');
b=iztrans((a+1)/2*1/(z−(1+a)/2))/a;
c=iztrans((a−1)/2*1/(z−(1−a)/2))/a;
f_n=b+c;
simplify(f_n)
```

And here is our output:

$-(5^{(1/2)}*((1/2 - 5^{(1/2)}/2)^{\wedge}n - (5^{(1/2)}/2 + 1/2)^{\wedge}n))/5$

We can simplify the above by multiplying through by the minus sign on the left and putting $5^{1/2}/5 = 1/\sqrt{5}$. We also extract the factor 2^n in the denominator. Our final answer is rather simple:

$$f(n) = \frac{1}{2^n \sqrt{5}}\left[\left(1+\sqrt{5}\right)^n - \left(1-\sqrt{5}\right)^n\right] \qquad (6.38)$$

We expect the Fibonacci numbers to be integers for n = 0, 1, 2, ..., and they are, despite the presence of the $\sqrt{5}$. Using the binomial theorem on each of the binomial expressions in the brackets, you should be able to prove that there are ultimately no terms in the brackets (after you do the indicated subtraction) that do not contain an integer times $\sqrt{5}$. This square root is canceled by the one outside the brackets in the denominator.

Here is a little program that prompts you to enter an integer. The output will be all the Fibonacci numbers up to and including that integer.

```
nmax=input('nmax =')
N=[0:nmax];
a=sqrt(5);
for j=1:length(N)
    n=N(j);
    nth_fibo(j)=1/a*((1+a)^n-(1-a)^n)/2^n;
end
'number=           fibonacci number='
disp([N' round(nth_fibo)'])
```

Notice the use of the function **round**. This ensures that we do not see any round-off errors arising from the use of the square root of 5 raised to various powers.

Here are the first 10 Fibonacci numbers from this program:

```
ans =
number = fibonacci number =
    0           0
    1           1
    2           1
    3           2
    4           3
    5           5
    6           8
    7          13
    8          21
    9          34
   10          55
```

Although as $n \to \infty$ the nth Fibonacci number tends to infinity, the ratio of the $n + 1$ number to the nth approaches a finite limit as n tends to infinity. From Equation 6.38, we have that

$$\frac{f(n+1)}{f(n)} = \frac{1+\sqrt{5}}{2}\left[\frac{1-p^{n+1}}{1-p^n}\right]$$

where

$$p = \frac{1-\sqrt{5}}{1+\sqrt{5}}$$

Notice that $|p| < 1$ and observe that as $n \to \infty$ the expression in the brackets has a limit of 1. Thus, the ratio of a Fibonacci number to the one preceding it has in the limit $n \to \infty$ the value $(1+\sqrt{5})/2$. This irrational expression is the *golden ratio* often written as φ and is approximately 1.618033988749895... There is a vast amount of literature on this number, and the reader may wish to begin with the Wikipedia entry on this subject to get started. A rectangle having these proportions has historically been said to be the most pleasing possible and to serve as the basis for classical Greek architecture.* The ratio of the 10th Fibonacci number to the 9th, which we obtain from the above list, is obtainable from MATLAB as follows:

>> vpa(55/34)
ans = 1.6176470588235294117647058823529

which looks almost like an irrational number (it cannot be, of course, and note the repeating decimal starting with 1764...) and which is remarkably close to the value φ. Note that we used **vpa** (i.e., variable precision arithmetic) so as to obtain as many decimal places as we could for our comparison. One can see that the use of Fibonacci numbers is a way to generate numbers whose numerical ratio (the $(n+1)$th divided by the nth), which unless you can find the repeating decimal, appears to be irrational. Of course, as n tends to infinity, the limit of this ratio *will be* irrational. The code that we used to generate the Fibonacci numbers, which is based on Equation 6.38, may not be the best way to find the Fibonacci numbers, as demonstrated in problem 12.

Here is another useful formula. Consider the Z transform:

$$Znf(n) = \frac{f(1)}{z} + \frac{2f(2)}{z^2} + \frac{3f(3)}{z^3} + \dots + \frac{nf(n)}{z^n}$$

Recall that

$$F(z) = \frac{f(0)}{z^0} + \frac{f(1)}{z^1} + \frac{f(2)}{z^2} + \dots + \frac{f(n)}{z^n} + \dots$$

---

* H. E. Huntley, *The Divine Proportion: A Study in Mathematical Beauty.* (Mineola, NY: Dover Books on Mathematics, 1970).

is the Z transform of $f(n)$. We differentiate both sides of this equation within the domain in which the given series is valid. We have

$$\frac{dF}{dz} = \frac{-f(1)}{z^2} - \frac{2f(2)}{z^3} - \frac{3f(2)}{z^4} \cdots - \frac{nf(n)}{z^{n+1}} - \cdots$$

Notice that the preceding right side can be turned into the Z transform of $nf(n)$ (see above) if we multiply it by $-z$. Thus,

$$Z(nf(n)) = -z\frac{dF}{dz} \tag{6.39}$$

where $F(z) = Zf(n)$.

Since the Z transform of $nf(n)$ is $-z$ times the derivative of $F(z)$, it must follow that the Z transform of the expression $n^2f(n)$, or $n \times nf(n)$, is $-z$ times the derivative of the Z transform of $nf(n)$ from which it follows:

$$Z(n^2 f(n)) = z\frac{dF}{dz} + z^2\frac{d^2F}{dz^2} \tag{6.40}$$

Here is a simple application of Equation 6.39. Suppose we want the Z transform of the function n. We regard n as the product of n and the number 1, and we saw at the start of this section that the Z transform of 1 is $\frac{z}{z-1}$. Thus, we have

$$Z(n \times 1) = -z\frac{d}{dz}\left(\frac{z}{z-1}\right) = \frac{z}{(z-1)^2}$$

We can check this result with MATLAB by either asking for **ztrans**(n) or **iztrans**$\left[\frac{z}{(z-1)^2}\right]$.

We can also take the inverse Z transform of $\frac{z}{(z-1)^2}$ with residues as described in Equation 6.32.

We have that the inverse Z transform of

$$\left[\frac{z}{(z-1)^2}\right]$$

is

$$\frac{1}{2\pi i}\oint z^{n-1}\frac{z}{(z-1)^2}dz$$

where we might use a circle of radius exceeding 1 and centered at the origin. Notice that the integrand has a pole of order 2 at $z = 1$. We use a standard formula for computing the residue at this pole (see W section 6.3 or S problem 7.3). Recall that if a function $f(z)$ has a pole of order 2 at a point z_0, then

$$\text{Res}[f(z), z_0] = \lim_{z \to z_0} \frac{d}{dz}((z - z_0)^2 f(z))$$

Employing this, we have that our inverse Z transform is $\frac{d}{dz} z^n$ evaluated at $z = 1$. The result is simply n, as required.

It is useful to have the Z transform of n^2 for a problem that is to follow. We use Equation 6.40, taking, as in the example just done, $f(n) = 1$ (whose Z transform is $\frac{z}{z-1}$). We obtain $\frac{z^2 + z}{(z-1)^3}$ as the required transform. To summarize,

$$Z(1) = \frac{z}{z - 1} \tag{6.41}$$

$$Z(n) = \frac{z}{(z - 1)^2} \tag{6.42}$$

$$Z(n^2) = \frac{z^2 + z}{(z - 1)^3} \tag{6.43}$$

Example 6.5

Problem
Find the sum of the squares of the first N integers (i.e., find a closed-form expression for $S = 1^2 + 2^2 + 3^2 + \dots N^2$).

Solution:
The reader is probably already aware of the simpler result

$$1 + 2 + 3 + \dots + N = \frac{1}{2}N(N + 1).$$

This result was known to the ancient Greeks (e.g., Pythagoras). In the early 18th century, a young Swiss boy, Leonhard Euler, who was to become a great mathematician, figured it out by writing down the sum of the terms on a horizontal line and then placing the same sum underneath in reverse order. He then added the two lines, term by term. The present problem is harder.

We begin with the difference equation

$$f(n + 1) = (n + 1)^2 + f(n),$$

where $f(0) = 0$ and $n = 0, 1, 2, ...$ We now take the Z transform of the right side and left side. Note that from Equations 6.35 and 6.43 that

$$Z(n+1)^2 = \frac{z^2(z+1)}{(z-1)^3}.$$

We use Equation 6.35 on the left as well. We obtain, after transformation and some simplification, this formula:

$$F(z) = \frac{z^2(z+1)}{(z-1)^4}.$$

In problem 8, we evaluate the inverse Z transform with residues. Here, we use MATLAB as follows:

```
syms z
r=iztrans(z^2*(z+1)/(z-1)^4)
```

The output is

r = 4*n + 5*nchoosek(n − 1, 2) + 2*nchoosek(n − 1, 3) − 3

The expression nchoose(a,b) means $\frac{a!}{b!(a-b)!}$, which is obviously a binomial coefficient.

We thus rewrite our answer as

$$4n + \frac{5(n-1)!}{(n-3)!2!} + 2\frac{(n-1)!}{(n-4)!3!} - 3 = 4n + \frac{5}{2}(n-1)(n-2) + \frac{2}{6}(n-1)(n-2)(n-3) - 3$$

After placing the last expression over the common denominator 6 and doing a little more simplification, we have the result that

$$f(n) = 1 + 2^2 + 3^2 + ... n^2 = \frac{n(n+1)(2n+1)}{6}$$

which you should check by putting $n = 0, 1, 2$.

We can skip the preceding by using the following in MATLAB. We add one more step to the code shown above:

```
expand(r)
```

This yields

ans = n^3/3 + n^2/2 + n/6

This is $\frac{n(2n^2 + 3n + 1)}{6}$ which agrees with $\frac{n(n+1)(2n+1)}{6}$.

6.4.1 The Z Transform of a Product

Suppose we have functions $f(n)$ and $g(n)$, both defined for the nonnegative integers. Consider now $h(n) = f(n)g(n)$ defined for these same integers. We have

$$Zh(n) = \frac{f(0)g(0)}{z^0} + \frac{f(1)g(1)}{z^1} + \frac{f(2)g(2)}{z^2} + \ldots + \frac{f(n)g(n)}{z^n} + \ldots \quad (6.44)$$

If we know the Z transforms of $f(n)$ and $g(n)$, is there some way we can obtain the Z transform of $h(n)$ without doing the calculation in Equation 6.44? The answer is yes, as we now show.

Now $f(n)$ has Z transform $F(z)$, and $g(n)$ has Z transform $G(z)$. Here,

$$F(z) = \sum_{m=0}^{\infty} c_m/z^m \qquad c_m = f(m) \quad (6.45)$$

$$G(z) = \sum_{n=0}^{\infty} d_n/z^n \qquad d_n = g(n) \quad (6.46)$$

We assume that both $F(z)$ and $G(z)$ are analytic in the domain $|z| > a > 0$.

Next we introduce a new variable w and work with $F(w)$ and $G(z/w)$. We place these two variables in the above series and have

$$F(w) = \sum_{m=0}^{\infty} c_m w^{-m} \quad (6.47)$$

and

$$G(z/w) = \sum_{n=0}^{\infty} d_n (z/w)^{-n} = \sum_{n=0}^{\infty} d_n w^n z^{-n} \quad (6.48)$$

Multiplying these functions and their corresponding series together, we have

$$F(w)G(z/w) = \sum_{m=0}^{\infty} \sum_{n=0}^{\infty} c_m d_n w^{n-m} z^{-n} \quad (6.49)$$

We now use a circle of radius b centered at the origin in the w plane. We assume that $b > a$. Thus, $F(w)$ is analytic on and outside this circle. We require that $|w| \geq b > a$. Now let us agree to choose $|z/w| \geq b$ or $|z| \geq b|w|$. With these constraints, both the series in Equations 6.47 and 6.48 are uniformly convergent in the region $|w| \geq b$. We now multiply both sides of Equations 6.47 and 6.48 by w^{-1} and integrate as follows around $|w| = b$. We can exchange the order of integration and summation because of the uniform convergence of both series, whose terms are continuous, and thus obtain

$$\frac{1}{2\pi i} \oint_{|w|=b} \frac{F(w)G(z/w)}{w} dw = \frac{1}{2\pi i} \sum_{m=0}^{\infty} \sum_{n=0}^{\infty} z^{-n} \oint_{|w|=b} \frac{c_m d_n w^{n-m}}{w} dw \qquad (6.50)$$

From the theory of residues, we know that the integral, which is the in variable w, will vanish except if $n = m$. Recalling that $\oint \frac{1}{w} dw = 2\pi i$ when the integral is taken around any simple closed contour enclosing the origin, we have that

$$\frac{1}{2\pi i} \oint_{|w|=b} \frac{F(w)G(z/w)}{w} dw = \sum_{n=0}^{\infty} c_n d_n z^{-n} = \sum_{n=0}^{\infty} f(n)g(n)z^{-n} \qquad (6.51)$$

Looking at Equation 6.44, we see that the right side of the above equation is the Z transform of the product $f(n)g(n)$. Thus, to summarize,

$$Z[f(n)g(n)] = \frac{1}{2\pi i} \oint_{|w|=b} \frac{F(w)G(z/w)}{w} dw \qquad (6.52)$$

where $F(z)$ and $G(z)$ are the Z transforms of $f(n)$ and $g(n)$, and the integration is taken around a circle centered at the origin such that $F(w)$ and $G(w)$ are analytic on and outside of it. With b as the radius of the circle, we require that $|z| > b|w|$. But since $|w| = b$ on the path of integration, this becomes simply $|z| > b^2$.

Example 6.6

Using MATLAB find the Z transforms of

$$f(n) = e^{in} \text{ and } g(n) = \frac{1}{n+1}.$$

With these results as well as Equation 6.52 and the calculus of residues, find the Z transform of $h(n) = \frac{e^{in}}{(n+1)}$. State any restrictions that must apply to the variable z in your answer. Check your result by using MATLAB to get $Zh(n)$.

Solution:
From MATLAB,

```
>> syms n
>> ztrans(exp(i*n))
ans = z/(z - exp(i))
>> ztrans(1/(n+1))
ans = -z*log(1 - 1/z)
```

Thus, $F(z) = \dfrac{z}{z - e^i}$ and $G(z) = -z\log(1 - 1/z) = z\log\dfrac{z}{z-1}$.

According to Equation 6.52, the Z transform of $h(n) = \dfrac{e^{in}}{n+1}$ is the integral

$$\frac{1}{2\pi i}\oint_{|w|=b}\frac{F(w)G(z/w)}{w}\,dw = \frac{1}{2\pi i}\oint_{|w|=b}\frac{1}{w-e^i}\frac{z}{w}\log\left[\frac{z/w}{z/w-1}\right]dw$$

$$= \frac{1}{2\pi i}\oint_{|w|=b}\frac{1}{w-e^i}\frac{z}{w}\log\left[\frac{z}{z-w}\right]dw$$

(6.53)

Notice that $F(w) = \dfrac{w}{w-e^i}$ is analytic on and outside any circular contour in the w plane given by $|w| = b > 1$. This is because the only singularity of this function is at $w = e^i$, which lies on the unit circle. Now

$$G(w) = w\log\frac{w}{w-1}.$$

We use the principal branch of this function. This is defined by means of a branch cut on the real axis of the w plane that is given by $0 \le w \le 1$. On this segment, the argument of the log will be negative real, 0, or infinity. Notice that $G(w)$ is also analytic for $|w| \ge b$. As discussed in our derivation, we take $|z| > b^2$ in Equation 6.53. Notice that

$$\log\frac{z}{z-w} = \log\frac{1}{1-w/z}$$

and that because $\left|\dfrac{w}{z}\right| < 1$ on the path of integration $|w| = b$, the path does not intersect the branch cut for $\log\dfrac{z}{z-w}$.

Now the reader should evaluate the integral in Equation 6.53, noting the apparent pole at $w = 0$. In fact, this is not a pole but *a removable singularity* (see W section 6.2 and S section 3.11), which has no residue. There is a residue to be found at $w = e^i$. Our final answer is

$$Z\frac{e^{in}}{n+1} = \frac{z}{e^i}\log\frac{z}{z-e^i} = ze^{-i}\log\frac{1}{1-e^i/z}$$

which is valid when $|z| > 1$.

We check this with MATLAB as follows:

```
>> syms n
>> ztrans(exp(i*n)/(n+1))
ans = -z*exp(-i)*log(1 - exp(i)/z)
```

The above is identical to $ze^{-i} \log \dfrac{1}{1 - e^i/z}$.

6.4.2 Inverse Z Transform of a Product and Convolutions

Suppose $f(n)$ has Z transform $F(z)$ and $g(n)$ has Z transform $G(z)$. Thus,

$$F(z) = c_0 + \frac{c_1}{z} + \frac{c_2}{z^2} + \dots \quad c_0 = f(0), c_1 = f(1), \dots c_n = f(n) \quad (6.54)$$

$$G(z) = d_0 + \frac{d_1}{z} + \frac{d_2}{z^2} + \dots \quad d_0 = g(0), d_1 = g(1), \dots d_n = g(n) \quad (6.55)$$

Let us multiply the two Laurent series together. We get the Laurent series for the product of the Z transforms:

$$F(z)G(z) = c_0 d_0 + \frac{c_0 d_1 + c_1 d_0}{z} + \frac{c_0 d_2 + c_1 d_1 + c_2 d_0}{z^2} + \dots \quad (6.56)$$

Note that the coefficient of $1/z^n$ in the above is

$$\sum_{k=0}^{k=n} c_k d_{n-k} = a_n \quad n = 0, 1, 2, \dots \quad (6.57)$$

and so

$$F(z)G(z) = a_0 + \frac{a_1}{z} + \frac{a_1}{z^2} + \dots$$

We say that the sequence of numbers $a_0, a_1, a_2, \dots$ results from the *convolution of the sequences* $c_0, c_1, \dots$ and $d_0, d_1, \dots$. Alternatively, we say that the function $f(n)$ is convolved with the function $g(n)$. We write this as $f(n) * g(n)$. The convolution is commutative—it is easy to see that $f(n) * g(n) = g(n) * f(n)$. This definition of the convolution, for functions defined only for discrete values of their independent variable ($n = 0, 1, 2, 3, \dots$), is the analog of the convolution for continuous functions that we discussed for the Laplace transform.

What we have discovered is that the Z transform of the convolution of two sequences is the product of the Z transforms of each sequence. But perhaps

just as important, we see that the inverse Z transforms of the product of two functions is the convolution of the sequences giving rise to each function.
  Here is an example.

Example 6.7

We saw in our work above that

$$Z(1) = \frac{z}{z-1} \text{ and } Z(e^{in}) = \frac{z}{z-e^i}.$$

Use convolution to find

$$Z^{-1} \frac{z^2}{(z-1)(z-e^i)}.$$

Check your result with residue calculus.

Solution:
Note that our result can be obtained through a convolution of $f(n) = 1$ and $g(n) = e^{in}$. Thus, our resulting sequence is $a_0 = 1 \times e^{i0}$, $a_1 = 1 \times e^{i1} + 1 \times e^{i0}$, $a_2 = 1 \times e^{i2} + 1 \times e^{i1} + 1 \times e^{i0} \dots$
  It should become apparent that

$$a_n = 1 + e^i + e^{2i} + \dots e^{in} = \left(e^i\right)^0 + \left(e^i\right)^1 + \dots + \left(e^i\right)^n$$

Recalling the sum of the finite geometric series:

$$1 + r + r^2 + \dots r^n = \frac{1 - r^{n+1}}{1-r}, \ r \neq 1$$

we see that

$$1 + e^i + e^{2i} + \dots e^{ni} = \frac{1 - e^{i(n+1)}}{1 - e^i}.$$

This is

$$Z^{-1} \frac{z^2}{(z-1)(z-e^i)} = \frac{1 - e^{i(n+1)}}{1 - e^i}.$$

We can check this result with Equation 6.32. We have

$$Z^{-1} \frac{z^2}{(z-1)(z-e^i)} = \frac{1}{2\pi i} \oint_C z^{n-1} \frac{z^2}{(z-1)(z-e^i)} dz, \quad n = 0, 1, 2, \dots$$

We will take the contour C as a circle centered at the origin and having radius greater than 1. Notice that there are simple poles of the integrand at $z = 1$ and $z = e^i$. Our result is

$$\text{Res} \frac{z^{n+1}}{(z-1)(z-e^i)} @1 + \text{Res} \frac{z^{n+1}}{(z-1)(z-e^i)} @e^i$$

This becomes

$$\frac{1}{1-e^i} - \frac{e^{i(n+1)}}{(1-e^i)}$$

which agrees with our answer obtained above. We can also check our answer by using the command **iztrans** in MATLAB, but there are some subtleties involved, as shown in exercise problem 13.

Exercises

1. a. Using MATLAB find the Z transform of $f(n) = e^{ian}$, where a is an arbitrary constant that in MATLAB you must treat as a symbol. Check this result by summing the series

$$\sum_{n=0}^{\infty} \left(\frac{e^{ia}}{z}\right)^n$$

   which is easily done as it is a geometric series.

   b. Use our result from part (a) to find the Z transform of $g(n) = e^{-ian}$.

   c. Use the two preceding results to find the Z transform of $h(n) = \sin(an)$.

   Check this result by using MATLAB.

2. a. Certain functions do not have a Z transform. Show that $f(n) = n!$ does not have such a transform. To do this apply the *ratio test* (see S theorem 6.11 or any standard calculus book) to the series whose nth term is $\frac{n!}{z^n}$ and see where this series converges in the complex plane.

   b. Verify your finding by trying to find the Z transform of $f(n)$ with MATLAB. Recall that in MATLAB we use **factorial** (n) to produce $n!$. What does MATLAB return to you?

   c. Can you think of another well-defined function of n, not involving factorials, that does not have a Z transform? Try to get this function's Z transform with MATLAB to see what happens.

3. a. Determine the Z transform of the function $f(n) = \frac{(-1)^n}{n!}$. Do this by writing down the defining Laurent series. The sum of this series should be familiar.

b. Check your answer by using MATLAB.

c. Provide another check by applying Equation 6.32 to the Z transform that you obtained above, and verify that you get back your $f(n)$.

4. We used MATLAB in this section to find the inverse Z transform of the function $F(z) = \dfrac{1}{z^2 - 1}$. Check that answer by finding the inverse Z transform using the calculus of residues.

5. a. In a certain sequence, each successive number is obtained as a weighted average consisting of two thirds of the previous number and one third of the number before that one. Suppose the initial number, call it $f(0)$, is zero and the next number $f(1)$ is one. Find a general formula for the nth number by writing down a difference equation that you will solve with Z transforms.

   b. Obtain a graph like that shown in Figure 6.8 that shows how quickly this weighted average procedure approaches its limit. What is the limit? Use the formula you derived for the nth term in the sequence.

6. a. Use MATLAB to find the Z transform of the function defined as $f(n) = 0, n \leq 1, f(n) = 1, n \geq 2$. It is helpful to use the Heaviside function here.

   b. We showed in this section that the Z transform of the function that is equal to unity for all $n \geq 0$ is $\dfrac{z}{z-1}$. Combine this result with the translation formula, Equation 6.34, to check your answer to part (a).

   c. Check your answer by recovering $f(n)$ by applying Equation 6.32 to the $F(z)$ you found in parts (a) and (b).

7. The formula $0 + 1 + 2 + 3 + \ldots + n = n(n + 1)$ is well known and can be derived by elementary school students if given some help. We derive it here with Z transforms. If you put $n = 0, 1, 2, 3, \ldots$ on the left in the preceding, you generate a sequence of numbers. Note that the elements in the sequence satisfy the difference equation $f(n + 1) - f(n) = n + 1$, where $f(0) = 0$.

   a. Take the Z transform of both sides of this equation and solve for $F(z)$.

   b. Find the inverse of this transform by using the calculus of residues (i.e., Equation 6.32).

   c. Check your answer by finding the inverse of the Z transform with MATLAB.

8. In solving the problem of obtaining the formula for the sum of the squares of the first n integers, in one of our worked examples, we had to find the inverse Z transform of

$$F(z) = \frac{z^2(z+1)}{(z-1)^4}.$$

We did this using MATLAB. Confirm your result by using residues in the complex plane.

9. a. Use MATLAB to obtain the Z transform of $f(n) = e^{in}$.

   b. Obtain the Z transform of $g(n) = n^2 e^{in}$ by using your result from part (a) as well as Equation 6.40.

   c. Check your result by using MATLAB to find the Z transform of $g(n)$.

10. a. We found the Z transform of the function $f(n) = n$ in this section. Use one of the translation formulas we derived to find the Z transform of $g(n) = n + 1$.

   b. Now use the transforms of n and $n + 1$ to find the Z transform of the function $h(n) = n(n + 1)$ by means of Equation 6.52. Evaluate the integral with residues.

   c. Check your answer to part (b) by asking MATLAB to compute the Z transform of $n(n + 1)$.

11. a. Use MATLAB to find the Z transform of the function given by

$$f(n) = \frac{1}{n}, \ n \geq 1 \text{ and } f(n) = 0, \ n \leq 0.$$

The Heaviside function is useful here. Notice that an expression such as heaviside$(x - 1/2)/x$ will, when evaluated at $x = 0$ by MATLAB, yield the result NaN (not a number) because of the ratio $0/0$. MATLAB skips this point when evaluating a Z transform.

   b. Check your answer to part (a) by recovering $f(n)$ by applying Equation 6.32 to the answer to part (a). Notice that the answer to part (a) requires the use of a branch cut. What are the requirements on this cut? The integral that you will get by applying Equation 6.32 cannot be evaluated with residue calculus because the interior of the contour you use contains branch cut singularities of the integrand. However, you should make the change of variables $w = 1/z$ and perform your integration in the w plane. Notice the branch cut is now outside the contour of integration, and you are now integrating in the clockwise (negative) direction. The series

$$\log(1-w) = -w - \frac{1}{2}w^2 - \frac{1}{3}w^3 - \frac{1}{4}w^4 - \dots \quad |w| < 1$$

is useful here.

12. We have seen that the Fibonacci numbers can be obtained as the coefficients in the Laurent series expansion

$$\frac{z}{z^2 - z - 1} = \sum_{n=0}^{\infty} \frac{c_n}{z^n}$$

Suppose we make a change of variables $w = \frac{1}{z}$ in both sides of the preceding equation. The right side becomes a Maclaurin series in the variable w, which converges to the rational function in w that now appears on the left. Now look up the MATLAB documentation for the function **taylor**. Recall that a Maclaurin series is a Taylor series expansion about the origin.

a. Use **taylor** to find the first 11 Fibonacci numbers (i.e., the first 10 nonzero ones). Note that MATLAB will give you them in reverse order.

b. Where does this infinite Taylor series converge in the complex w plane?

13. a. Try to use MATLAB to obtain the inverse Z transform of the function

$$F(z) = \frac{z^2}{z - 1}\frac{1}{z - e^i}$$

   Notice that you will get a confusing and messy answer because MATLAB will convert e^i to a numerical value.

b. Repeat part (a) but use the symbolic expression exp(sym(i)) in place of e^i in MATLAB. Be sure to apply the operation **simplify** to your answer. Which is the more satisfying expression, the one from part (a) or part (b)?

14. Find a closed-form expression for the sum of the series:

$$\frac{e^0 \cos 0}{z^0} + \frac{e^0 \cos 1 + e^1 \cos 0}{z} + \frac{e^0 \cos 2 + e^1 \cos 1 + e^2 \cos 0}{z^2} + \ldots + \sum_{k=0}^{n} \frac{e^k \cos(n - k)}{z^n} + \ldots$$

by taking the Z transform of two simple functions of n that you can find with MATLAB. State the conditions on z for this series to converge.

15. a. Use Equation 6.40 to find the Z transform of the function $g(n) = n^3$ by starting with the transform of n, which we have already derived. You may use MATLAB to take the needed derivatives. Use the function **diff**.

Check your answer by finding the transform with MATLAB. Without using MATLAB, find the Z transform of $h(n) = (n + 1)^3$.

b. Suppose $f(0) = 0$, $f(1) = 1$, and that $f(n + 1) = (n + 1)^3 + f(n)$. Starting with $n = 0$, the terms generated are 0^3, 1^3, $1^3 + 2^3$, $1^3 + 2^3 + 3^3$, ..., $1^3 + 2^3 + 3^3 + \ldots n^3$, ... Take the Z transform of both sides of the difference equation, and use the result derived in part (a) to show that

$$F(z) = \frac{z(z^2 + 4z + 1)}{(z - 1)^5}$$

c. Use either MATLAB (the expand function is useful here) or residues as in Equation 6.32 to show that

$$f(n) = \frac{n^2}{4}(n + 1)^2$$

How can you be sure this will always produce an integer?

If you elect to use Equation 6.32, you have to compute a fourth derivative. It is best to do this with MATLAB to avoid mistakes. Check your $f(n)$ by putting in $n = 0, 1, 2, 3, \ldots$.

6.5 The Hilbert Transform

Unlike the Laplace and Fourier transforms, which emerged in the eighteenth and nineteenth centuries, respectively, the Hilbert transform is relatively recent and came into use in the twentieth century. It ultimately proved itself useful in the latter twentieth century in analysis of electrical systems and in the study of modulated electrical signals such as radio waves. Students of complex variable theory will see it as a means of relating the real and imaginary parts of many analytic functions, while practitioners of electrical engineering will see its value in relating the real and imaginary parts of the system function of a broad class of electrical systems. In contrast to the Laplace and Fourier transforms, the present version of MATLAB (R2015a) does not yield the Hilbert transform of a continuous function. It will perform a Hilbert transformation of a function that is defined for uniformly spaced discrete time intervals, but this takes us too far from the world of functions of a complex variable and will not be discussed here.

Some earlier versions of MATLAB would perform the Hilbert transform of a function that is defined over the real axis by a mathematical expression. This capability was contained in the Signal Processing Toolbox. Because the present version of MATLAB does not contain this feature, we write our own software to do the Hilbert transform of a broad class of functions of a real variable. This is a useful exercise in MATLAB programming.

6.5.1 Definition

Suppose $g(t)$ is a function of the real variable t. In most applications this is a real function, but this is not a requirement. It will, however, simplify our discussion if we make this assumption. We call the function's Hilbert transform $\hat{g}(x)$ and define it by the following integral, assuming that x is real and that the integral exists:

$$\hat{g}(x) = \frac{1}{\pi} \int_{-\infty}^{\infty} \frac{g(t)}{x-t} dt \qquad (6.58)$$

We also write

$$\hat{g}(x) = \mathbf{H}g(t) \qquad (6.59)$$

Here, $\mathbf{H}$ indicates that we are performing the Hilbert transform (transformation). The integral in Equation 6.58 is to be thought of as a Cauchy principal value both with respect to how its infinite limits are treated and with how we deal with the singularity at $t = x$.

The inverse of the above equation is

$$g(t) = \mathbf{H}^{-1}\hat{g}(x)$$

where $\mathbf{H}^{-1}$ indicates that we are to take the inverse Hilbert transform. (See W section 6.10.) Let us use the calculus of residues to find $\mathbf{H}\sin t$. We must do the integration

$$\frac{1}{\pi} \int_{-\infty}^{\infty} \frac{\sin t}{x-t} dt$$

This is a familiar integral from residue calculus which is performed using the method of indented contours and is evaluated as

$$\mathrm{Im} \frac{1}{\pi} \int_{-\infty}^{\infty} \frac{e^{it}}{x-t} dt$$

We use the contour shown in Figure 6.9, which is in the complex τ plane, not the z plane. The real axis in the τ plane is the t axis. The integral

$$\frac{1}{\pi} \oint \frac{e^{i\tau}}{x-\tau} d\tau$$

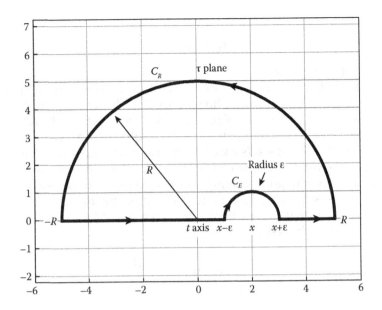

FIGURE 6.9
An indented contour to compute our Hilbert transform.

taken around this contour is zero because the integrand is analytic on and in the contour.

We pass to the limits $R \to \infty$ and $\varepsilon \to 0+$. Notice, from Jordan's lemma, that

$$\lim R \to \infty \frac{1}{\pi} \int \frac{e^{i\tau}}{x - \tau} d\tau = 0$$

over the arc C_R, while over the arc C_ε, we have

$$\lim \varepsilon \to 0 + \frac{1}{\pi} \int \frac{e^{i\tau}}{x - \tau} d\tau = -\frac{\pi i}{\pi} \operatorname{Res} \frac{e^{i\tau}}{x - \tau} @ \tau = x$$

After evaluating this, we have

$$\frac{1}{\pi} \int \frac{e^{i\tau}}{x - \tau} d\tau = ie^{ix}$$

over the indentation. Thus, with a little rearrangement, we have

$$\operatorname{Im} \frac{1}{\pi} \int_{-\infty}^{\infty} \frac{e^{it}}{x - t} dt = -\operatorname{Im} ie^{ix} = -\cos x$$

or

$$\text{H}\sin t = -\cos x$$

for our Hilbert transform.

Once a function, say $g(t)$, has been Hilbert transformed into $\hat{g}(x)$, how do we recover $g(t)$? In other words, how do we find the inverse transform? The answer is that for a wide class of functions, we can take the Hilbert transform of $\hat{g}(x)$ and put a minus sign in front of the result. There are two proofs in W (section 6.10). The simplest is based on Fourier transforms and requires that we accept the validity of the Fourier transform inversion formula. A more complicated proof requires complex variable theory and the use of indented contours and residues. The Fourier transform proof has the advantage that it is less restrictive as to what functions can be transformed (e.g., delta functions).

To summarize the results, if

$$\hat{g}(x) = \text{H}g(t) = \frac{1}{\pi} \int_{-\infty}^{\infty} \frac{g(t)}{x-t} dt$$

then

$$g(t) = \frac{-1}{\pi} \int_{-\infty}^{\infty} \frac{\hat{g}(x)}{t-x} dx = \text{H}^{-1}\hat{g}(x) \tag{6.60}$$

The middle expression in the above equation is the negative of the Hilbert transform of $g(t)$.

Of course, the preceding assumes that both integrals exist as Cauchy principal values.

How can we use MATLAB to evaluate a Hilbert transform and its inverse? We do *not* use the MATLAB command **hilbert** for the problems we have been doing, as it is intended to be used with functions defined for discrete uniformly spaced values of some independent variable.

Study Equation 6.58 for a moment. If

$$\hat{g}(x) = \frac{1}{\pi} \int_{-\infty}^{\infty} \frac{g(t)}{x-t} dt$$

then the function $\hat{g}(x)$ is seen to be the convolution of the functions $g(t)$ and $\frac{1}{\pi t}$. See Equation 6.21 as a reminder, and recall that the Fourier transform of a convolution of two functions is the product of the Fourier transform of each.

Let us now work with $\hat{g}(t)$ instead of $\hat{g}(x)$. Thus,

$$\text{F}\hat{g}(t) = \text{F}g(t)\text{F}\frac{1}{\pi t}$$

The Fourier transform of $\hat{g}(t)$ is the product of the Fourier transforms of $g(t)$ and $\dfrac{1}{\pi t}$.

Suppose we take the inverse Fourier transform of both sides. The left side is simply $\hat{g}(t)$, and we have $\hat{g}(t) = $ inverse Fourier transform $\left[\mathbf{F}g(t)\mathbf{F}\dfrac{1}{\pi t} \right]$.

In the language of MATLAB, we have

$$\hat{g}(t) = \mathbf{ifourier}\left[\mathbf{fourier}g(t) \times \mathbf{fourier}\dfrac{1}{\pi t} \right] \tag{6.61}$$

The Fourier transform of $\dfrac{1}{\pi t}$ is easily calculated with either residues or MATLAB, and we have

```
>> syms t
>> Fourier(1/(pi*t))
ans = heaviside(-w)*2*i - i
```

This function is equal to $-i$ for $w > 0$ and i for $w < 0$. It is zero for $w = 0$. The following MATLAB script M file will determine the Hilbert transform of a function that we supply. We have elected to give this M file the name **hiltrial**. After creating the file and saving it, you can type this word in the command window. You will be prompted to produce the function of t that you want to be transformed.

If you elect to use this M file, you may save it with any other name that MATLAB does not use for a function or file. Do not use the name **hilbert**—it has been taken.

Here is the M file:

```
syms t
g=input ('g is equal to ')
G=fourier (g);
syms pi
h=1/(pi*t);
H=fourier (h);
C=ifourier (G*H);
assume (x,'real')
The_Hilbert=simplify (C)
```

Notice that we treat pi as a symbol so that it will not be reduced to a numerical value in our calculation. We did choose to have MATLAB find the Fourier transform of $\dfrac{1}{\pi t}$, although we might have used the transform we found above; in a later example we do that.

We use the M file, as follows, in the command window. We demonstrate its use in obtaining the Hilbert transform of $\sin t$:

>> hiltrial
g is equal to sin(t)
g = sin(t)

The_Hilbert = −cos(x)

We were given the prompt *g is equal to,* and we then supplied sin(t). Our result The_Hilbert confirms our calculation by residues: namely, −cos(x).

Another way to evaluate Hilbert transforms in MATLAB is by creating a *function* to do the job. We employ this function in the same way that we employ any MATLAB function, such as the **sin, log,** or int. We have chosen to call our function **hlbtrn,** and we write it in the following code:

```
function [output ] = hlbtrn( func )
% this gives you the hilbert transform of the
%function that is the argument of hlbtrn.
%the input is a function of t, the output is a function of x.
  syms t w h x
    H=@(f)fourier(f(t),t,w);%this creates a nested in line
      %function% of the function f(t)
f(t)=func; % this establishes that f(t) is the argument you
  %fed in %when calling hlbtrn
h=H(f);% this evaluates the Fourier transform
% of the input f(t)or func
%(2*heaviside(-w) - 1)*i is the Fourier transform of 1/(pi*t)
output=simplify(ifourier(h*(2*heaviside(-w) - 1)*i));
end
% the end is needed because you have a nested function within
%the one we created
```

The function will evaluate the Hilbert transform of $\sin t$ if we proceed as follows in the command window:

>> syms t
>> hlbtrn(sin(t))
ans = −cos(x)

The reader may wonder why we did not use the MATLAB function **int** to perform the integration in Equation 6.58. This should directly yield the Hilbert transform without our taking a route through Fourier transforms. Try getting the Hilbert transform on your computer of the function $\sin t$ by using the following:

>> syms x t pi
>> assume(x,'real')
>> H=1/pi*int(sin(t)/(x−t),t,−inf,inf,'PrincipalValue',true,'IgnoreAnalyticCons
   traints',true)

You will be surprised at how complicated and clumsy the answer looks. It does not resemble the correct answer of $-\cos t$. MATLAB is much more sophisticated in handling Fourier and inverse Fourier transforms than in performing integrations using **int**. This may change in later versions.

Suppose we wish to use either **hiltrial** or **hlbtrn** to find the Hilbert transform of the number 1. Here we try both. Notice that both **hiltrial** and **hlbtrn** expect to be fed a symbolic function, not a number. Thus, we create a symbol to stand for the number 1, and then summon either the M file or the function:

```
>> a=sym('1')
a = 1
>> hiltrial
g is equal to a
g = 1
The_Hilbert = 0
```

Now we try out the function that we created called **hlbtrn**

```
>> hlbtrn(a)
ans = 0.
```

In both cases, we obtain zero. In fact, if you try any constant, you will find that its Hilbert transform is equal to zero. You may confirm this by using real variable calculus to evaluate Equation 6.58, taking the integrand as a real constant. This is alarming as it tells us that the Hilbert transform is not unique and therefore neither is its inverse transform. Thus, you could argue that the inverse Hilbert transform of zero is any number you wish.

How can we avoid this situation? From W (see pages 420–421) we learn the following: Consider the function $f(t) = g(t) + i\hat{g}(t)$. Here the imaginary part is the Hilbert transform of the real part. Now consider the function $f(\tau) = g(\tau) + i\hat{g}(\tau)$, where τ is a complex variable whose real part is t.

Now assume that

$$\lim R \to \infty \frac{1}{\pi} \int \frac{f(\tau)}{x - \tau} d\tau = 0,$$

where the integration is performed over a semicircle of radius R in the complex τ plane, and x has any real value. The center of the circle is at the origin of the τ plane, and the semicircle lies in either the upper or lower half plane. The function $f(\tau)$ must be analytic on and inside at least one of these semicircles. If this behavior is satisfied, the transform is unique. A function $g(t)$ equal to a constant will not satisfy this condition as $f(\tau)$ will not satisfy the requirement on the arc of the semicircle when $R \to \infty$.

Because MATLAB has more sophistication in handling Fourier and inverse Fourier transforms than it does in ordinary symbolic integration, a third method—probably the best—is presented here for finding Hilbert transforms. Recall the definition of the Fourier transform:

$$\mathbb{F}f(t) = F(w) = \int_{-\infty}^{\infty} f(t)e^{-iwt}dt$$

Suppose we take

$$f(t) = \frac{g(t)}{x-t} \tag{6.62}$$

If we evaluate the Fourier transform of the above and then set $w = 0$, we have arrived at the Hilbert transform of $g(t)$. In other words,

$$\hat{g}(x) = \lim w \to 0 \int_{-\infty}^{\infty} \frac{g(t)}{\pi(x-t)}e^{-iwt}dt \tag{6.63}$$

Typically, the Fourier transform of a real function of t is a complex function. But the Hilbert transform resulting from the above calculation must be real.

We have created a MATLAB M file called **hilfor** that will prompt the user to supply a function $g(t)$ and return its Hilbert transform. Here is the code that you might wish to experiment with by trying various functions, for example, $\frac{1}{t^2+1}$:

```
% the script file hilfor
syms t x w
g=input('g(t) is equal to ')
syms pi
h=(1/pi)/(x-t);
G=@(w)fourier (g*h,t,w);
assume(w,'real')
The_hilbert=simplify(real(G(0)))
```

Exercises

1. Do the following problems by using either one or the script files or functions described above, like **hiltrial**, or **hilfor** or **hlbtrn**, or a program of your own devising that does Hilbert transforms. Find the Hilbert transforms of these functions, $g(t)$. Check your answers by doing contour integration directly.

   a. $\cos 2t$

   b. $\sin(t + 1)$

   c. $\dfrac{1-\cos 2t}{t}$

2. We asserted that if you take the Hilbert transform of a function $g(t)$, then take the Hilbert transform of the resulting function, and multiply the result by -1, you should recover the original $g(t)$. Verify this in the case of the function $g(t) = \sin t$ by using one of the programs described above in Problem 1.

3. a. Find the Hilbert transform of the function $g(t) = \dfrac{\sin t}{t}$ by using residues.

   b. Find the Hilbert transform of this same function by using **hiltrial, hlbtrn,** and **hilfor** given above. Which gives the most useful result? Does it agree with the result of part (a)?

   c. Using **hilfor,** find the Hilbert transform of the *result* of part (a)— that is, transform $\hat{g}(t)$—and verify that you get the negative of the function $g(t)$ supplied in part (a) except that the independent variable is now x.

4. The Hilbert transform can be used to solve certain integral equations. Consider the equation

$$\frac{1}{\pi}\int_{-\infty}^{\infty}\frac{f(t)}{x-t}dt + \frac{\sin x}{x} = 2\cos x$$

   Notice that the integral on the far left is the Hilbert transform of the function $f(t)$. Thus, if we take the inverse Hilbert transform of the expression, we recover $f(t)$. Take the inverse Hilbert transform of both sides of the above equation (using the linearity property of the Hilbert transformation and its inverse) and find $f(t)$. Check your answer by using either residue calculus to find the Hilbert transform of your answer or by using the program **hilfor** developed in this section.

7

Coda: Fractals and the Mandelbrot Set

A FRACTAL GIRAFFE from *The New Yorker* magazine

In this book, I have sought to create a bridge between complex variable theory and the MATLAB programming language. It seems fitting to end with a branch of mathematics, which, although dating in its origins to the early 1900s, really did not come into flower until the widespread use of high-speed digital computers in the 1980s and cannot really be appreciated without a programming language such as MATLAB®. This is the subject known as *fractals*, which is a branch of a larger field in mathematics called *chaos theory*. The public is exposed to chaos theory in newspaper articles that speak of "the butterfly effect." The term was coined in 1969 by Edward Lorenz, an MIT Professor of Earth Sciences, who is famous for his research in weather prediction. He is usually credited with the observation that if a butterfly flaps its wings, it might trigger a storm appearing many hundreds of miles away. To a mathematician, this is saying that the solution of a differential equation, or some iterative process, is acutely sensitive to the *initial conditions* imposed on the equations. Lorenz was at the time working on solutions of the Navier–Stokes equations, which are nonlinear differential equations that are at the heart of the theory of the behavior of fluids and therefore of the weather.

We will not be studying these equations, or in fact any differential equations. The reader might unknowingly already be familiar with chaos theory, which is about the acute sensitivity of the outcome of a real or computer experiment to initial conditions. Go to the top of a hill on a windless day and release, one at a time, a series of seemingly identical newly bought rubber balls from a particular spot. You will find, of course, that the balls take different paths, and that the paths differ increasingly as the balls make their ways down the hill. The differences are due to small errors arising because you did not place each ball in exactly the same spot, or because the balls have subtle differences in their manufacture, or because you inadvertently imparted a slightly different velocity to each ball as you set it adrift.

Our glimpse of chaos theory will come from some computer experiments in the complex plane. We perform the experiments using MATLAB. We will be using MATLAB programs to generate what are called *fractals*. The term was coined in 1975 by the Polish-born mathematician, Benoit Mandelbrot, who spent most of his professional life in the United States. It has the same Latin root as the words *fractured* and *fraction* and suggests something that is not whole (i.e., broken).

What is a *fractal*? There is no neat universally accepted definition of the term. We describe here some of the attributes of those fractals that can, in theory, be drawn on a piece of paper, provided you have a pen with an infinitely fine nib. With a real pen or a computer's printer, you can construct a fair approximation to many fractals. There are fractals that you might build out of clay and hold in the palm of your hand, but these fractals possessing mass will not concern us.

Look at a map of England or the state of Maine and follow the coastline depicted there. With the aid of a ruler, or perhaps a piece of thread and some pins, you can, if you know the scale of the map, estimate the length of the coast. But suppose you decided to walk along the actual coast, following the high water mark of the tides. You would come up with a greater measure for the coastline's length. An ant following the high water mark would walk an even greater distance than you, while an insect smaller than the ant, and more sensitive to the fluctuations of the boundary, would log an even greater trip. This procedure of using a finer and finer measure could proceed indefinitely (we stop short of the molecular scale) so that we might think that the coastline is of infinite length. Moreover, we might conclude that the curve describing the coastline could never be said to possess at any point a definable tangent, since at some scale of scrutiny the curve would be seen as rough, not smooth. A fractal is typically composed of an infinite set of points. If they are committed to a piece of paper, plotted with our fine-pointed pen, they will create a figure that like the coastline has no definable tangent. The language of fractals is used by some mathematicians to describe coastlines of countries and states.

You and I are accustomed to speaking of one-, two-, and three-dimensional objects. A fractal, by contrast, often, but not always, has a dimensionality that is not a whole number—possibly fractional or irrational. We do not have the space here to explain what a fractional dimension is, so the reader might want to refer to other books for additional information.[*]

We explore a number of fractal sets in the exercises. In what follows we study the fractal set that has received the most publicity and which is the most interesting one known to this writer. This is called the *Mandelbrot set*. It is obtained from an iterative process. Mandelbrot has taken credit for his eponymous set, which he claimed to discover in 1979, but there is some fascinating controversy about the set's origins which the reader might wish to pursue at this web address http://www.scientificamerican.com/article/mandelbrot-set-1990-horgan/

An iterative process can take the form

$$z_{n+1} = f(z_n) \tag{7.1}$$

where n is a nonnegative integer and $f(z)$ is a suitably defined function. We begin with an initial value, typically z_0, and then obtain z_1 from the above equation, and from this z_2, and so on. By way of illustration, we use the function used in creating the Mandelbrot set:

$$z_{n+1} = z_n^2 + c \tag{7.2}$$

This does look very simple, but you are in for a surprise. Before showing how this formula defines the Mandelbrot set, we need to review the concept of a sequence of complex numbers that remains bounded.

Suppose the elements of the sequence are the set of numbers $z_0, z_1, z_2, \ldots z_n, z_{n+1}, \ldots$. This sequence is *bounded* if and only if the following is true: given any positive real number M there exists a real number N such

$$|z_n| < M \text{ for all } n > N \tag{7.3}$$

Thus, the sequence $i, i^2, i^3, \ldots i^n, \ldots$ is bounded (the magnitude of each term is one), while the sequence $1+i, (1+i)^2, (1+i)^3, \ldots (1+i)^n, \ldots$, whose magnitude for the nth term is $\left(\sqrt{2}\right)^n$, is not bounded. You should be able to show that the sequence consisting of principal values $1^i, 2^i, 3^i, \ldots n^i, \ldots$ is bounded, but if we replace i with $1 + i$, we do get a sequence that is not bounded.

Here is how we find the elements of the Mandelbrot set, the most famous fractal. We choose a numerical value for c (any number, real or complex) and choose for our "seed" the number $z_0 = 0$. We plug this into Equation 7.2 and compute z_1, which is of course equal to c. Knowing z_1, we return to this

---

[*] *Fractals, Chaos, Power Laws: Minutes from an Infinite Pradise* by Manfred R. Schroeder, New York: WH Freeman, 1990.

equation and see that $z_2 = c^2 + c$, and in the same way, we have $z_3 = (c^2 + c)^2 + c = c^4 + 2c^3 + c^2 + c$. The expressions become increasingly complicated. The Mandelbrot set is defined as follows:

> It is the set of values of c for which the sequence of values obtained from Equation 7.2 is bounded when we use as our starting value $z_0 = 0$.

Notice that we said nothing about the convergence of the sequence. The sequence might very well diverge, but c could still be in the Mandelbrot set. The sequence whose nth term is $i\sin(n\pi/4)$ does not converge (you might want to look up the definition of convergence), but if you found such a sequence in the course of using Equation 7.2, you would say that the value of c in use does belong to the Mandelbrot set.

It should be obvious that certain values of c do or do not belong to the Mandelbrot set. For example, $c = 1$ results in the sequence 0, 1, 2, 5, 26, ..., which you should verify by means of Equation 7.2. The terms are getting bigger and bigger. Thus, $c = 1$ is not in the set.

What about $c = -1$? You should verify that the sequence looks like 0, –1, 0, –1, 0, ..., which although a divergent sequence, is bounded. Thus, $c = -1$ does lie in the set.

And what about $c = i$? This is quite interesting. The sequence is easily found to be 0, i, –1+i, –i, –1+i, –i, –1+i, ... Once we have passed the first two terms, every term is either –i or –1+i. Again, we have a divergent sequence but not an unbounded one. The reader should see that if we took $c = -i$, our sequence would consist of the conjugate of every term found for the case $c = i$. Thus, $c = \pm i$ are both in the Mandelbrot set. You should notice that if a value of c is found to be in the Mandelbrot set, then so is the conjugate of that value, since the sequences $z_0, z_1, z_2, ... z_n, z_{n+1}, ...$ obtained in each case are conjugates of each other, and their being bounded (or not) is identical in each case. And of course, if a value of c is not in the set, then neither is its conjugate. You should also observe the truth of the following: in trying to determine whether a certain value of c belongs to the Mandelbrot set, you encounter two values z_p and z_q that are identical ($z_p = z_q$, $p \neq q$), then the sequence will not be unbounded but will cycle between these two values. Thus, c is in the Mandelbrot set.

How can we tell if a number c belongs to the Mandelbrot set without a lot of tedious numerical calculation? At this point you should read the important results contained in the appendixes.

From the first we see that a *necessary* condition (but not sufficient condition) is that $|c| \leq 2$. Thus, the Mandelbrot set is *contained entirely on and inside a circle of radius 2* centered at the origin in the complex plane. From the second appendix, we find that if we are testing to see whether c is in the Mandelbrot set (first assuring ourselves that $|c| \leq 2$) and discover a value of $|z_n| > 2$, then we can conclude that the value of c is not in the set. Indeed, we can conclude that the sequence $|z_n|, |z_{n+1}|, |z_{n+2}|, ...$ is monotonically increasing without bound.

Imagine that we begin, by means of a MATLAB program, to compute the values $z_1, z_2, z_3, \ldots$. Suppose after computing 10 or 20 such numbers we find none whose magnitude exceeds the number 2. Can we conclude that we have found a value of c that lies in the Mandelbrot set, or might we be worried that the next number we compute might be one whose magnitude exceeds 2, thus indicating that c is not in the set.

This is a real concern that mathematicians have thought about. In the 1980s, when pure and applied mathematicians began using computers to explore fractals and especially the Mandelbrot set, a number of popular articles and books assured the reader that he or she could stop the computation employing Equation 7.2 if after reaching the 200th term a value of $|z_n|$ exceeding 2 had not been reached. They asserted that it was highly unlikely that further iteration would result in a term whose magnitude exceeds 2.* In 2015, as I write this, computers are many orders of magnitude faster, and we will typically go to a much higher level of iteration (e.g., 10 million iterations). We will see why this is sometimes necessary.

It is easy to write a MATLAB program that prompts us to supply a value of c to be tested and that will go through 10^7, namely 10 million, iterations of Equation 7.2 and then tell us if c lies in the Mandelbrot set.

Here is the code, which we call **mandelcheck**:

```
q=1;
while q>0
c=input('the complex value of c to be checked')
tic
nmax=10e6;
%this is the max number of iterations, but
%the iterations will stop if |zn|>2
n=1;
z=0;
while(abs(z)<=2)
  z=z^2+c;
      n=n+1;
      if n>nmax
              break
      end
end
if n<=nmax
disp('the value of zn that causes early termination if n<nmax')
    z
end
  disp('number of iterations used')
      n-1
      if n-1==nmax
```

---

* See A. J. Crilly, R. A. Earnshaw, and H. Jones. *Fractals and Chaos* (New York: Springer, 1991): 36.

```
        disp('the number c is in the set')
    else
        disp('the number c is not in the set')
    end
toc
    q=input('q is neg to stop')
end
```

Let us try the program out by seeing if the point .2+.5*i is in the Mandelbrot set. Here is what occurs on our computer screen when we run the above program:

>> mandelcheck
the complex value of c to be checked .2+.5*i
c = 0.2000 + 0.5000i
number of iterations used
ans = 10000000
the number c is in the set
Elapsed time is 8.165984 seconds.

The calculation took around 8 seconds. If the point chosen lies outside the set, the calculation takes less time because the iteration procedure terminates before the 10 millionth value of z_n is found. We see that in determining whether $c = .2 +.7*i$ lies in the set—it does not—we require a lot less time. Here are our results:

>> mandelcheck
the complex value of c to be checked .2+.7*i
c = 0.2000 + 0.7000i
the value of zn that causes early termination if n<nmax
z = −1.0957 + 1.7956i
number of iterations used
ans = 6
the number c is not in the set
Elapsed time is 0.003120 seconds.

We see that by the time we reach z_6, whose magnitude exceeds one, we can conclude that .2+.7*i is not in the Mandelbrot set.

Even though we are allowing for 10 million possible iterations, the program is not perfect. It can be shown that if c is positive real, then it must satisfy $c \leq 1/4$. This is derived in problem 3. Thus, we might wish to test our program by choosing c just to the right of one quarter. Here is an instance where we try both c = .25 + 1000*eps and also .25+100*eps. In the second case, we got an incorrect answer. The number should have been found to *not* lie in the set. Recall that eps is the smallest positive real that MATLAB can handle. For my version of MATLAB, this was found as follows:

>> eps
ans = 2.2204e–16

If you type *eps* you will find the value in use on your computer. Our results for these two values of c are as follows:

>> mandelcheck
the complex value of c to be checked .25+1000*eps
c = 0.2500
the value of zn that causes early termination if n<nmax
z = 2.3790
number of iterations used
ans = 6666992
the number c is not in the set
Elapsed time is 0.136802 seconds.
q = negative to stop, positive to keep going 2
q = 2
the complex value of c to be checked .25+100*eps
c = 0.2500
number of iterations used
ans = 10000000
the number c is in the set
Elapsed time is 0.202285 seconds.

As noted, the second answer is erroneous. *But*, if you change the number of iterations to something larger (e.g., 100 million), the computer determines that the number is not in the set.

At this point, we might recall Lorenz's statement about a butterfly flapping its wings. Here is an analogous finding from the Mandelbrot set. Suppose $c = i$. We saw above that $c = i$ does lie in the Mandelbrot set. If you try out our program Mandelcheck with any of these four values $i\pm10^{-6}$ or $i\pm i10^{-6}$, you will find that none of them lies in the Mandelbrot set. In other words, if we use the iterative formula Equation 7.2 with any of these four values of c, we develop sequences the magnitudes of whose elements grow without bound. Yet these points seem to lie "close" to $z = i$. These small deviations from i might make us think of the flapping of the butterfly's wings causing a storm.

The reader might be thinking, "What's so special about that? After all, the series $1 + z + z^2 + z^3 + \dots$ will converge if $z = .99999999$ but will diverge to infinity if $z = 1$." The difference is that we can state quite simply where this series will converge, namely, it converges if and only if $|z| < 1$. If asked to find all points in a small neighborhood of $c = i$ that lie in the Mandelbrot set, we are facing a very difficult problem. In the exercises at the end of this section, we see how to find some such points by numerical, not analytic, means.

At this juncture, we have found a way to determine whether a point is in the Mandelbrot set but we have not created a graphical picture of the set, which would be our first example of a fractal. What is a fractal? There is no uniformly agreed-upon on definition. The reader might wish to refer to the book by Falconer in the footnotes for an overview of the definition and might also want to look in a complex variables text to review the meaning of such words as *open set, closed set, boundary point, interior point, connected set,* and *bounded set.*[*] The following is a good start at a working definition of a fractal.

1. A fractal exhibits some degree of *self-similarity* at all scales of magnification. This means that if you examine a portion of a fractal through a magnifying glass, it will look much, or exactly, like, some other portion. A familiar example might be the hourly change of the Dow Jones stock average as reported in the newspaper. Without the axes scales that were provided, you might think that this shows the same quantity over a year. If you look at the photograph of a cloud and there is nothing else in the picture, you have no way of knowing how big the cloud is. The ears of the giraffe at the start of this Coda look like the giraffe. And if you could study the ears on the ears you might see another giraffe, although I doubt if the cartoonist bothered with such a small scale.

2. A fractal exhibits complexity at every degree of magnification. This reminds one of the coastline of Maine, which looks complicated on a map and which will look complicated if you study the line left by the high water mark while you are walking. Look at the photograph of the edge of a cloud and then look at the photograph through a magnifying class. It still looks complicated.

3. The classical methods of calculus, used to determine the slope of a curve (or its tangent), do not apply. We cannot say in general what the local tangent is to a curve generated from fractals.

4. Mathematical fractals (as opposed to fractals in nature, like clouds) are usually obtained from a fairly simple recursive formula. We have just seen the one that Mandelbrot used, Equation 7.2.

It is easy and fun to use MATLAB to generate a curve exhibiting self-similarity. Here is an example of such a program and its output:

```
x=[0:100];%x assumes the integer values from 0 to 100
y=(1-(-1).^x)/2;%at even x values this is zero,at odd x is one
r=(.9).^((x-1)/2);%at x=1,3,5,...gives values of
    %1,.9,.9^2,,.9^3....
```

* Kenneth Falconer, *Fractals: A Very Short Introduction* (New York: Oxford University Press, 2013).

```
y=y.*r; %multiplies y at the peaks by 1,.9,.9^2,.9^3 etc
xx=(1-sqrt(.9).^(x+1))/(1-sqrt(.9));% this is the partial sum
   %of the series
%1+a+a^2+a^2...,a=sqrt(.9) each term in the sum is sqrt(.9)
   %times the one preceding it,
 plot(xx,y,'linewidth',2);grid
```

This program produces the output as shown in Figure 7.1.

Notice that the height of each peak is 9/10, the height of the peak just to the left of it. The base of each triangular peak is 9/10, the width of the peak just to the left of it. Magnifying any portion of the curve the right amount can make it look like a portion of the curve to the left of it. This is the essence of self-similarity. But this is *not* a fractal—it lacks the necessary complexity; except where the curve abruptly changes slope, it has a well-defined tangent.

To illustrate what a fractal can look like, we now write and use the code that will illustrate the Mandelbrot set. We call this program simply mandel_one.

```
q=1;
while q>0
xo=input('x coord center of box=')
yo=input('y coord center of box=')
dx=input('half width of box, x direction=')
dy=input('half width of box, y direction=')
nx=input('number of x divisions')
ny=input('number of y divisions')
tic
```

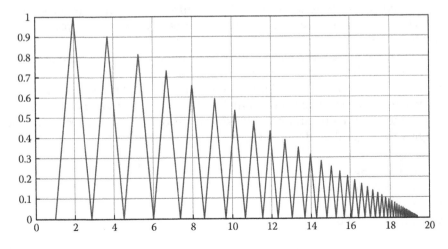

FIGURE 7.1
Self similarity.

```
cr=linspace(xo-dx,xo+dx,nx);
ci=linspace(yo-dy,yo+dy,ny);
[Cr,Ci]=meshgrid(cr,ci);
c=Cr+i*Ci;%this creates a grid of complex numbers for c
nmax=200;% we use 200 iterations of the recursion relation
j=1;
z=zeros(size(c));%this starts off z, for each value of c,
   %at the value zero
while j<=nmax;%iterates the expression below nmax times;
   %200 here;
z=z.*z+c;
j=j+1;
end
ck=abs(z)<=2;%puts a symbol of 1 in the matrix where |z|<=2;
   %otherwise puts %0
dk=1.0*ck;%converts symbolic elements to numerical in above
   %matrix.
p=pcolor(cr,ci,dk);
set(p,'EdgeColor','none');colormap(gray);grid;%use gray for a
   %black and white %picture
set(gca,'layer','top'); %required if you hope to see the grid
axis([xo-dx xo+dx yo-dy yo+dy])
q=input('negative q to stop')
end
```

In this program, we are asked for the center and dimensions of a rectangular box. The program then creates a plot on your screen where points in the box that lie in the Mandelbrot set are given the color white, while the other points are given the color black. Recall that we introduced the function **pcolor** at the end of Chapter 2. If you wish to reverse this convention, use the line of code **p=pcolor(cr,ci,-dk)**. Notice the minus sign in front of the symbol dk. However, there is then the risk that a spot of dark dirt on your screen might be erroneously taken for an element of the set.

We know that the Mandelbrot set must lie on and within a circle of radius 2 centered at the origin of the complex c plane. We generate our set inside a square box centered at the origin. The sides are parallel to the real or imaginary axes, and each side is of length 4. Thus, we are guaranteed that the set will be inside the box and no part will be left out (Figure 7.2). We will use slightly over 1 million date points, as follows:

```
>> mandel_one
x coord center of box=0
xo = 0
y coord center of box=0
yo = 0
half width of box, x direction=2
dx = 2
```

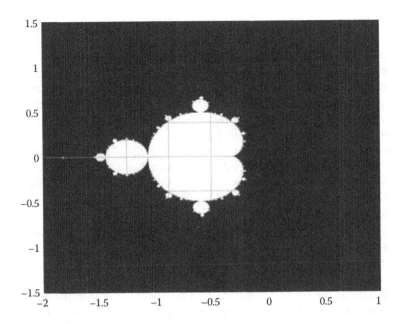

FIGURE 7.2
The Mandelbrot set.

half width of box, y direction=2
dy = 2
number of x divisions1001
nx = 1001
number of y divisions1001
ny = 1001
negative q to stop (−1)
q = −1

We chose odd numbers (namely 1001) for nx and ny, the number of x and y subdivisions of our box in the x and y directions, to ensure that there would be elements lying on the horizontal and vertical lines passing through the center of the box. This is especially useful if the center of the box lies on the x axis.

By choosing ny as odd, we ensure that we are choosing elements that lie on the x axis, which is an axis of symmetry for the Mandelbrot set.

In 1981–1982, two mathematicians, A. Douady (a Frenchman) and J.H. Hubbard (an American) proved that the Mandelbrot set is simply connected and closed. The set has no holes and contains all its boundary points. Choosing any two points in the set, we should be able to connect them by a chain of curved and straight segments all of whose points lie in the Mandelbrot set. We know that $z = i$ is in the Mandelbrot set.

And so is its conjugate $z = -i$. Yet neither of these points appears as a white dot in the above plot, and there is no path going from these points to the points depicted in white in the figure. There is in fact a slender thread going from the points $z = \pm i$ to the main body of the set, but the number of points that we might discover in any section of the thread are so few that if they could be reduced to pixels they cannot be seen on the screen. There are several ways out of this dilemma. We present one here.

If we apply the iterative formula Equation 7.2 to test whether a value of c lies in the Mandelbrot set and if that value of c does not belong to the set, then the application of Equation 7.2 will eventually result in a value of $|z_{n+1}| > 2$. It has been found that the further this value of c lies from a boundary point of the Mandelbrot set, the sooner the condition $|z_{n+1}| > 2$ is encountered. Similarly, the closer this value lies to a boundary point, the later this condition is encountered. Thus, a plot showing not only the Mandelbrot set but showing points that *narrowly escape* being in the set can help us see the set, especially close to its boundaries. This procedure is written in a code employing what is called "an escape time algorithm." Such an algorithm assigns a color to not only points in the set but to points that nearly made it into the set. Because for a narrow filament connecting the main (cardioid shaped) body of the Mandelbrot set to the point $z = i$, there are numerous nearby points that do not quite fall into the Mandelbrot set but do surround the filament, a plot of them suggests the path of the filament. Such a technique can help us study the complicated boundary of the Mandelbrot set that contains innumerable such filaments.

Here is an example of a program that will plot not only the Mandelbrot set but nearby points that did not quite qualify for being in the set. We call this program *Mandelescape*:

```
q=1;
while q>0
xo=input('x coord center of box=')
dx=input('half width of box, x direction=')
yo=input('y coord center of box=')
dy=input('half width of box, y direction=')
nx=input('number of x divisions')
ny=input('number of y divisions, preferably use odd=')
tic
cr=linspace(xo-dx,xo+dx,nx);
ci=linspace(yo-dy,yo+dy,ny);
[Cr,Ci]=meshgrid(cr,ci);
c=Cr+i*Ci;
figure(1)
z=zeros(size(c));
for j=1:20
    z=z.*z+c;
end
```

```
D=abs(z);
d=(D<=2);%this gives a matrix having ones for those values of c
%that are apparently in the Mandelbrot set after 20 iterations
grid;axis equal;
w=z;
for j=21:1000
    w=w.*w+c;
end
 DD=abs(w);
    dd=(DD<=2);%this gives a matrix having ones for those
        %values of c
    %that are taken to be in the Mandelbrot set after 1000
        %iterations
    p=pcolor(cr,ci,d/2+dd/2);%the matrix d/2+dd/2 will have
        %ones at those
    %values of c that are in the matrix after 1000 iterations
    % it will have the value 1/2 at those values of c that are
        %in the
    % matrix after 20 iterations but not after 1000; these
        %values are not
    %in the Mandelbrot set;values of c that are eliminated
        %after just 20
    %iterations are represented by zeros
    set(p,'EdgeColor','none')
    colormap(hot);hold on
    grid on;axis equal
    xlim([xo-dx xo+dx])
    ylim([yo-dy yo+dy])
    set(gca,'layer','top');% this allows you to see the grid
    toc
    q=input('choose a negative q to stop')
end
```

From our previous plot, we have some idea of the rough outlines of the set. Thus, we will run the program so that xo=−.5, yo=0, dx=1.5, dy=1.5. We have then a box centered at [−.5,0]. The sides of the box are each of length 3. We choose nx=ny=501 so that the grid on which we determine the Mandelbrot set has slightly over a quarter million data points. We used a "hot" colormap whose middle tones are orange. The reader might wish to study the topic "colormap" in MATLAB help. We have chosen to place points in the set if after 1,000 iterations of Equation 7.2 we have not achieved $|z_n| > 2$. These are portrayed with a white dot. Points that appeared to be in the set after 20 iterations but are found not to be in the set (they escaped from it) after we were somewhere between 21 and 1,000 iterations are portrayed by this program with an orange dot. All the points that were found to be outside the set in the first 20 iterations appear in black. The reader might wish to experiment with these settings.

The following plot is a gray-scale version of the color plot generated by the code given above. The plot gives us some idea of the narrow filaments surrounding the central cardioid set—they are embedded within the set of points that escaped. It can be shown that all the points on the interval $y = 0$, $-2 \leq x \leq 1/4$ lie in the Mandelbrot set, and this is clearly indicated. Notice, too, that we can begin to see the shape of the filaments that extend outward from the cardioid to the points $z = \pm i$, which we already knew to be in the set.

Our result is presented in Figure 7.3.

Compare this figure with Figure 7.2. Notice that we have some points here that are neither black nor white but are shown in gray. These points did not make the cut—they are not in the Mandelbrot set—but after only 20 iterations they appeared to be. However, after more iterations, we realized they were not. Their presence helps us to locate the positions of points in the Mandelbrot set which exist on such a narrow thread that they cannot be shown on the screen of your computer. The gray points surround these. If you run the code provided above, you will not obtain a gray-scale plot but a more dramatic color plot where there is orange instead of gray.

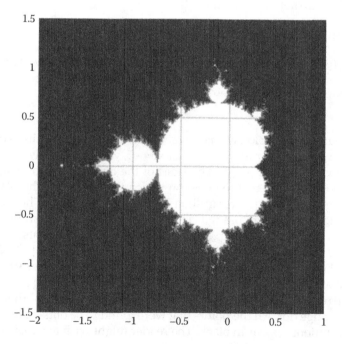

FIGURE 7.3
The Mandelbrot set enhanced with escape points.

7.1 Julia Sets

Suppose we just assign any number for the value of c in Equation 7.2. We begin the iteration progress in this equation not with $z_0 = 0$ (the value used to determine the Mandelbrot set), but with some arbitrary value, real or complex. Starting the iteration process we will find, as in our previous work, that we obtain a sequence of values for $|z_1|, |z_2| \ldots$ whose terms either become unbounded or do not become unbounded. If the sequence remains bounded (i.e., does not diverge to infinity), we take note of the value of z_0. Now suppose we go throughout the complex z plane, continuing to use the same value of c, and take each point in turn as a possible value for z_0.

We note which values of z_0 do not result in a sequence whose elements become unbounded. If we plot these numbers as points in the complex plane, we obtain a certain set pertaining to c. It has been shown that there are two possibilities for our set:

(a) The set is connected, bounded, and closed

   or

(b) The set is composed of points, none of which can be connected to any other point by a contour composed of elements lying entirely in the set.

The set described in (b) may be thought of as mathematical "dust particles" comprising the elements of the set.

The set of points in (b) is composed entirely of its own boundary points. Every neighborhood of every element contains points not belonging to the set. And a neighborhood of every element can be found such that there are no other points (of the set) in the neighborhood except the given element. The set described in (b) is called a *Julia set* and is named after a French mathematician Gaston Julia who discovered it in the early decades of the twentieth century.

The set of points in (a) has a subset—its boundary points. These boundary points are also called a *Julia set*. The *entire* set in (a) is called a *"filled-in Julia set."* The term *Julia set* (without *filled- in*) refers to the set of boundary points. This Julia language can be applied to iterative processes other than the one used in Equation 7.2. Julia sets are often but not always fractals, as we later see (e.g., problem 6).

We do not provide the software for generating Julia or filled-in Julia sets. The MATLAB code is sufficiently similar to that given for generating the Mandelbrot set that readers should be able to write their own code. Note that in Appendix C we prove that no value of z_0 whose magnitude exceeds the larger of $|c|$ and 2 can belong to any kind of Julia set for c. We also assert that if we try any value of z_0 (irrespective of its magnitude) and if using the iterative formula (Equation 7.2), we encounter a value of $|z_n|$ that exceeds the larger of $|c|$ and 2, then z_0 is not in a Julia set, filled-in or not.

Shown below is the filled-in Julia set for the case $c = .25 + .45i$. Note that $|c| < 2$. The points in the plane are to be interpreted as values of z_0. We have used 1,000 iterations in the formula (Equation 7.2) and have decided that if the magnitude of the value of z_n has not exceeded 2 after this many iterations, then the starting value z_0 is in a Julia set.

The filled-in Julia set is in white. The border consisting of the boundary points is complicated. This is the Julia set for $c = .25 + .45i$. The filled-in Julia set—the interior plus boundary—looks like the map of a country. Think of France, for example. The number of data points used in creating this image is $(1001)^2$—slightly over a million.

Here is another example of a Julia set (see Figure 7.4). This one uses $c = -1 + .287i$. We obtain the Figure 7.5. It looks like a constellation of stars in the sky. Note that although there appear to be portions of the set containing points that are not isolated (i.e., interior points whose every neighborhood contain elements other than the point in question), all members of this set are indeed isolated; the computer is not capable of showing the true dust-like nature of this Julia set. How do we know that all the elements are isolated? We answer this in a moment.

The image was created using a grid for possible values of z_0 equal to $(501)^2$ data points. There is an art to choosing the fineness of the grid. If you make the grid too fine, the number of pixels assigned to each value of z_0 will be so few that a white spot will not appear on the computer screen or be reproducible from your printer.

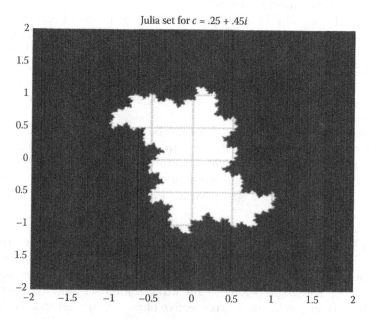

FIGURE 7.4
An example of a Julia set. This one uses $c = .25 + .45i$.

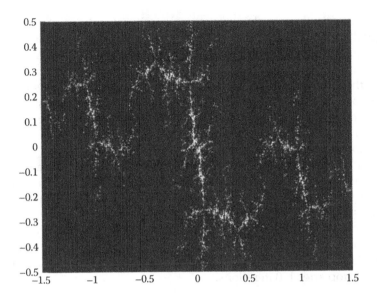

FIGURE 7.5
Julia set for $c = -1 + .287i$.

The use of "escape time algorithms" which enhanced our pictures of the Mandelbrot set can also be advantageously applied also in the code used to display Julia sets.

When does a value of c yield a connected set, and when does it yield "dust"—a set consisting entirely of boundary points? The answer to this is sometimes called *The Fundamental Theorem of the Mandelbrot Set*. The Julia set is connected (and this applies to the filled-in set as well) if c lies in the Mandelbrot set. The Julia set is not connected (nor is any subset of it connected) if c lies outside the Mandelbrot set. The proof was made independently, circa 1919, by the two French mathematicians Gaston Julia and Pierre Fatou. The theorem was proved long before the Mandelbrot set received its name—the naming did not take place until the early 1980s and originated with the mathematicians John Hubbard and Andrew Douady, who wished to honor Mandelbrot for his contribution to our understanding the set.

The point $c = -1 + .287i$ does not lie in the Mandelbrot set. You can verify this with our program called Mandelcheck. Thus, Figure 7.5 shows a Julia set—every point is boundary point and every point is an isolated member of the set. The figure preceding that shows a closed connected set. This is because $.25 + .45i$ lies in the Mandelbrot set, as can be verified with our program Mandelcheck.

Here is a hint for writing the software to generate Julia sets and filled-in Julia sets. We use Equation 7.2, $z_{n+1} = z_n^2 + c$; our Julia plot is for some chosen value of c. We then explore the complex plane; each point of interest is assigned the value z_0, and Equation 7.2 is used to iterate this formula. If c is known to be in the Mandelbrot set, then you may stop the procedure as soon as you encounter a value of z_{n+1} whose magnitude exceeds 2. The z_0 in question is thus not in a Julia set. If you do 1,000 iterations and this magnitude has not been achieved, it is highly likely that the z_0 chosen does lie in a Julia set. We make a mark in the complex plane (we chose white in the above plots) having a distinctive color at the point we called z_0. We chose black for points found not to be in a Julia set.

If you know that c is not in the Mandelbrot set, or if you are not sure, then you perform the operation described above. As soon as you encounter a value of z_{n+1} whose magnitude exceeds the larger of 2 and $|c|$, you can argue that the starting value z_0 is not included in a Julia set. If you reach 1,000 iterations and this situation has not occurred, then it is highly likely that this z_0 does lie in a Julia set. It should be noted that the number 1,000 quoted here and in the previous paragraph is very conservative. Many authors say that it is safe to stop after 100 iterations.

Appendices to Coda

Appendix A

We prove here that if $|c| > 2$, then c is not in the Mandelbrot set.

We make considerable use in this discussion of a version of the triangle inequality (see W section 1.3 or S section 1.5).

If $u = v + w$ with $|v| > |w|$, then we have

$$|u| > |v| - |w| > 0 \tag{7.4}$$

Let us assume that $|c| > 2$.
Now from our iterative formula,

$$z_{n+1} = z_n^2 + c \tag{7.5}$$

and with our taking $z_0 = 0$ we have that $z_1 = c$; and so

$$|z_1| = |c| > 2 \tag{7.6}$$

Consider now $z_2 = z_1^2 + c = z_1(z_1 + c/z_1)$. Thus, $|z_2| = |z_1||(z_1 + c/z_1)|$. Now $|z_1| > \left|\dfrac{c}{z_1}\right|$ from Equation 7.6. Thus, from our triangle inequality, $|z_2| > |z_1|(|z_1| - |c|/|z_1|)$.

With Equation 7.6, this reduces to

$$|z_2| > |c|(|c| - 1) \qquad (7.7)$$

And, since $|c| > 2$, we have from Equation 7.7,

$$|z_2| > 2 \text{ and } |z_2| > |c| \qquad (7.8)$$

Also, from Equation 7.7 and with $|c| > 2$, we obtain

$$\left|\frac{c}{z_2}\right| < \frac{1}{(|c| - 1)} < 1 \qquad (7.9)$$

Continuing, $z_3 = z_2^2 + c = z_2(z_2 + c/z_2)$, so that $|z_3| = |z_2||z_2 + c/z_2|$. Recognizing from Equation 7.9 that $|c|/|z_2| < 1$ and $|z_2| > 2$ we have with our triangle inequality $|z_3| = |z_2||z_2 + c/z_2| > |z_2|(|z_2| - |c|/|z_2|)$, so that

$$|z_3| > |z_2|(|z_2| - 1) \qquad (7.10)$$

Since $|z_2| > |c|(|c| - 1)$ (see Equation 7.7 and $|z_2| > |c|$), we can rewrite Equation 7.10 as

$$|z_3| > |c|(|c| - 1)^2$$

One should see that a pattern is emerging. We now make the following proof by the method of induction. Our goal is to prove if $|c| > 2$ and if for *some* integer $n > 0$ that

$$|z_n| > |c|(|c| - 1)^{n-1} \qquad (7.11)$$

These conditions will ensure that

$$|z_{n+1}| > |c|(|c| - 1)^n \qquad (7.12)$$

In other words, we will have proved that Equation 7.12 remains valid for all values of n larger than the one where we already knew it to be valid.

Notice by the way that we have already proved that Equation 7.11 is true for $n = 2,3$.

From our iterative formula, taking magnitudes:

$$|z_{n+1}| = |z_n^2 + c| = |z_n||(z_n + c/z_n)| \qquad (7.13)$$

Going back to Equation 7.11, which we assume to be true, for some n, and recalling that $|c| > 2$, it should be apparent that $|z_n| > |c| > 2$ and also from Equation 7.11 that

$$|c| / |z_n| < \frac{1}{\left(|c|-1\right)^{n-1}} < 1 < |z_n|$$

Using our triangle inequality, we have from Equation 7.13,

$$|z_{n+1}| > |z_n|(|z_n|-|c|/|z_n|) > |z_n|(|c|-1)$$

Replacing $|z_n|$ on the far right with our assumption in Equation 7.11, we have finally our desired result $|z_{n+1}| > |c|(|c|-1)^n$. We can repeat this process indefinitely, each time using the latest result, so that $|z_{n+k+1}| > |c|(|c|-1)^{n+k}$, where $k \geq 0$ is any integer. Notice that as $n \to \infty$ in Equation 7.12, we find that $|z_{n+1}|$ grows without bound, in other words, the sequence *diverges to infinity*, a subject we touched on in Chapter 3. Thus, if $|c| > 2$, then c is not in the Mandelbrot set.

Appendix B

In this appendix, we prove the following. If we assume that $|c| \leq 2$ (which means that c *might* lie in the Mandelbrot set), and if we perform our iterative procedure (Equation 7.2) and discover a value of $|z_n| > 2$, then $|z_{n+1}| > |z_n|$, $|z_{n+2}| > |z_{n+1}|$, and so on, so that the magnitude of each term in the sequence exceeds that of any term preceding it. This, of course, does not prove that the sequence becomes unbounded. But we will in fact prove that the sequence $|z_0|$, $|z_1|$, $|z_2|$,... does grow without bound and thus c is not in the Mandelbrot set.

From Equation 7.5, $z_{n+1} = z_n^2 + c$, which gives us the familiar

$$|z_{n+1}| = |z_n^2 + c| = |z_n|\left|z_n + \frac{c}{z_n}\right| > |z_n|\left(|z_n| - \left|\frac{c}{z_n}\right|\right) > |z_n|(|z_n|-1) \qquad (7.14)$$

Since $|z_{n+1}| > |z_n|(|z_n| - 1)$, we have that $|z_{n+1}| > |z_n|$. And since $|z_n| > 2$, we can argue from the preceding that $|z_{n+1}| > 2$. Notice that for this reason it is now permissible to change the index from n to $n + 1$ in Equation 7.14 to get

$$|z_{n+2}| > |z_{n+1}|(|z_{n+1}| - 1) \qquad (7.15)$$

We now use Equation 7.14 and the fact that $|z_{n+1}| > |z_n|$ to rewrite Equation 7.14 as

$$|z_{n+2}| > |z_n|(|z_n| - 1)^2 \qquad (7.16)$$

Continuing along in this way, we have the generalization that

$$|z_{n+k}| > |z_n|(|z_n| - 1)^k \tag{7.17}$$

provided $|z_n| > 2$ and $k \geq 0$. Since $|z_n| > 2$, the limit of $|z_{n+k}|$ is infinity as $k \to \infty$.

Thus, as soon as we find a $|z_n| > 2$, we know that the value c we are using cannot belong in the Mandelbrot set.

The equation $|z_{n+1}| > |z_n|(|z_n|-1)$ can be applied whenever $|z_n| > 2$. Thus, it can be applied to the sequence $z_1, z_2, ..., z_n, ...,$ provided you choose a term z_n whose magnitude exceeds 2. Hence, for this term or any term to the right of it, you can say that the succeeding term is larger in magnitude than the one before it. As soon as you find a term whose magnitude exceeds 2, you will know that from then on the terms will get bigger and bigger in magnitude and grow without bound.

Appendix C

Here we prove that if a point z_0 satisfies the condition that $|z_0|$ is greater than the larger of 2 and $|c|$, then this point is not in a Julia set.

We have that $z_1 = z_0^2 + c$, so that

$$|z_1| = |z_0^2 + c| = |z_0|\left|z_0 + \frac{c}{z_0}\right| > |z_0|\left(|z_0| - \frac{|c|}{|z_0|}\right) \tag{7.18}$$

Recall that $\left|\dfrac{c}{z_0}\right| < 1$ by assumption. Also $|z_0| > 2$ by assumption. Thus, it must be true that $|z_1| > |z_0|$ and that $|z_1|$ is greater than the larger of 2 and $|c|$, since z_0 is. Continuing in this way we could argue that

$$|z_2| > |z_1|\left(|z_1| - \frac{|c|}{|z_1|}\right) \tag{7.19}$$

and that $|z_2|$ is greater than the larger of 2 and $|c|$. Combining our two inequalities, we have

$$|z_2| > |z_0||z_1|\left(|z_0| - \frac{|c|}{|z_0|}\right)\left(|z_1| - \frac{|c|}{|z_1|}\right)$$

and continuing on, we have

$$|z_n| > |z_0||z_1| \ldots |z_{n-1}|\left(|z_0| - \frac{|c|}{|z_0|}\right)\left(|z_1| - \frac{|c|}{|z_1|}\right) \cdots \left(|z_{n-1}| - \frac{|c|}{|z_{n-1}|}\right)$$

Suppose we pass to the limit as $n \to \infty$. The product $|z_0||z_1| \cdots |z_{n-1}|$ consists of terms each of which exceeds 2 and thus it grows without bound. Each of the other terms in the product

$$\left(|z_0| - \frac{|c|}{|z_0|}\right)\left(|z_1| - \frac{|c|}{|z_1|}\right) \cdots \left(|z_n| - \frac{|c|}{|z_n|}\right)$$

exceeds one. Thus, our limit goes to infinity—the product is unbounded and the iterative process leads to an unbounded result. As in the above two appendixes, we obtain a sequence diverging to infinity.

Notice that had we begun with a value of z_0 whose magnitude was *not* greater than the larger of 2 and $|c|$ but, through a process of iteration, encountered a value of z_n whose magnitude did exceed these two quantities, we could readily modify the preceding proof to show that the resulting sequence did become unbounded, and we could then conclude that z_0 is not in a Julia set.

Exercises

1. Consider the sequence $|z_0|, |z_1|, |z_2|, \ldots$ For the expressions z_n given below, explain whether this sequence is bounded. If the expression is a function of z, state the domain or set of points where the sequence becomes unbounded.

   a. $z_n = n\cos(n\pi/2)$

   b. $z_n = e^{nz}$

   c. $z_n = n^z$

2. a. We established that $z = i$ is in the Mandelbrot set. Since the Mandelbrot set is connected, every neighborhood of this point must contain at least one other point in the set. Consider $z = x + iy$ satisfying $-.1 \le x \le .1, .9 \le y \le 1.1$. Using our program Mandelcheck, try to find a point in this region *other than* $z = i$ that lies in the Mandelbrot set. This could be quite tedious. It is doubtful that you will succeed by proceeding by trial and error.

   b. A systematic way to solve the preceding problem is with the following code. In order to understand it, read the MATLAB documentation for the command **find**. Try running the code for the region described below using nx=ny=1001 subdivisions for the square region. Find at least one point in the Mandelbrot set (the program should yield many). Test that this point lies in the Mandelbrot set by using the code Mandelcheck given earlier in this chapter.

```
q=1
 format long
 while q>0
xo=input('x coord center of box=')
yo=input('y coord center of box=')
dx=input('half width of box, x direction=')
dy=input('half width of box, y direction=')
%below use odd numbers for nx and ny to include center of box
%in your calculations
nx=input('number of x divisions')
ny=nx; %this is number of y divisions
% the two lines below give the real and imaginary parts of c.
tic
cr=linspace(xo-dx,xo+dx,nx);
ci=linspace(yo-dy,yo+dy,ny);
[Cr,Ci]=meshgrid(cr,ci);
c=Cr+i*Ci;%this creates a grid of complex numbers for c
nmax=1000;% we use 1000 iterations of the recursion relation
j=1;
z=zeros(size(c));%this starts off z, for each value of c,
   %at the value zero
while j<=nmax;%iterates the expression below nmax times;
z=z.*z+c;
j=j+1;
end
ck=abs(z)<=2;%puts a symbol of 1 in the matrix where |z|<=2 ;
   %otherwise puts zero
dk=1.0*ck;%converts symbolic elements to numerical in above
   %matrix.
[rows,cols,vals] = find(dk);%this finds the nonzero elements
   %in dk
for k=1:length(rows)
    locations=c(rows(k),cols(k))
    figure(1)
    plot(locations,'k.'); hold on
    %the above is optional and will plot the points
    %you just found as dots
    %the "locations" are points in the complex plane lying in
       the % mandelbrot set
    end
toc
q=input('negative q to stop')
  end
```

3. We can prove that if c is real and $c > 1/4$, then c is not in the Mandelbrot set.

   a. Why is $z_{n+1} - z_n = z_n^2 - z_n + c$?

      Note that all values of z_n obtained through iteration are real if c is real.

   b. Using the result of (a), argue that $z_{n+1} - z_n$ is always greater than or equal to some positive constant that is *independent* of n.

      Hint: Look at the equation in the x–y plane $y = x^2 - x + c$, $c > 1/4$ and argue that this is always positive. One way to do this is to find where the first derivative vanishes and show that the second derivative is positive there. Thus, the function displays a minimum. It is easy to show that the function is positive at the minimum. What is its value at the minimum? It is convenient to take $c = \frac{1}{4} + \varepsilon$, where $\varepsilon > 0$.

   c. Since $z_{n+1} - z_n \geq \varepsilon$, how does this show that as $m \to \infty$ z_m grows positive without bound? This completes the proof. Notice that we already know that if c is real and $c < -2$, then c is not in the Mandelbrot set because $|c| > 2$.

4. Our code **Mandelescape** shows points that are presumed to be in the Mandelbrot set when we have made 1,000 iterations with our working iterative formula (Equation 7.2). It also shows points that are presumed to be in the set after 20 iterations but are found not to be in the set when we have made between 21 and 1,000 iterations. It also shows points found to not be in the set when we have made 1 to 20 iterations.

   Improve upon the code as follows. It should show points that are presumed to be in the Mandelbrot set when we have made 1,000 iterations with our working iterative formula (Equation 7.2). Also show points with a distinctive color that are presumed to be in the set after 5 iterations but are found not to be in the set when we have made between 6 and 15 iterations. Similarly, show points found to not be in the set when we have made 16 to 1,000 iterations. And show points found not to be in the set when we have made 1 to 5 iterations.

5. Show that no matter what value is chosen for c, the corresponding Julia set or filled Julia set must contain the two points given by $z_0 = \frac{1}{2} + \frac{1}{2}(1 - 4c)^{1/2}$, $c \neq \frac{1}{4}$. If $c = \frac{1}{4}$, this reduces to the fact that the filled Julia set contains $z_0 = \frac{1}{2}$.

   Hint: Study Equation 7.2 and look for *fixed points* (i.e., points which the iterative procedure transforms back into themselves).

6. a. Write code similar to our code called **Mandelbrot**, above, to generate a Julia set. The code will prompt you for the value of c

to be considered and will iterate (Equation 7.2) throughout a grid placed in a rectangular box in the complex z plane. We take z_0 as the points lying at the intersections of the grid. Check your program by generating the Julia sets for $.25+.45i$ and $-1+.287i$ and compare them with the results given in this section. Notice that in the previous problem (number 5) we saw that if $c = \dfrac{1}{4}$, then the Julia set must contain $z = \dfrac{1}{2}$. Verify that this is so by using your program to generate the filled Julia set for this value of c.

b. Not every Julia set obtained from Equation 7.2 is a fractal. If $c = 0$, find mathematically the corresponding filled Julia set and the Julia set as well. Why is the latter not a fractal set? Consider issues of self-similarity and tangents. Now verify the Julia sets (filled and ordinary) by running the program found in part (a) for $c = 0$.

7. a. Show that no matter what value is chosen for c, the corresponding Julia set or filled Julia set must contain the values of z_0 that are solutions of the equation: $z_0^4 + 2cz_0^2 - z_0 + c^2 + c = 0$.

   Hint: Suppose we have a value of z_0 in Equation 7.2 such that $z_0 = z_2$. Why does this guarantee that z_0 belongs to a Julia set? How does the preceding lead to the quartic equation?

   b. There is a closed-form solution available for quartic equations; it is quite complicated. One is generally better served by the MATLAB function **roots** (which you might wish to review) into which one enters the numerical values of the coefficients in the preceding equations. Using MATLAB, find the values of z_0 in the above quartic corresponding to $c = i/2$.

   c. Run the code you wrote in problem 6, to generate Julia sets, choosing $c = i/2$ to see if the four values found above appear to be in the set. Put them on your plot. Note that some appear to be boundary points.

8. There are other iterative schemes besides Equation 7.2 that generate fractal shapes. Here are some to experiment with. In the absence of other information, you may assume that a point being studied does not belong to the set being sought if after 200 iterations we have $|f_n| > 10$. Obtain fractal plots for the following, finding the regions of the complex c plane where the sequence does not become unbounded, provided we begin with $z_0 = 0$.

   a. $z_{n+1} = z_n^3 + c$
   b. $z_{n+1} = \cos(z_n + c)$
   c. $z_{n+1} = \cos(cz_n)$
   d. $z_{n+1} = e^{icz_n}$. Plot in the complex plane for $0 \le \mathrm{Re}(c) \le 6$, $0 \le \mathrm{Im}(c) \le 6$.

9. Write a program to create a curve that has self-similarity but is different from the one using the sequence of diminishing positive triangles that we used earlier in this section. Try, for example, a damped triangular wave having peaks and troughs: $1, -.9, .9^2, -.9^3, .9^4, -.9^5, \ldots$.

Index

Note: Page numbers in *italics* refer to figures.